THE CHALLENGE OF DEVELOPING STATISTICAL LITERACY, REASONING AND THINKING

The Challenge of Developing Statistical Literacy, Reasoning and Thinking

Edited by

Dani Ben-Zvi
University of Haifa,
Haifa, Israel

and

Joan Garfield
University of Minnesota,
Minneapolis, U.S.A.

KLUWER ACADEMIC PUBLISHERS
DORDRECHT / BOSTON / LONDON

A C.I.P. Catalogue record for this book is available from the Library of Congress.

ISBN 1-4020-2277-8 (HB)
ISBN 1-4020-2278-6 (e-book)

Published by Kluwer Academic Publishers,
P.O. Box 17, 3300 AA Dordrecht, The Netherlands.

Sold and distributed in North, Central and South America
by Kluwer Academic Publishers,
101 Philip Drive, Norwell, MA 02061, U.S.A.

In all other countries, sold and distributed
by Kluwer Academic Publishers,
P.O. Box 322, 3300 AH Dordrecht, The Netherlands.

Printed on acid-free paper

Printed in the Netherlands.

Contents

List of Authors

Bakker, Arthur
Freudenthal Institute, Utrecht University, the Netherlands
A.Bakker@fi.uu.nl

Batanero, Carmen
Universidad de Granada, Spain
batanero@ugr.es

Ben-Zvi, Dani
University of Haifa, Israel
dbenzvi@univ.haifa.ac.il

Chance, Beth
California Polytechnic State University, USA
bchance@calpoly.edu

Cobb, Paul
Vanderbilt University, USA
paul.cobb@vanderbilt.edu

Confrey, Jere
Washington University at St. Louis, USA
jconfrey@wustl.edu

delMas, Robert C.
University of Minnesota, USA
delma001@umn.edu

Gal, Iddo
University of Haifa, Israel
iddo@research.haifa.ac.il

Garfield, Joan
University of Minnesota, USA
jbg@umn.edu

Gravemeijer, Koeno P. E.
Freudenthal Institute, Utrecht University, the Netherlands
K.Gravemeijer@fi.uu.nl

Heaton, Ruth M.
University of Nebraska-Lincoln, USA
Rheaton1@unl.edu

Jones, Graham A.
Griffith University, Gold Coast Campus, Australia
g.jones@griffith.edu.au

Konold, Clifford
University of Massachusetts, Amherst, USA
konold@srri.umass.edu

Langrall, Cynthia W.
Illinois State University, USA
langrall@ilstu.edu

Makar, Katie
University of Texas at Austin, USA
kmakar@mail.utexas.edu

McClain, Kay
Vanderbilt University, USA
kay.mcclain@vanderbilt.edu

Mickelson, William T.
University of Nebraska-Lincoln, USA
wmickelson2@unl.edu

Mooney, Edward S.
Illinois State University, USA
mooney@ilstu.edu

Moore, David S.
Purdue University, USA
dsmoore@stat.purdue.edu

Moritz, Jonathan
University of Tasmania, Australia
jonathan.moritz@utas.edu.au

Pfannkuch, Maxine
The University of Auckland, New Zealand
m.pfannkuch@auckland.ac.nz

Pollatsek, Alexander
University of Massachusetts, Amherst, USA
pollatsek@psych.umass.edu

Reading, Chris
University of New England, Australia
creading@metz.une.edu.au

Sánchez, Victoria
Universidad de Sevilla, Spain
mvsanchez@cica.es

Shaughnessy, J. Michael
Portland State University, USA
mike@mth.pdx.edu

Tauber, Liliana Mabel
Universidad Nacional del Litoral, Santa Fe, Argentina
lilianatauber@gigared.com

Thornton, Carol A.
Illinois State University, USA
thornton@ilstu.edu

Watson, Jane M.
University of Tasmania, Australia
Jane.Watson@utas.edu.au

Wild, Chris
The University of Auckland, New Zealand
c.wild@auckland.ac.nz

Foreword

David S. Moore
Purdue University

This unique book is very welcome, for at least three reasons. The first is that teachers of statistics have much to learn from those whose primary expertise is the study of learning. We statisticians tend to insist that we teach first of all from the base of our knowledge of statistics, and this is true. Teachers at all levels must understand their subject matter, and at a depth at least somewhat greater than that of the content they actually teach. But teachers must also understand how students learn, be aware of specific difficulties, and consider means to guide students toward understanding. Unaided, we gain skill intuitively, by observing our own teachers and by experience. Teachers below the university level receive specific instruction in teaching—this is, after all, their profession—and this book will improve that instruction where statistics is concerned. Teachers at the university level consider themselves first of all mathematicians or psychologists or statisticians and are sometimes slow to welcome pedagogical wisdom from other sources. This is folly, though a folly typical of professionals everywhere. I have learned a great deal from some of the editors and authors of this book in the past, and yet more from reading this volume.

Second, this book is timely because data-oriented statistics has at last moved into the mainstream of mathematics instruction. In the United States, working with data is now an accepted strand in school mathematics curricula, a popular upper-secondary Advanced Placement syllabus adds a full treatment of inference, and enrollment in university statistics courses continues to increase. (Indeed, statistics is almost the only subject taught by university mathematics departments that is growing.) Similar trends, particularly in school mathematics, are evident in other nations. The title of this volume, with its emphasis on "statistical literacy, reasoning, and thinking" reflects the acceptance of statistics as a mainstream subject rather than a technical specialty. If some degree of statistical literacy is now part of the equipment of all educated people, then more teachers, and teachers of more varied backgrounds, must be prepared to help students learn to think statistically. Here at last is a single source that can inform our preparation.

Finally, statisticians in particular should welcome this book because it is based on the recognition that statistics, while it is a mathematical science, is not a subfield of mathematics. Statistics applies mathematical tools to illuminate its own subject matter. There are core statistical ideas—think of strategies for exploratory data analysis and the distinction between observational and experimental studies with the related issue of establishing causation—that are not mathematical in nature. Speaking broadly, as long as "statistics education" as a professional field was considered a subfield of "mathematics education," it was in fact primarily the study of learning probability ideas. Understanding that statistics is not just mathematics is giving rise to a new field of study, closely related to mathematics education but not identical to it. The editors and authors of this volume are leaders in this new field. It is striking that the chapters in this book concern reasoning about data more than about chance. Data analysis, variation in data and its description by distributions, sampling, and the difficult notion of a sampling distribution are among the topics receiving detailed study.

It is not often that a book serves to synthesize an emerging field of study while at the same time meeting clear practical needs. I am confident that The Challenge of Developing Statistical Literacy, Reasoning, and Thinking will be seen as a classic.

Preface

Over the past decade there has been an increasingly strong call for statistics education to focus on statistical literacy, statistical reasoning, and statistical thinking. Our goal in creating this book is to provide a useful resource for educators and researchers interested in helping students at all educational levels to develop these cognitive processes and learning outcomes. This book includes cutting-edge research on teaching and learning statistics, along with specific pedagogical implications. We designed the book for academic audiences interested in statistics education as well as for teachers, curriculum writers, and technology developers.

The events leading to the writing of this book began at the Fifth International Conferences on Teaching Statistics (ICOTS-5), held in 1998 in Singapore. We realized then that there are no consistent definitions for the often stated learning goals of statistical reasoning, thinking, and literacy. In light of the rapid growth of statistics education at all levels, and the increasing use of these terms, we realized that it was important to clearly define and distinguish between them in order to facilitate communication as well as the development of instructional materials and educational research.

A small, focused conference bringing together an international group of researchers interested in these topics appeared to be an important next step in clarifying the terms, connecting researchers working in this area, and identifying ways to move the field forward together. The first International Research Forum on Statistical Reasoning, Thinking, and Literacy (SRTL-1) was held in Israel in 1999 to address these needs. Due to the success of SRTL-1 and the strong feeling that this type of forum should be repeated, SRTL-2 was held in 2001 in Australia, this time with a focus on different types of statistical reasoning. Many of the papers from these first two forums have led to chapters in this book. The forums continue to be offered every two years, with SRTL-3 held in the USA in 2003, as interest and research in this area steadily increase.

To get the most out of this book, readers may find the following points useful:

- Chapter 1 may be a good starting point. It offers preliminary definitions and distinctions for statistical literacy, reasoning, and thinking. It also describes some of the unique issues addressed by each chapter to help readers in their journey within the book.
- The first part of this book (Chapters 2 through 5) is a comprehensive overview of statistical literacy, reasoning, and thinking from historical, psychological, and educational perspectives. In addition, cognitive models of development in statistical reasoning are examined. Readers who wish to examine the theoretical foundations upon which the individual studies in subsequent parts are based are referred to these chapters.
- Many chapters that focus on a particular type of statistical reasoning follow a unified structure, starting with a description of the type of reasoning studied and ending with key practical implications related to instruction, assessment, and research. Readers can examine these sections to quickly determine the chapter's contents.
- The closing chapter (Chapter 17) describes the current state of statistics education research and its uniqueness as a discipline. It offers a summary of issues and challenges raised by chapters in this book and presents implications for teaching and assessing students.

The seventeen chapters in this volume by no means exhaust all issues related to the development of statistical literacy, reasoning, and thinking. Yet, taken as a whole, the chapters constitute a rich resource summarizing current research, theory, and practical suggestions related to these topics. We hope that this volume will contribute to and stimulate the scholarly discourse within the statistics education community, and that in coming years additional publications will more closely examine the many issues and challenges raised.

A project of this magnitude would have been impossible without the help of numerous individuals and organizations. First and most importantly, we would like to thank our many contributors, who remained focused on the goal of sharing their experiences and insights with the educational community while enduring multiple review cycles and editing demands. Their enthusiasm, support, and friendship are valuable to us and have made this long process easier to complete.

Many thanks go to Kibbutz Be'eri (Israel), the University of New England (Australia), and the University of Nebraska–Lincoln (USA) for hosting and supporting the three SRTL Research Forums in 1999, 2001, and 2003. These meetings, which we co-chaired, informed our work as well as the writings of some of many contributors to this volume. In addition, numerous organizations and institutions helped sponsor these forums: the University of Minnesota (USA), the Weizmann Institute of Science (Israel), the International Association on Statistics Education (IASE), the American Statistical Association (ASA) Section on Statistics Education, and Vanderbilt University. This funding has been pivotal in enabling us to sustain our extended effort through the years it took to complete this project.

We are grateful to Kluwer Academic Publishers for providing a publishing venue for this book, and to Michel Lokhorst, the humanities and social sciences publishing editor, who skillfully managed the publication on their behalf. We appreciate the support received from the University of Minnesota (USA) and the University of Haifa (Israel) for copyediting and formatting this volume. We are especially grateful for the contributions of our copy editor, Christianne Thillen, as well as Ann Ooms and Michelle Everson, who under a tight production schedule diligently and ably worked to prepare this book for publication.

Lastly, many thanks go to our spouses, Hava Ben-Zvi and Michael Luxenberg, and to our children—Noa, Nir, Dagan, and Michal Ben-Zvi, and Harlan and Rebecca Luxenberg. They have been our primary sources of energy and support.

Dani Ben-Zvi[1] and Joan Garfield[2]
University of Haifa, Israel[1] and University of Minnesota, USA[2]

PART I

INTRODUCTION TO STATISTICAL LITERACY, REASONING, AND THINKING

Chapter 1

STATISTICAL LITERACY, REASONING, AND THINKING: GOALS, DEFINITIONS, AND CHALLENGES

Dani Ben-Zvi[1] and Joan Garfield[2]
University of Haifa, Israel[1]; University of Minnesota, USA[2]

INTRODUCTION

Over the past decade there has been an increasingly strong call for statistics education to focus more on statistical literacy, reasoning, and thinking. One of the main arguments presented is that traditional approaches to teaching statistics focus on skills, procedures, and computations, which do not lead students to reason or think statistically. This book explores the challenge posed to educators at all levels—how to develop the desired learning goals for students by focusing on current research studies that examine the nature and development of statistical literacy, reasoning, and thinking. We begin this introductory chapter with an overview of the reform movement in statistics education that has led to the focus on these learning outcomes. Next, we offer some preliminary definitions and distinctions for these often poorly defined and overlapping terms. We then describe some of the unique issues addressed by each chapter and conclude with some summary comments and implications.

THE GROWING IMPORTANCE OF STATISTICS IN TODAY'S WORLD

Quantitative information is everywhere, and statistics are increasingly presented as a way to add credibility to advertisements, arguments, or advice. Being able to properly evaluate evidence (data) and claims based on data is an important skill that all students should learn as part of their educational programs. The study of statistics provides tools that informed citizens need in order to react intelligently to quantitative information in the world around them. Yet many research studies indicate that adults in mainstream society cannot think statistically about important issues that affect their lives.

D. Ben-Zvi and J. Garfield (eds.),
The Challenge of Developing Statistical Literacy, Reasoning and Thinking, 3–15.

As former president of the American Statistical Association, David Moore (1990) wrote, "Statistics has some claim to being a fundamental method of inquiry, a general way of thinking that is more important than any of the specific techniques that make up the discipline" (p. 134). It is not surprising, given the importance of statistics, that there has been an increase in the amount of statistical content included in the elementary and secondary mathematics curriculum (NCTM, 2000) and an ever-increasing number of introductory statistics courses taught at the college level.

THE CHALLENGE OF TEACHING STATISTICS

Despite the increasing need for statistics instruction, historically statistics education has been viewed by many students as difficult and unpleasant to learn, and by many instructors as frustrating and unrewarding to teach. As more and more students enroll in introductory statistics courses, instructors are faced with many challenges in helping these students succeed in the course and learn statistics. Some of these challenges include

- Many statistical ideas and rules are complex, difficult, and/or counterintuitive. It is difficult to motivate students to engage in the hard work of learning statistics.
- Many students have difficulty with the underlying mathematics (such as fractions, decimals, algebraic formulas), and that interferes with learning the related statistical content.
- The context in many statistical problems may mislead the students, causing them to rely on their experiences and often faulty intuitions to produce an answer, rather than select an appropriate statistical procedure.
- Students equate statistics with mathematics and expect the focus to be on numbers, computations, formulas, and one right answer. They are uncomfortable with the messiness of data, the different possible interpretations based on different assumptions, and the extensive use of writing and communication skills.

Amidst the challenges of dealing with students' poor mathematics skills, low motivation to learn a difficult subject, expectations about what the course should be, and reliance on faulty intuitions and misconceptions, many instructors strive to enable students to develop statistical literacy, reasoning, and thinking. There appears to be a consensus that these are the most important goals for students enrolled in statistics classes, and that these goals are not currently being achieved. The dissatisfaction with students' ability to think and reason statistically, even after formally studying statistics at the college and graduate level, has led to a reexamination of the field of statistics.

EFFORTS TO CHANGE THE TEACHING OF STATISTICS

Today's leading statisticians see statistics as a distinct discipline, and one that is separate from mathematics (see Chapter 4). Some suggest that statistics should in fact be considered one of the liberal arts (e.g., Moore, 1998). The liberal arts image emphasizes that statistics involves distinctive and powerful ways of thinking: "Statistics is a general intellectual method that applies wherever data, variation, and chance appear. It is a fundamental method because data, variation, and chance are omnipresent in modern life. It is an independent discipline with its own core ideas rather than, for example, a branch of mathematics" (Moore, 1998, p. 1254).

As the discipline has evolved and become more distinct, changes have been called for in the teaching of statistics. Dissatisfaction with the introductory college course has led to a reform movement that includes focusing statistics instruction more on data and less on theory (Cobb, 1992). Moore (1997) describes the reform in terms of changes in content (more data analysis, less probability), pedagogy (fewer lectures, more active learning), and technology (for data analysis and simulations).

At the elementary and secondary level, there is an effort to help students develop an understanding and familiarity with data analysis (see Chapter 6) rather than teaching them a set of separate skills and procedures. New K–12 curricular programs set ambitious goals for statistics education, including developing students' statistical reasoning and understanding (e.g., Australia—Australian Education Council, 1991, 1994; England—Department for Education and Employment, 1999; New Zealand—Ministry of Education, 1992; USA—National Council of Teachers for Mathematics, 2000; and Project 2061's Benchmarks for Science Literacy, American Association for the Advancement of Science, 1993).

Several factors have led to these current efforts to change the teaching of statistics at all educational levels. These factors include

- Changes in the field of statistics, including new techniques of data exploration
- Changes and increases in the use of technology in the practice of statistics, and its growing availability in schools and at home
- Increased awareness of students' inability to think or reason statistically, despite good performance in statistics courses
- Concerns about the preparation of teachers of statistics at the K–12 and college level, many of whom have never studied applied statistics nor engaged in data analysis activities.

Many recommendations have been given for how statistics courses should be taught, as part of the general reform movement. Some of these recommendations are as follows:

- Incorporate more data and concepts.
- Rely heavily on real (not merely realistic) data.
- Focus on developing statistical literacy, reasoning, and thinking.

- Wherever possible, automate computations and graphics by relying on technological tools.
- Foster active learning, through various alternatives to lecturing.
- Encourage a broader range of attitudes, including appreciation of the power of statistical processes, chance, randomness, and investigative rigor, and a propensity to become a critical evaluator of statistical claims.
- Use alternative assessment methods to better understand and document student learning.

There appears to have been some impact on teaching practices from these recommendations at the college level (Garfield, Hogg, Schau, & Whittinghill, 2002). However, despite reform efforts, many statistics courses still teach the same progression of content and emphasize the development of skills and procedures. Although students and instructors appear to be happier with reformed courses, many students still leave the course perceiving statistics as a set of tools and techniques that are soon forgotten. Pfannkuch and Wild (Chapter 2) discuss how current methods of teaching have often focused on the development of skills and have failed to instill the ability to think statistically.

STATISTICAL LITERACY, REASONING, AND THINKING: DEFINITIONS AND DISTINCTIONS

It is apparent, when reading articles about recommendations to reform the teaching of statistics, that there are no consistent definitions for the often stated learning goals of literacy, reasoning, and thinking. Statistical literacy is used interchangeably with quantitative literacy, while statistical thinking and reasoning are used to define the same capabilities. This confusion of terms was especially evident at the Fifth International Conference on Teaching Statistics, held in Singapore in 1998. It became apparent that when statistics educators or researchers talk about or assess statistical reasoning, thinking, or literacy, they may all be using different definitions and understandings of these cognitive processes.

The similarities and differences among these processes are important to consider when formulating learning goals for students, designing instructional activities, and evaluating learning by using appropriate assessment instruments. A small, focused conference consisting of researchers interested in these topics appeared to be an important next step in clarifying the issues, connecting researchers and their studies, and generating some common definitions, goals, and assessment procedures. The first International Research Forum on Statistical Reasoning, Thinking, and Literacy (SRTL-1) was held in Israel in 1999 to address these needs. At this first conference some preliminary definitions were presented and discussed. A second forum (SRTL-2) was held in 2001 in Australia, with a focus on different types of statistical reasoning. Many of the papers from these first two forums have led to chapters in this book. The forums continue to be offered every two years (SRTL-3 in USA, 2003) as interest and research in this area steadily increase.

Although no formal agreement has been made regarding the definitions and distinctions of statistical literacy, reasoning, and thinking, the following list summarizes our current thoughts (Garfield, delMas, & Chance, 2003):

- ***Statistical literacy*** includes basic and important skills that may be used in understanding statistical information or research results. These skills include being able to organize data, construct and display tables, and work with different representations of data. Statistical literacy also includes an understanding of concepts, vocabulary, and symbols, and includes an understanding of probability as a measure of uncertainty.
- ***Statistical reasoning*** may be defined as the way people reason with statistical ideas and make sense of statistical information. This involves making interpretations based on sets of data, representations of data, or statistical summaries of data. Statistical reasoning may involve connecting one concept to another (e.g., center and spread), or it may combine ideas about data and chance. Reasoning means understanding and being able to explain statistical processes and being able to fully interpret statistical results.
- ***Statistical thinking*** involves an understanding of why and how statistical investigations are conducted and the "big ideas" that underlie statistical investigations. These ideas include the omnipresent nature of variation and when and how to use appropriate methods of data analysis such as numerical summaries and visual displays of data. Statistical thinking involves an understanding of the nature of sampling, how we make inferences from samples to populations, and why designed experiments are needed in order to establish causation. It includes an understanding of how models are used to simulate random phenomena, how data are produced to estimate probabilities, and how, when, and why existing inferential tools can be used to aid an investigative process. Statistical thinking also includes being able to understand and utilize the context of a problem in forming investigations and drawing conclusions, and recognizing and understanding the entire process (from question posing to data collection to choosing analyses to testing assumptions, etc.). Finally, statistical thinkers are able to critique and evaluate results of a problem solved or a statistical study.

For more discussion of these definitions and distinction, see papers by Chance (2002), delMas (2002), Garfield (2002), Rumsey (2002), and Chapters 2 through 4 in this book.

RATIONALE AND GOALS FOR THIS BOOK

With the increasing attention given to the need to develop students' statistical literacy, reasoning, and thinking at all levels, it has become apparent that these educational outcomes were not being adequately addressed in the research literature and, therefore, not used as the foundation for curricular programs. In fact, research studies on statistical reasoning are still evolving, and are just beginning to suggest ways to help students develop these outcomes.

Our goal in creating this book is to provide a useful resource for educators and researchers interested in helping students at all educational levels to develop statistical literacy, statistical reasoning, and statistical thinking. Given the increased attention being paid worldwide to the need for statistically literate citizens, the broad inclusion of statistics in the K–12 mathematics curriculum, the increasing numbers of students taking statistics at the secondary level (e.g., Advanced Placement Statistics courses in high school in the USA), and the increasing numbers of students required to take introductory statistics courses in postsecondary programs, it is crucial that the cutting-edge research being conducted on teaching and learning statistics be collected and disseminated along with specific pedagogical implications.

This book offers a synthesis of an emerging field of study, while at the same time responding to clear practical needs in the following ways:

- It establishes a research base for statistics education by focusing on and distinguishing between different outcomes of statistics instruction.
- It raises awareness of unique issues related to teaching and learning statistics, and it distinguishes statistical literacy, reasoning, and thinking from both general and mathematical literacy, reasoning, and thinking.
- It provides a bridge between educational research and practice, by offering research-based guidelines and suggestions to educators and researchers.

Although the word *statistics* is often used to represent both probability and statistical analysis, the authors and editors of this book focus on reasoning and thinking exclusively on the statistical analysis area, rather than on probability. Although statistics as a discipline uses mathematics and probability, probability is actually a field of mathematics. Since most of the early work in statistics education focused on the teaching and learning of probability, we wanted to move away and look at how students come to reason and think about data and data analysis. However, because the two subjects are so interrelated, several chapters mention issues related to learning probability as they relate to the focus of a particular chapter.

AUDIENCE FOR THIS BOOK

This book was designed to appeal to a diverse group of readers. The primary audience for this book is current or future researchers in statistics education (e.g., graduate students). However, we encourage others who do not identify themselves as researchers to read the chapters in this book as a way to understand the current issues and challenges in teaching and learning statistics. By asking authors to specifically address implications for teaching and assessing students, we hope that teachers of students at all levels will find the research results directly applicable to working with students.

SUGGESTED WAYS TO USE THIS BOOK

Given the different audiences for this book, we suggest several different ways to use this book for researchers, teachers, curriculum writers, and technology developers.

- **Researchers** Each chapter includes a detailed review of the literature related to a particular topic (e.g., reasoning about variability, statistical literacy, statistics teachers' development), which will be helpful to researchers studying one of these areas. The chapters also provide examples of current research methodologies used in this area, and present implications for teaching practice as well as suggestions for future research studies. By providing cutting-edge research on statistical literacy, reasoning, and thinking, the book as a whole outlines the state of the art for the statistics education research community. In addition, the contributing authors may be regarded as useful human resources for researchers who are interested in pursuing studies in these areas.
- **Curriculum writers** By reading this book, people designing statistics instructional activities and curricula may learn about current research results in statistics education. Curriculum development involves tightly integrated cycles of reviewing related research, instructional design, and analysis of students' learning, which all feed back to inform the revision of the design. Many chapters in this book also give recommendations for appropriate ways to assess learning outcomes.
- **Technology** Many chapters in this book offer discussion on the role of technology in developing statistical reasoning. Types of technologies used are presented and assessed in relation to their impact on students' reasoning.

Given the different uses just listed, we believe that this book can be used in a variety of graduate courses. Such courses include those preparing mathematics teachers at the K–12 level; courses preparing teachers of statistics at the high secondary and tertiary level; and research seminars in mathematics, statistics education, or psychology.

We advise readers focused on students at one level (e.g., secondary) not to skip over chapters describing students at other levels. We are convinced that students who are introduced to statistical ideas and procedures learn much the same material and concepts (e.g., creating graphical displays of data, describing the center and dispersion of data, inference from data, etc.) regardless of their grade level. Furthermore, reasoning processes develop along extended periods of time, beginning at early encounters with data in elementary grades and continuing through high school and postsecondary education. Therefore, we believe that discussions of reasoning issues couched in the reality of one age group will be of interest to those working with students of other ages and abilities.

OVERVIEW OF CHAPTERS

All of the chapters in this book discuss issues pertaining to statistical literacy, reasoning, or thinking. Some chapters focus on general topics (e.g., statistical literacy) while others focus on the context of a specific educational level or setting (e.g., teaching middle school students to reason about distribution). Whenever possible, the chapter authors outline challenges facing educators, statisticians, and other stakeholders. The chapters present many examples (or references to resources) of activities, data sets, and assessment tasks suitable for a range of instructional levels. This emphasis of connection to practice is a result of our strong belief that researchers are responsible for translating their findings to practical settings.

All the chapters that focus on a particular type of student or teacher statistical reasoning (Chapters 6 through 15) follow a unified and familiar structure to facilitate their effective use by the readers. These chapters typically start with a section introducing the key area of reasoning explored in the chapter. This is followed by clear and informative descriptions of the *problem* (a description of the type of reasoning studied, why it is important, and how this type of reasoning fits into the curriculum); *literature and background* (prior and related work and relevant theoretical background); *methods* (the subjects, methods used, data gathered, and activities or interventions used); *analysis and results* (description of how the data were analyzed, and the results and findings of the study); and *discussion* (lessons learned from the study, new questions raised, limitations found). Finally, in the *implications* section, each chapter highlights key practical implications related to teaching and assessing students as well as implications for research.

The chapters have been grouped into three parts, each of which is summarized here.

Part I. Introduction to Statistical Literacy, Reasoning, and Thinking (Chapters 2 through 5)

The first part of this book is a comprehensive overview of the three interrelated but distinct cognitive processes (or learning outcomes) of statistical literacy, reasoning, and thinking from historical, psychological, and educational perspectives. This part is therefore the basis upon which the individual studies in subsequent parts are built.

In the first chapter of this part (Chapter 2), Pfannkuch and Wild present their paradigm on statistical thinking (part of their four-dimensional framework for statistical thinking in empirical enquiry; Wild & Pfannkuch, 1999). The authors identify five types of thinking, considered to be fundamental to statistics. They follow the origins of statistical thinking through to an explication of what is currently understood to be statistical thinking. They begin their historical exploration with the early developers of statistics; move on to more recent contributions from epidemiology, psychology, and quality management; and conclude with a discussion of recent writings of statistics education researchers and statisticians influential in the movement of pedagogy from methods toward thinking.

Gal proposes in Chapter 3 a conceptualization of statistical literacy and its main components. Statistical literacy is described as a key ability expected of citizens in information-laden societies, an expected outcome of schooling, and a necessary component of adults' numeracy and literacy. Statistical literacy is portrayed as the ability to interpret, critically evaluate, and communicate about statistical information and messages. Gal suggests that statistically literate behavior is predicated on the joint activation of both a *knowledge* component (comprised of five cognitive elements: literacy skills, statistical knowledge, mathematical knowledge, context knowledge, and critical questions) and a *dispositional* component (comprised of two elements: critical stance, and beliefs and attitudes).

The focus of delMas's chapter (Chapter 4) is on the nature of mathematical and statistical reasoning. The author first outlines the general nature of human reasoning, which he follows with an account of mathematical reasoning as described by mathematicians along with recommendations by mathematics educators regarding educational experiences to improve mathematical reasoning. He reviews the literature on statistical reasoning and uses findings from the general literature on reasoning to identify areas of statistical reasoning that students find most challenging. Finally, he compares and contrasts statistical reasoning and mathematical reasoning.

The last chapter in this part (Chapter 5) is a joint work by Jones, Langrall, Mooney, and Thornton that examines cognitive models of development in statistical reasoning and the role they can play in statistical education. The authors consider models of development from a psychological perspective, and then describe how models of statistical reasoning have evolved historically from models of development in probability. The authors describe and analyze comprehensive models of cognitive development that deal with multiple processes in statistical reasoning as well as models of cognitive development that characterize students' statistical reasoning as they deal with specific areas of statistics and data exploration. The authors suggest that school students' statistical reasoning passes through a number of hierarchical levels and cycles.

Part II. Studies of Statistical Reasoning (Chapters 6 through 13)

The chapters in this part focus on how students reason about specific areas of statistics. The topics of these chapters include data analysis, distributions, measures of center, variation, covariation, normal distribution, samples, and sampling distributions. These studies represent the current efforts in the statistics education community to focus statistical instruction and research on the big ideas of statistics (Chapter 17) and on developing students' statistical reasoning at all levels of education.

In the first chapter of this part (Chapter 6), Ben-Zvi describes and analyzes the ways in which middle school students begin to reason about data and come to understand exploratory data analysis (EDA). He describes the process of developing reasoning about data while learning skills, procedures, and concepts. In addition, the author observes the students as they begin to adopt and exercise some of the habits and points of view that are associated with statistical thinking. Ben-Zvi offers two case studies focusing on the development of a global view of data and data representations, and on design of a meaningful EDA learning environment that promotes statistical reasoning about data analysis. In light of the analysis, the author proposes a description of what it may mean to learn to reason about data analysis.

Bakker and Gravemeijer explore (Chapter 7) how informal reasoning about distribution can be developed in a technological learning environment. They describe the development of reasoning about distribution in seventh-grade classes in three stages as students reason about different representations. The authors show how specially designed software tools, students' created graphs, and prediction tasks supported the learning of different aspects of distribution. In this process, several students came to reason about the shape of a distribution using the term *bump* along with statistical notions such as outliers and sample size.

Chapter 8 presents an article by Konold and Pollatsek originally published in a research journal; therefore, it does not follow the same format as the other chapters in this part. Their chapter offers a conceptualization of averages as a stable feature of a noisy process. To explore the challenges of learning to think about data as signal and noise, the authors examine that metaphor in the context of three different types of statistical processes. For each process, they evaluate the conceptual difficulty of regarding data from that process as a combination of signal and noise. The authors contrast this interpretation of averages with various other interpretations of averages (e.g., summaries of groups of values) that are frequently encountered in curriculum materials. They offer several recommendations about how to develop and extend the idea of central tendency as well as possible directions for research on student thinking and learning.

Understanding the nature of variability and its omnipresence is a fundamental component of statistical reasoning. In Chapter 9, Reading and Shaughnessy bring together findings from a number of different studies, conducted in three different countries, designed to investigate students' conceptions of variability. The focus of the chapter is on details of one recent study that investigates reasoning about variation in a sampling situation for students aged 9 to 18.

In Chapter 10, Moritz investigates three skills of reasoning about covariation: (a) speculative data generation, demonstrated by drawing a graph to represent a verbal statement of covariation; (b) verbal graph interpretation, demonstrated by describing a scatterplot in a verbal statement and by judging a given statement; and (c) numerical graph interpretation, demonstrated by reading a value and interpolating a value. The authors describe survey responses from students in grades 3, 5, 7, and 9 in four levels of reasoning about covariation.

Batanero, Tauber, and Sánchez present (Chapter 11) the results of a study on students' learning of the normal distribution in a computer-assisted, university-level introductory course. The authors suggest a classification of different aspects of students' correct and incorrect reasoning about the normal distribution as well as giving examples of students' reasoning in the different categories.

Chapter 12, written by Watson, extends previous research on students' reasoning about samples and sampling by considering longitudinal interviews with students 3 or 4 years after they first discussed their understanding of what a sample was, how samples should be collected, and the representing power of a sample based on its size. Of the six categories of response observed at the time of the initial interviews, all were confirmed after 3 or 4 years, and one additional preliminary level was observed.

Reasoning about sampling distributions is the focus of Chance, delMas, and Garfield in the last chapter of this part (Chapter 13). In this chapter, the authors present a series of research studies focused on the difficulties students experience when learning about sampling distributions. In particular, the authors trace the 7-year history of an ongoing collaborative classroom-based research project investigating the impact of students' interaction with computer software tools to improve their reasoning about sampling distributions. The authors describe the complexities involved in building a deep understanding of sampling distributions, and formulate models to explain the development of students' reasoning.

Part III. Curricular, Instructional, and Research Issues (Chapters 14 through 16)

The third and final part of this book deals with important educational issues related to the development of students' statistical reasoning: (a) teachers' knowledge and understanding of statistics, and (b) instructional design issues.

Mickelson and Heaton (Chapter 14) explore the complexity of teaching and learning statistics, and offer insight into the role and interplay of teachers' statistical knowledge and context. Their study presents an analysis of one third-grade teacher's statistical reasoning about data and distribution in the context of classroom-based statistical investigation. In this context, the teacher's statistical reasoning plays a central role in the planning and orchestration of the class investigation.

Makar and Confrey also discuss (Chapter 15) teachers' statistical reasoning. They focus on the statistical reasoning about comparing two distributions of four secondary teachers addressing the research question: "How do you decide whether two groups are different?" The study was conducted at the end of a 6-month professional development sequence designed to assist secondary teachers in making sense of their students' results on a state-mandated academic test. The authors provide qualitative and quantitative analyses to examine the teachers' reasoning.

In Chapter 16, Cobb and McClain propose design principles for developing statistical reasoning about data in the contexts of EDA and data generation in elementary school. They present a short overview of a classroom design experiment, and then frame it as a paradigm case in which to tease out design principles addressing five aspects of the classroom environment that proved critical in supporting the students' statistical learning: The focus on central statistical ideas, the instructional activities, the classroom activity structure, the computer-based tools the students used, and the classroom discourse.

Summary and Implications
(Chapter 17)

In the closing chapter (Chapter 17) the editors summarize issues, challenges, and implications for teaching and assessing students emerging from the collection of studies in this book. We begin with some comments on statistics education as an emerging research area, and then concentrate on the need to focus research, instruction, and assessment on the big ideas of statistics. We address the role of technology in developing statistical reasoning as well as the diversity of various statistics learners (e.g., students at different educational levels as well as their teachers). Next we present a summary of research methodologies used to study statistical reasoning, along with comments on the extensive use of qualitative methods and the lack of traditional experimental designs. Finally, we consider some implications for teaching and assessing students and suggest future research directions.

We hope that the articulated, coherent body of knowledge on statistical literacy, reasoning, and thinking presented in this book will contribute to the pedagogical effectiveness of statistics teachers and educators at all levels; to the expansion of research studies on statistical literacy, reasoning and thinking; and to growth of the statistics education community.

REFERENCES

American Association for the Advancement of Science (Project 2061). (1993). *Benchmarks for science literacy.* New York: Oxford University Press.

Australian Education Council (1991). *A national statement on mathematics for Australian schools.* Carlton, Vic.: Author.

Australian Education Council (1994). *Mathematics—a curriculum profile for Australian schools.* Carlton, Vic.: Curriculum Corporation.

Chance, B. L. (2002). Components of statistical thinking and implications for instruction and assessment. *Journal of Statistics Education* [Online], *10*(3). Retrieved June 24, 2003, from http://www.amstat.org/publications/jse/

Cobb, G. W. (1992). Report of the joint ASA/MAA committee on undergraduate statistics. In *American Statistical Association 1992 Proceedings of the Section on Statistical Education*, (pp. 281–283). Alexandria, VA: ASA.

delMas, R. C. (2002). Statistical literacy, reasoning, and learning: A commentary. *Journal of Statistics Education* [Online], *10*(3). Retrieved June 24, 2003, from http://www.amstat.org/publications/jse/

Department for Education and Employment (1999). *Mathematics: The national curriculum for England.* London: Author and Qualifications and Curriculum Authority.

Garfield, J. (2002). The challenge of developing statistical reasoning. *Journal of Statistics Education* [Online], *10*(3). Retrieved June 24, 2003, from http://www.amstat.org/publications/jse/

Garfield, J., delMas, R., & Chance, B. (2003). *Web-based assessment resource tools for improving statistical thinking.* Paper presented at the annual meeting of the American Educational Research Association, Chicago.

Garfield, J., Hogg, B., Schau, C., & Whittinghill, D. (2002). First courses in statistical science: The status of educational reform efforts. *Journal of Statistics Education* [Online], *10*(2). Retrieved June 24, 2003, from http://www.amstat.org/publications/jse/

Ministry of Education (1992). *Mathematics in the New Zealand curriculum.* Wellington, NZ: Author.

Moore, D. S. (1990). Uncertainty. In Lynn Steen (Ed.), *On the shoulders of giants: A new approach to numeracy* (pp. 95–137). National Academy of Sciences.

Moore, D. (1997). New pedagogy and new content: The case of statistics. *International Statistical Review, 65*, 123–137.

Moore, D. (1998). Statistics among the liberal arts. *Journal of the American Statistical Association, 93*, 1253–1259.

National Council of Teachers of Mathematics (2000). *Principles and standards for school mathematics.* Reston, VA: Author.

Rumsey, D. J. (2002). Statistical literacy as a goal for introductory statistics courses. *Journal of Statistics Education* [Online], *10*(3). Retrieved June 24, 2003, from http://www.amstat.org/publications/jse/

Wild, C. J., & Pfannkuch, M. (1999). Statistical thinking in empirical enquiry. *International Statistical Review 67*(3), 223–265.

Chapter 2

TOWARDS AN UNDERSTANDING OF STATISTICAL THINKING

Maxine Pfannkuch and Chris Wild
The University of Auckland, New Zealand

INTRODUCTION

There has been an increasingly strong call from practicing statisticians for statistical education to focus more on statistical thinking (e.g., Bailar, 1988; Snee, 1993; Moore, 1998). They maintain that the traditional approach of teaching, which has focused on the development of skills, has failed to produce an ability to think statistically: "Typically people learn methods, but not how to apply them or how to interpret the results" (Mallows, 1998, p. 2).

Solutions offered for changing this situation include employing a greater variety of learning methods at undergraduate level and compelling students to experience statistical thinking by dealing with real-world problems and issues. A major obstacle, as Bailar (1988) points out, is teacher inexperience. We believe this is greatly compounded by the lack of an articulated, coherent body of knowledge on statistical thinking that limits the pedagogical effectiveness even of teachers who are experienced statisticians. Mallows (1998) based his 1997 Fisher Memorial lecture on the need for effort to be put into developing a theory for understanding how to think about applied statistics, since the enunciation of these principles would be useful for teaching.

This chapter focuses on thinking in statistics rather than probability. Although statistics as a discipline uses mathematics and probability, as Moore (1992b) states, probability is a field of mathematics, whereas statistics is not. Statistics did not originate within mathematics. It is a unified logic of empirical science that has largely developed as a new discipline since the beginning of the 20th century. We will follow the origins of statistical thinking through to an explication of what we currently understand to be statistical thinking from the writings of statisticians and statistics educationists.

D. Ben-Zvi and J. Garfield (eds.),
The Challenge of Developing Statistical Literacy, Reasoning and Thinking, 17–46.

Model for Interpretation of Literature

We will be interpreting the literature from our own paradigm (Figure 1) on statistical thinking (Wild & Pfannkuch, 1999). The model was developed by interviewing statisticians and tertiary students about statistical projects they had been involved in; interviewing tertiary students as they performed statistical tasks; and analyzing the literature below (see "Discussion and Summary" for more detail). In our model we identified the types of thinking we consider to be fundamental to statistics (Figure 1b). These five fundamental thinking types are now elaborated upon.

Recognition of the Need for Data

The foundations of statistical enquiry rest on the assumption that many real situations cannot be judged without the gathering and analysis of properly collected data. Anecdotal evidence or one's own experience may be unreliable and misleading for judgments and decision making. Therefore, properly collected data are considered a prime requirement for reliable judgments about real situations.

Transnumeration

For this type of thinking we coined the word *transnumeration,* which means "changing representations to engender understanding." Transnumeration occurs in three specific instances. If one thinks of the real system and statistical system from a modeling perspective, then transnumeration thinking occurs when (1) measures that "capture" qualities or characteristics of the real situation are found; (2) the data that have been collected are transformed from raw data into multiple graphical representations, statistical summaries, and so forth, in a search to obtain meaning from the data; and (3) the meaning from the data, the judgment, has to be communicated in a form that can be understood in terms of the real situation by others.

Consideration of Variation

Adequate data collection and the making of sound judgments from data require an understanding of how variation arises and is transmitted through data, and the uncertainty caused by unexplained variation. It is a type of thinking that starts from noticing variation in a real situation, and then influences the strategies we adopt in the design and data management stages when, for example, we attempt to eliminate or reduce known sources of variability. It further continues in the analysis and conclusion stages through determining how we act in the presence of variation, which may be to either ignore, plan for, or control variation. Applied statistics is about making predictions, seeking explanations, finding causes, and learning in the context sphere. Therefore we will be looking for and characterizing patterns in the variation, and trying to understand these in terms of the context in an attempt to solve the problem. Consideration of the effects of variation influences all thinking through every stage of the investigative cycle.

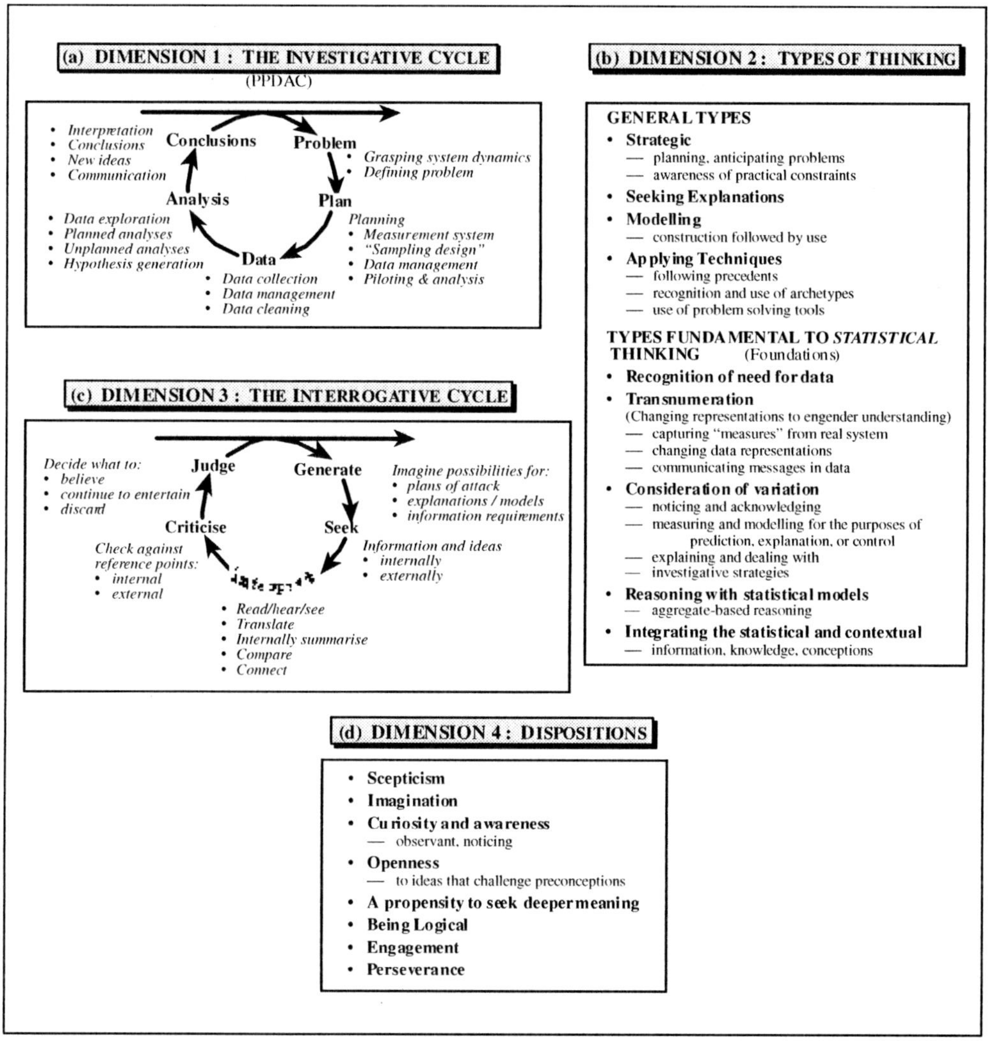

Figure 1. A four-dimensional framework for statistical thinking in empirical enquiry. (From "Statistical Thinking in Empirical Enquiry," by C. J. Wild and M. Pfannkuch, 1999, International Statistical Review, 67, p. 226. Copyright 1999 by International Statistical Institute. Reprinted with permission.)

Reasoning with Statistical Models

The predominant statistical models are those developed for the analysis of data. When we talk about "statistical models," most people interpret the term as meaning, for example, regression models or time-series models. Even much simpler tools such as statistical graphs can be thought of as statistical models since they are statistical ways of representing and thinking about reality. When we use statistical models to reason with, the focus is more on aggregate-based rather than individual-based reasoning, although both types are used. Proper individual-based reasoning concentrates on single data points with little attempt to relate them to the wider data set, whereas aggregate-based reasoning is concerned with patterns and relationships in the data set as a whole. A dialogue is set up between the data and statistical models. The models may allow us to find patterns in the data, find group propensities, and see variation about these patterns via the idea of distribution. The models enable us to summarize data in multiple ways depending on the nature of the data. For example, graphs, centers, spreads, clusters, outliers, residuals, confidence intervals, and *p*-values are read, interpreted, and reasoned with in an attempt to find evidence on which to base a judgment. Different types of statistical models based on the idea of "process" are starting to be used for reasoning with in the other stages of the investigative cycle (e.g., see Joiner, 1994; Wild & Pfannkuch, 1999, Section 4).

Integrating the Statistical and Contextual

Although the above types of thinking are linked to contextual knowledge, the integration of statistical knowledge and contextual knowledge is an identifiable fundamental element of statistical thinking. The statistical model must capture elements of the real situation; thus the resultant data will carry their own literature base (Cobb & Moore, 1997), or more generally, their own body of context knowledge. Because information about the real situation is contained in the statistical summaries, a synthesis of statistical and contextual knowledge must operate to draw out what can be learned from the data about the context sphere.

These ideas will be used to analyze and interpret the perspectives of different fields on statistical thinking.

CONTRIBUTIONS FROM DIFFERENT FIELDS

Statistics has been like a tiny settlement taking root and steadily growing into a large, rich country through continual two-way trade with the many neighbors on its borders. Tracing all the contributions from all the fields that have fed and enriched statistics would be an impossibly large undertaking; see, for example, the three volumes of Kotz and Johnson (1992). We will just concentrate on some high points in the development of statistical ways of thinking, and more recently of pedagogy aimed at enhancing statistical thinking (see Scheaffer, 2001). Our account stresses thinking that led to new ways of perceiving a world reality. We do not, for example, discuss how different schools of inference use probability models to draw inferences from data. The big developmental step, as we see it, was to begin *to use* probability models to draw inferences from data.

We begin this section with the early developers of statistics; move on to much more recent contributions from epidemiology, psychology, and quality management; and conclude the section with a discussion of recent writings of statistics education researchers and statisticians influential in the movement of pedagogy from methods toward thinking.

Origins

Statistical thinking permeates the way we operate and function in everyday life. Yet, it remains an enigma as to why even the most basic of the statistical perspectives on the world—namely, reasoning from data—is less than 350 years old (Davis & Hersh, 1986). Many have put forward explanations for the delay. The current theory (Hacking, 1975) is that in the Renaissance two significant shifts in thinking occurred about what was considered to be the nature of knowledge and the nature of evidence. First, the concept of knowledge shifted from an absolute truth toward a knowledge based on opinion, resulting in the thinking shifting toward a probabilistic perspective. This required a skeptical attitude and inductive thinking. Second, the nature of evidence shifted away from the pronouncements of those in authority and toward making inferences from observations, resulting in the thinking shifting toward reasoning from data. Both of these shifts initiated a new paradigm for viewing and learning about the world.

Drawing Inferences from Data

The roots of such statistical thinking can be traced to John Graunt (David, 1962; Kendall, 1970; Greenwood, 1970), who in 1662 published the book *Natural and Political Observations*. Previously, official statistics had lain dormant as stored data. Graunt's new way of thinking is best illustrated with a centuries-old controversy about whether the plague was carried by infection from person to person or carried through infectious air. Most people believed both methods were true. They believed sick people caused the air to be infectious. They also knew that the plague could start at the dock since a ship from overseas brought with it foul and infectious air. The practice and advice were to flee such sources of infection. But when Graunt looked at the number of plague cases, he reasoned (Hacking, 1975, p. 105):

> The contagion of the plagues depends more on the disposition of the air than upon the effluvia from the bodies of men. Which we also prove by the sudden jumpings which the plague hath made, leaping in one week from 118 to 927, and back again from 993 to 258, and from thence again the very next week to 852.

If the plague was passed from one person to another, then these statistics could not be explained; but they could be explained by the infectious air theory. In this graphic example, we see Graunt taking mankind's first steps in making inferences from data. He uses some fundamental statistical thinking, such as noticing and seeking to explain the differences in the numbers using his context knowledge. Graunt also gave the "first reasoned estimate of the population of London" (Hacking, 1975, p. 106) using arithmetical calculations. From knowing the number of births, he inferred the number of women of childbearing age and hence estimated the total number of families and the mean size of a family to produce an estimate of the population. In his time Graunt was regarded by some of his peers as the "Columbus" who discovered how to think and reason with data and hence opened up a new world in which old and new demographic reports could be surveyed.

Similar ways of thinking arose independently in many parts of Western Europe in the same decade. Other pioneers were Petty, King, Halley, Hudde, Huyghens, and Davenant. According to Kendall (1970, p. 46), these political arithmeticians had an inferential approach to data and "thought as we think today" since "they *reasoned* about their data." Their approach was to estimate and predict and then learn from the data, not to describe or collect facts.

Recognition of the Need for Data

Graunt and these other political arithmeticians, besides calculating insurance rates—which involved much discussion among them on producing realistic mortality tables—were also promoting the notion that state policy should be informed by the use of data rather than by the authority of church and nobility (Porter, 1986). In these ideas we see fundamental statistical thinking operating—there is a recognition that data are needed in order to make a judgment on a situation. This notion was not a part of the mainstream consciousness until the late 1800s (Cline Cohen, 1982), when politicians were urged to argue for a policy based on quantitative evidence since "without numbers legislation is ill-informed or haphazard" (Porter, 1986, p. 37).

The Beginnings of Statistical Modeling

Even though the foundations of probability were laid down, by Pascal (1623–1662) and later by Bernoulli (1654–1705) at the same time and in parallel with the foundations of statistics, probability ideas were not incorporated into empirical data or statistical analyses. There appeared to be stumbling blocks in (1) relating urn-device problems to real-world problems; (2) a lack of equiprobability in the real-world problems; (3) the notion that prediction is impossible when there is a multitude of causes; (4) thinking tools such as graphs not being available; and (5) the inevitable time lags in drawing disparate and newly developed strands together into a coherent whole. According to Stigler (1986), the chief conceptual step toward the application of probability to quantitative inference involved the inversion of the probability analyses of Bernoulli and de Moivre (1667–1754).

This ground-breaking inference work of Bayes in 1764 was encouraged by two critical key ideas. The first key idea was not to think in terms of games of chance.

That is, instead of thinking of drawing balls from an urn, Bayes thought in terms of a square table upon which two balls were thrown. This new thinking tool allowed for continuous random quantities to be described as areas and for the problem to assume a symmetric character. The second key idea was from Simpson, who in 1755 had a conceptual breakthrough in an astronomical problem. Rather than calculating the arithmetic mean of the observations, Simpson focused on the errors of the observations (the difference between the recorded observation and the actual position of the body being observed) and assumed a specific hypothesis for the distribution of the measurement errors. These critical changes in thinking opened the door to an applicable quantification of uncertainty. Lightner (1991, p. 628) describes this as a transition phase as "many concepts from probability could not be separated from statistics, for statisticians must consider probabilistic models to infer properties from observed data."

Thus in astronomy and geodesy (surveying) the use of probability to assess uncertainty and make inferences from data employing the mathematical methods of Laplace (1749–1827) and Gauss (1777–1855) such as the normal distribution for measurement errors and the method of least squares became commonplace. At this stage we see the beginning of some more fundamental statistical thinking; there is a movement from reasoning with arithmetic to reasoning with statistical models and to the measuring and modeling of *error*. It is important to note that there was still no concept of variation in nature. This concept and some other major conceptual barriers had to be overcome before this thinking could spread to the social sciences.

Social Data and Reasoning with the Aggregate

At the beginning of the 19th century a new sense of dynamism in society, after the French Revolution, produced a subtle shift in thinking when statistics was seen as a science of the state. The statists, as they were known, conducted surveys of trade, industrial progress, labor, poverty, education, sanitation, and crime (Porter, 1986). "The idea of using statistics for such a purpose—to analyze social conditions and the effectiveness of public policy—is commonplace today, but at that time it was not" (Cohen, 1984, p. 102). Into this milieu a pioneer in social statistics, Quetelet (1796–1874), arrived. Quetelet argued that the foundations of statistics had been established by mathematicians and astronomers. He looked at suicide rates and crime rates and was amazed to find large-scale regularity. Through realizing that general effects in society are produced by general causes and that chance could not influence events when considered collectively, he was able to recast Bernoulli's law of large numbers as a fundamental axiom of social physics. Porter (1986, p. 55) suggested that Quetelet's major contribution was in: "persuading some illustrious successors of the advantage that could be gained in certain cases by turning attention away from the concrete causes of individual phenomena and concentrating instead on the statistical information presented by the larger whole." The effect of Quetelet's findings reverberated. Debates raged about the "free will of man." Politicians and writers such as Buckle and Dickens were impressed; they wrote about these constant statistical laws that seemed to govern the moral and physical condition of people. For instance, the argument was promoted that if a particular individual did not commit a crime, others would be impelled until the annual quota of crime had been reached. Thus this new way of processing information was catalyzing a new awareness of reality and a reevaluation of determinism.

Quetelet's other major contribution occurred in 1844, when he announced that the astronomer's error law, or error curve, also applied to human traits such as height. He viewed the "average man" (his findings about the average man became so well known in his day that the phrase is still part of our language today) as the perfect archetype. All men were designed according to this specification; but because of nutrition, climate, and so forth failed to achieve the average man's measurements (Porter, 1986). He believed, therefore, that such human measurements were indeed errors. Although too many of his data sets revealed evidence of normality, he succeeded in creating a climate of awareness that empirical social observations could be modeled by theoretical distributions. His work provided "evidence" that there appeared to be an averaging of random causes and "that nature could be counted on to obey the laws of probability" (Stigler, 1986, p. 220). Quetelet started to shift the interest within probability from measurement error to variation and began the process by which the "error curve" became a distribution governing variation.

Variation as a Concept

Galton in the late 19th century provided the major conceptual breakthrough (Stigler, 1986) for rationalizing variation in nature to the normal curve. To him the curve stood as a denial of the possibility of inheritance. In other words, why did population variability in height not increase from year to year, since tall parents should have taller children and short parents should have shorter children? His pondering on the size of pears (large, moderate, and small) in a garden and his development of the quincunx as an analogy "demonstrated" that the resulting mixture of approximately normal conditional distributions was itself approximately normal. This empirical theory, coupled with his work on reversion in sweet pea experiments and his study of hereditary stature, eventually led to the theory of regression to the mean. For the first time, statistical thinking had incorporated the notion of variation rather than error.

Debates about the use of statistics in the social sciences continued. An argument promoted was that statistical regularities proved nothing about the causes of things. When Einstein declared in his famous quotation that "God did not play dice," he was stating the viewpoint of the late 19th century that scientific laws were based on causal assumptions and reflected a causal reality. The defense of human freedom inspired a wide-ranging reevaluation of statistical thought. Variation and chance were recognized as fundamental aspects of the world in a way that they had not been before. This acceptance of indeterminism constituted one of the noteworthy intellectual developments of the time. According to Porter (1986, p. 319) the evolvement of statistical thinking from 1662 to 1900 "has been not just to bring out the chance character of certain individual phenomena, but to establish regularities and causal relationships that can be shown to prevail nonetheless."

New Tools and Transnumeration Thinking

The use of abstract, nonrepresentational pictures to show numbers, rather than tables of data, was not thought of until 1750–1800. Statistical graphics such as time-series and scatter plots were invented long after the use of Cartesian coordinates in mathematics. "William Playfair (1759–1823) developed or improved upon nearly all the fundamental graphical designs, seeking to replace conventional tables of numbers with the systematic visual representations of his 'linear arithmetic'" (Tufte, 1983, p. 9). Another pioneer, Florence Nightingale (1820–1910), also developed new graphical representations (Cohen, 1984). The representation of her tables of data into new graph forms, for example, revealed the extent to which deaths in the Crimea War had been preventable. This changing of data representation in order to trigger new understandings from the data or to communicate the messages in the data illustrates some fundamental statistical thinking.

Emergence of a New Discipline

Porter (1986, p. 315) states that "the intellectual character of statistics" had been crystallized by 1900, and that modern statisticians perceived "the history of their field as beginning with Galton, [(1822–1911)] if not Pearson [(Karl Pearson, 1857–1936)]." The emergence of statistical thinking appears to have been based on four main factors. The first factor is a fundamental realization that the analysis of data will give knowledge about a situation. The basis to this factor is recognition that knowledge acquisition can be based on investigation. The second factor is a recognition that mathematical probability models can be used to model and predict group (e.g., human group) behavior. Thus an interplay between the mathematical probability model and the real situation resulted in a shift of thinking to include a nondeterministic view of reality. The third factor is the application of mathematical probability models to a variety of domains, resulting in new ways of thinking, perceiving, and interpreting in the statistics discipline. For example, these new ways of thinking occurred when mathematical error models were used by Quetelet in the social science field, and by Galton in the biological sciences, and consequently became reinterpreted in fundamentally different ways as variation or chance statistical models. The fourth factor is the development of new tools for analysis, arising from the new situations where statistics was being applied. These new tools helped to aid the development of statistical thinking. Statistical thinking appears to have arisen from a context-knowledge base interacting with a statistical-knowledge base, with the resultant synthesis producing new ways of modeling and perceiving the world.

At the beginning of the 20th century people such as Karl Pearson, Ronald A. Fisher (1890–1962), Jerzy Neyman (1894–1981) and Egon Pearson (1885–1980) built the foundations of modern statistics (see Salsburg, 2001). Their particular insights into principles such as randomization in experiments and surveys, coupled with the development of theoretical statistics, promoted new ways of thinking in many fields. In particular, Fisher's work is regarded as providing the conceptual underpinnings not only for the academic discipline of statistics but also for fields such as plant and animal breeding, evolutionary biology, and epidemiology. Krishnan (1997) believes that Fisher's most important contribution to statistics and science was his formulation of the basics of experimental design—randomization, replication, and local control. Consideration of variation (e.g., variation in the growing conditions for plants) is a core element in the thinking behind such experimental design.

Fisher's famous thought experiment on "the lady and the cup of tea," on which he based his discussion on experimental designs, was never undertaken. The idea arose from an actual incident 12 years earlier, when a Dr. Muriel Bristol declined a cup of tea on the grounds that the milk had not been poured in first. Fisher and her fiancé immediately set out to test whether she could tell the difference. Her fiancé declared she was able to prove her case. Box (1978, p. 134), however, thinks that Fisher pondered on questions such as: "How many cups should be used in the test? . . . What should be done about chance variations in the temperature, sweetness and so on? What conclusions could be drawn from a perfect score or from one with one or more errors?" Therefore, Fisher initiated his groundbreaking work by considering questions relevant to designing an experiment for the following situation:

> A lady declares that by tasting a cup of tea made with milk she can discriminate whether the milk or the tea infusion was first added to the cup. We will consider the problem of designing an experiment by means of which this can be asserted. (Fisher, 1935, cited in Box, 1978, p. 135)

Fisher's two main innovations for the design of experiments were the introduction of analysis of variance and randomization. According to Box (1997, p. 102), Fisher elucidated "the underlying theory and provided the statistical methods that research workers urgently needed to deal with the ubiquitous variation encountered in biological experimentation." Fisher also played a pivotal role in the actual use of randomization in controlled agricultural experiments (Fienberg and Tanur, 1996). Randomization was described by Fisher as a method that was necessary for the validity of any test of significance, since it "affords the means, in respect of any particular body of data, of examining the wider hypothesis in which no normality of distribution is implied" (1935; cited in Box, 1978, p. 151). Without randomization, confounding factors would give biased estimates. Fisher's work contributed to the recognition that uncertainty could be captured by quantifiable measures that led to a deeper appreciation and understanding of its nature (Box, 1997). Porter (1986) also observed that Fisher's integration of statistics with experimental design essentially changed the character of statistics by moving it beyond observing patterns in data to demonstrating the existence of causal relationships.

Some Contributions from Epidemiology

Variation and Randomization

In accounts of statistical thinking in medicine, variation is never mentioned; yet it is at the heart of the methodology and the thinking. Perhaps it is because medicine has only recently accepted the quantification and objectification of its practice (Porter, 1995). This awareness of the importance of statistical thinking and methods in epidemiology can largely be attributed to the work of three statisticians—Austin Bradford Hill, Jerome Cornfield, and Richard Doll (Gail, 1996)—during the mid-20th century. They were the main statisticians behind the general acceptance by the medical profession of (1) the randomized comparative clinical trial, starting with Hill's pioneering work with the whooping-cough vaccine in the 1940s; and (2) acceptance of a code of practice for observational studies, through their data analyses on the association between smoking and lung cancer. Before the technique of randomized comparative trials could be applied to humans, however, there were ethical issues to be overcome, as well as a largely innumerate profession (Gail, 1996). Another reason for this recent acceptance of randomized comparative clinical trials is that the statistical methods for comparison were only invented in the 1920s by Fisher (in the context of agricultural experiments). It is noteworthy that what are now common practices and ways of thinking about what constitutes evidence only began to be accepted by the medical profession during the 1960s.

Causal Inference

Fisher was a significant protagonist in the prolonged debate on whether smoking causes lung cancer (Box, 1978). However, his insistence on raising other possible causes for lung cancer—together with Cornfield, Doll, and Hill's careful, logical analyses of data—markedly increased awareness of the importance of statistical thinking in medicine (Gail, 1996). Alternative explanations for the association between lung cancer and smoking suggested by Fisher and others were systematically refuted by Cornfield, Doll, and Hill until there could be no other plausible interpretation of the data. The debate on the association between smoking and lung cancer, which began in 1928, culminated in the 1964 publication of the U.S. Surgeon General's report, a landmark in the setting of standards of evidence for inference of a causal relationship from observational studies.

Thus in epidemiology it was recognized that purely statistical methods applied to observational data cannot prove a causal relationship. Causal significance was therefore based on "expert" judgment utilizing a number of causal criteria such as consistency of association in study after study, strength of association, temporal pattern, and coherence of the causal hypothesis with a large body of evidence (Gail, 1996). It should be noted that whether the study is experimental or observational, the researcher always has the obligation to seek out and evaluate alternative explanations and possible biases before drawing causal inference.

Causal inference is at the heart of epidemiology. Epidemiology laid down the foundations for causal criteria as first enunciated by Hill (1965). According to Porter (1995), these agreed-upon rules and conventions are paramount for trusted communication globally. Thalidomide was originally considered safe according to expert judgment. The resulting disaster led to more criteria being laid down for scientific procedure and quantification of new knowledge. Gail (1996, p. 1) believes that:

> Statistical thinking, data collection and analysis were crucial to understanding the strengths and weaknesses of the scientific evidence ... [and] gave rise to new methodological insights and constructive debate on criteria needed to infer a causal relationship. These ideas form the foundation for much of current epidemiologic practice.

The statistical thinking that would seem to permeate epidemiology is a synthesizing of contextual knowledge with statistical knowledge and the consideration of variation at all stages of the investigative cycle for experimental and observation studies. Statistical thinking in this context is about seeking causes with a knowledge and understanding of variation.

Some Contributions from Psychology

The centrality of variation in statistical thinking was being recognized in experimental design and in observational studies. In psychology in the late 1960s, however, a link was recognized between statistics and how people think in everyday situations.

Recognizing Statistical Thinking as a Way of Perceiving the World

In the early 1970s Kahneman and Tversky began publishing important work on decision making under uncertainty (see Tversky and Kahneman, 1982). They discovered that statistical thinking is extraordinarily difficult for people. These researchers' particular insights transformed the idea of statistical thinking from making inferences from purposefully collected data, to making inferences from everyday data that are not collected for any purpose nor seen as data. To illustrate this concept, the story of how this field was started is related. According to McKean (1985), Kahneman mentioned, in a psychology course to flight instructors, that from research with pigeons there was evidence that reward was a more effective teaching strategy than punishment. The flight instructors disagreed vehemently that this research was applicable to humans. They knew from their experience that if they praised a person for a good maneuver then invariably the next maneuver would be worse, and that if they yelled at the person for a badly executed maneuver then the next one would more than likely be an improvement. At that instant, Kahneman made an insightful connection with Galton's statistical principle of regression to the mean.

We can explain the idea as follows (Figure 2). If you look at a time-series plot of data points independently sampled from a random distribution, say the normal as in the figure, you will see that the observation that follows a fairly small value tends to be larger, and the observation that follows a fairly large value tends to be smaller. It tends to go back, or "regress," toward the mean.

Thus if flight performance was a random process and praise for good performance and censure for poor performance had absolutely no effect at all, flight instructors would tend to have experienced students performing better after censure and worse after praise. They would then come to exactly the same conclusion—that censure was effective and praise was, if anything, counterproductive:

The student pilots, Kahneman explained, were improving their skills so slowly that the difference in performance from one maneuver to the next was largely a matter of luck. Regression dictated that a student who made a perfect three-point landing today would make a bumpier one tomorrow—regardless of praise or blame. But the flight instructors, failing to realize this, had underestimated the effect of reward and overestimated the effect of punishment. (McKean, 1985, p. 25)

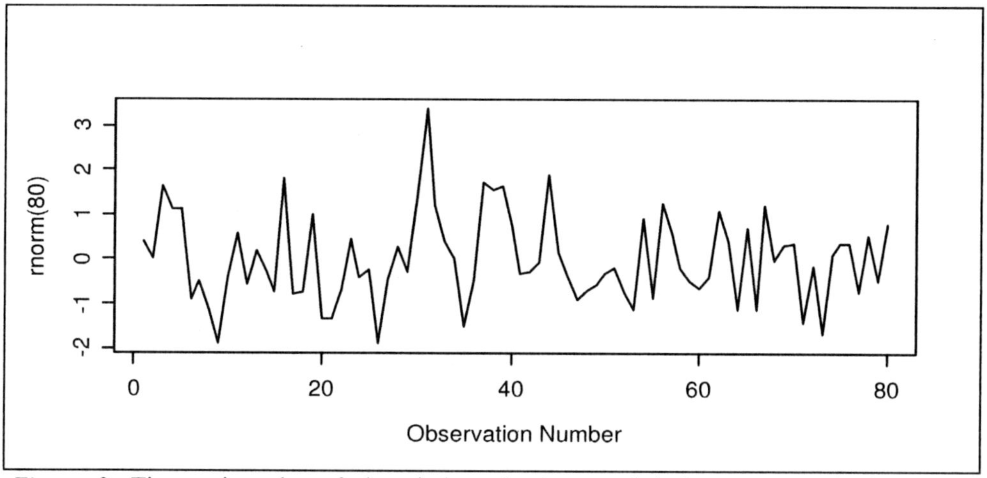

Figure 2. Time-series plot of data independently sampled from a normal distribution (μ=0, σ=1).

This type of statistical thinking we will label as understanding the behavior of variation, though Kahneman and Tversky do not explicitly write in these terms. It requires admitting the possibility of indeterminism. People were mistakenly attributing each change to a cause rather than perceiving the students' performance as a random process with an underlying mean. Kahneman and Tversky became sensitized to seeing regression to the mean everywhere. They developed a long list of phenomena that people have found surprising that can be explained in terms of regression to the mean.

This insight led to the two men thinking of other statistical principles that were counterintuitive. One of these was that people believe that a small sample is a representative sample, or that a small sample should reflect the characteristics of the population (Tversky & Kahneman, 1982). From a variation perspective there is more variation in a small sample than in a large sample. Consequently, people often put too much faith in the results of small samples. But there is an ambivalence here. At other times people severely doubt results from small (i.e., small in proportion to the size of the population) randomly selected samples (Bartholomew, 1995) and do not believe that the sample will reflect the population.

Tversky and Kahneman's work has revealed that statistical thinking is not embedded in how people act and operate in the world, which is not surprising given its youth. In fact the psychologists Falk and Konold (1992, p. 151) believe people must undergo their own 'probabilistic revolution' and shift their perception of the world from a deterministic view to one "in which probabilistic ideas have become central and indispensable." A complementary but very differently expressed view is shared within the quality management field, where there is a belief that peoples' conception of statistical thinking will alter their understanding of reality (Provost & Norman, 1990).

Some Contributions from Quality Management

Statistical thinking is at the forefront of the quality management literature. Snee (1999) believes that the development of statistical thinking will be the next step in the evolution of the statistics discipline, while Provost and Norman (1990, p. 43) state "the 21st century will place even greater demands on society for statistical thinking throughout industry, government, and education." Such strong beliefs about the value of statistical thinking pervade the quality management field, which focuses on systematic approaches to process improvement. At the heart of these approaches is learning from and about processes so that changes can be made to improve them. This has led to a literature and to a large numbers of courses in statistical thinking, many of them concerned with the skill sets required of managers (e.g., Joiner, 1994). What stands out immediately in their definitions of statistical thinking is the role of variation. Process improvement, in large part, consists of controlling and minimizing variation.

Controlling Variation

Hare, Hoerl, Hromi, and Snee (1995) state that statistical thinking has its roots in the work of Shewhart, who in 1925 published a paper about maintaining the quality of a manufactured product. This led to the development of the quality control field, of which Deming was also at the forefront (Shewhart & Deming, 1939). The basis of Shewhart and Deming's work was that there are two sources of variation in a process: special-cause variation and common-cause variation, or chance variation. For quality control the prevailing wisdom for a long time had been to identify, fix, and eliminate the special causes (thus bringing the process to ever-improved levels of statistical stability) and to accept the inherent variability within a process (i.e., the common cause or chance variation). So long as the observations fell within the three-sigma limits, the rule was to leave the process alone. This attitude to variation has been changing due to a climate of continually shifting standards and higher expectations. It is no longer quality control but continuous quality improvement that is the focus of management.

Minimizing Variation

Pyzdek's (1990, p. 104) approach to thinking about variation is summarized as:

- All variation is caused.
- Unexplained variation in a process is a measure of the level of ignorance about the process.
- It is always possible to improve understanding (reduce ignorance) of the process.
- As the causes of process variation are understood and controlled variation will be reduced.

This understanding of variation enables not only the reduction of process variation but also the changing of the average level of the process (Snee, 1999). Thus in quality improvement it is believed that to truly minimize variability, the sources of variation must be identified and eliminated (or at least reduced). The first task, however, is to distinguish common-cause and special-cause variation. It is recognized that variation from special causes should be investigated at once, while variation from common causes should be reduced via structural changes to the system and long-term management programs. The method for dealing with common causes is to investigate cause and effect relationships using such tools as cause and effect diagrams, stratification analysis, pareto analysis, designed experiments, pattern analysis, and modeling procedures. In-depth knowledge of the process is essential. Patterns in the data must be looked for, and depending on the question asked, data must be aggregated, re-aggregated, stratified, or re-stratified. There is a need to look at the data in many ways in the search for knowledge about common causes. The context must also be known in order to ask good questions of the data.

Pyzdek (1990) gives a graphic example of how viewing "chance" as being explicable and reducible rather than unexplainable but controllable in a system can lead to improvements. In a manufacturing process the average number of defects in solder-wave boards declined from 40 to 20 per 1,000 leads, through running the least dense circuit pattern across the wave first. Another two changes to the system later on reduced the average number of defects to 5 per 1,000 leads. Therefore Pyzdek (1990, p. 108) repudiates the "outdated belief that chance causes should be left to chance and instead presents the viewpoint that all variation is caused and that many, perhaps most processes can be improved economically." His perspective is on the marketplace with its increasing emphasis on continuous improvement. Although this may be considered a deterministic outlook, there is still an acceptance of indeterminism—it is more about reducing the level of indeterminism by acquiring more knowledge.

In reality, variation is ever present. If patterns cannot be found in the data, then the extent of the variability can be estimated and allowed for in the process. If patterns are found, but the cause is not manipulable (e.g., gender), then the identification of the cause enables better prediction for individuals and processes can be designed to allow for the variation. If the cause is manipulable, then the process can be changed to increase the "desirable" outcomes (Wild & Pfannkuch, 1999). Therefore the thinking is to search for causes, for all possible explanations, but to recognize that variation will be present. Coupled with this thinking is the cognition that what may appear to be a pattern may in reality be random or unexplained variation.

Variation as a Way of Perceiving the World

In the quality management area, common consensus is being developed on the characteristics of the statistical thinking required for improving systems. As people in the quality field have moved from quality control to quality management, the nature of the thinking required has developed from an emphasis on stable variability in manufactured products toward an emphasis on the way managers (from any environment) should operate and think.

Snee (1990, p. 116) believes there is a need to acquire a greater understanding of statistical thinking and the key is to focus on statistical thinking at the conceptual level or from a "systems" perspective rather than focusing on the statistical tools:

> I define statistical thinking as thought processes, which recognize that variation is all around us and present in everything we do, all work is a series of interconnected processes, and identifying, characterizing, quantifying, controlling and reducing variation provide opportunities for improvement. This definition integrates the ideas of processes, variation, analysis, developing knowledge, taking action and quality improvement.

According to Hare et al. (1995, p. 55), "Statistical thinking is a mind-set. Understanding and using statistical thinking requires changing existing mind-sets." They state that the key components of statistical thinking for managers are "(1) process thinking; (2) understanding variation; (3) using data whenever possible to guide actions." In particular, they reinforce ideas like these: improvement comes from reducing variation; managers must focus on the system, not on individual people; and data are the key to improving processes. Kettenring (1997, p. 153) supports this view when he states that managers need to have an "appreciation for what it means to manage by data."

Snee (1999, p. 257), however, contends that while data should be used for effective statistical thinking, data are not essential to the use of statistical thinking. He observes variation is present in processes without data being available. For example, it is generally known that "decreasing the variation of process inputs decreases the variation of process outputs." Hence, without data, statistical thinking would suggest, for example, that companies should significantly reduce their number of suppliers. Britz, Emerling, Hare, Hoerl, and Shade (1997, p. 68) sum up this ability to use statistical thinking without data as follows: "the uniqueness of statistical thinking is that it consists of thought processes rather than numerical techniques. These thought processes affect how people take in, process, and react to information."

Tversky and Kahneman's insights about how regression to the mean affects people's beliefs about the effects of reward and punishment are widely promulgated in quality management as part of "understanding the theory of variation." The setting used to illustrate this is typically the reactions of sales managers to the highs and lows in sales figures of their staff. According to Joiner and Gaudard (1990), many managers fail to recognize, interpret, and react appropriately to variation over time in employee performance data. These statisticians are attempting to get managers to understand that looking at single time-interval changes and meting out praise and censure is not conducive to improving performance. The way to improve performance is to make some system change that will increase the average level of performance. Managers need to recognize that there will always be variation, and that unless there is a system change there will be regression to the mean. This suggests that managers are being asked to take on a world view that allows for indeterminism.

Statistical thinking in quality management is now seen not only as necessary for gleaning information from data but also as a way of perceiving the world reality. From quality management we learn that statistical thinking is, first and foremost, about thought processes that consider variation, about seeking explanations to explain the variation, about recognizing the need for data to guide actions, and about reasoning with data by thinking about the system or process as a whole. Implicit in their concepts about variation is that system (not people or individual) causal thinking is paramount. Once the type of variation has been categorized as special-cause or common-cause, then there are appropriate strategies for identifying the causes of that variation. The quality management thinking approach is not to leave variation to chance, but to reduce it in an attempt to improve processes and performance.

Some Contributions from Statistics Education Researchers

The quality management approach to statistical thinking arose from the confluence of a focus on empirical data and the need to improve processes. In contrast, the statistics education field tended to have its origins in mathematics education and in a deductive rather than inductive culture.

Statistics education research emerged in the late 1970s and focused mainly on probability (e.g., Fischbein, 1975; Tversky & Kahneman, 1982). It has really only been in the last decade that statistical thinking has begun to be addressed. We will now discuss some of these developments.

Integrating the Statistical and the Contextual

It emerged from the research of Biehler and Steinbring (1991) that the interplay between data and context was essential for the generation and interpretation of graphical representations. They used the term *statistical detective work* to describe this process of questioning the data through to a judgment or decision about the original situation. Shaughnessy, Garfield, and Greer (1996, p. 206) also suggested that students need to set up a dialogue with data with the mind-set of a detective and to "look behind the data" since data arise from a specific context.

> Data are often gathered and presented by someone who has a particular agenda. The beliefs and attitudes lying behind the data are just as important to include in the treatment of data handling as are the methods of organizing and analyzing the data … it is mathematical detective work in a context … relevance, applicability, multiple representations and interpretations of data are lauded in a data handling environment. Discussion and decision-making under uncertainty are major goals … so too are connections with other disciplines.

Transnumeration and Context Knowledge

From their research on students involved in statistical projects using technology, Ben-Zvi and Friedlander (1997) emphasized, in their hierarchy of thinking modes, the role of representation and implicitly the role of context. Students who were handling multiple representations in a meaningful and creative way, and were using graphs to search for patterns and to convey ideas—coupled with a critical attitude—were considered to be thinking statistically. One of the main notions identified in this hierarchy is the fundamental type of statistical thinking that we call transnumeration.

Hancock, Kaput, and Goldsmith (1992, p. 339) view statistics from a modeling perspective encapsulating the idea that data are a model of a real-world situation. They identified data creation and data analysis as making up the domain of data modeling. "Like any model it is a partial representation and its validity must be judged in the context of the uses to which it will be put. The practical understanding of this idea is the key to critical thinking about data-based arguments." They state that data creation has been neglected and includes:

> Deciding what data to collect, designing a structure for organizing the data and establishing systematic ways of measuring and categorizing … data creation informs data analysis because any conclusion reached through analysis can only be as reliable and relevant as the data on which it is based. The most interesting criticisms of a data-based argument come not from scrutinizing graphs for misplotted points … but from considering some important aspect of the situation that has been neglected, obscured or biased in the data collection.

This is a good example of (1) transnumeration at the beginning of the problem when relevant "measures" need to be captured from the real system and (2) bringing to the problem context knowledge of the situation and integrating it with statistical knowledge to challenge the interpretation of the data.

Reasoning with Statistical Models

Hancock et al. (1992) and Konold, Pollatsek, Well, and Gagnon (1997) conclude, from their research on students, that reasoning about group propensities rather than individual cases is fundamental in developing statistical thinking. But, according to research by Konold et al. (1997), students dealing with data find it very difficult to make the transition from thinking about and comparing individual cases to aggregate-based reasoning. For example, in mathematics one counterexample disproves a conjecture or claim, whereas in statistics a counterexample (an individual case) does not disprove a theory concerning group propensities. Furthermore, for students to reason with a statistical graph they must "see" patterns in the data set as a whole, with the proviso that patterns can be seen in randomness and that individual-based reasoning may be required in some situations.

Recognition of the Need for Data

Hancock et al. (1992), Konold et al. (1997), and Watson et al. (1995) have observed in their research that it was not unusual to find students who expected that the collection and analysis of data would confirm their personal knowledge of the situation. In fact, the students often ignored the graphs they had constructed and wrote their conclusions based on their own beliefs. This fundamental statistical thinking element, which some students seem to lack, is the recognition that data are needed to judge a situation. This facet includes the recognition that personal experience and opinions may be inadequate or possibly biased, and furthermore that opinions may need to be revised in light of the evidence gained.

Statistical Thinking and Interacting with Statistically Based Information

Many mathematics curricula (e.g., Ministry of Education, 1992) have incorporated the interpretation and critical evaluation of media and other statistically based reports as a desirable outcome in a statistics course. This is not surprising given the high level of statistical information present in the media (Knight et al., 1993) and that the general aim of education programs is to produce literate citizens.

The ability to question claims in the media and to critically evaluate such reports requires high-level thinking skills (Watson, 1997). When students are confronted with having to form a judgment on a report, they have to weigh up what they are willing to believe, what else should be done, or what should be presented to them to convince them further. Gal (1997) suggests that evaluation of a report requires students to have a critical list of "worry" questions in their heads, coupled with a critical disposition. This list of worry questions is based on critiquing the investigative cycle stages. This underlying thinking requires the students to place themselves in the position of being the investigators and thereby determining the considerations that an investigator should give to such aspects as the measures, design, alternative explanations, inference space, and so forth. In doing so, the student checks for possible flaws in the design and reasoning. This evaluation process requires the students to use not only their statistical knowledge, but their contextual knowledge. Often when thinking of, for example, alternative explanations for the meaning of findings, students must "consider other information about the problem context or consult world knowledge they may have to help in ascribing meaning to the data" (Gal, 1997, p. 50).

Gal and Watson, through their research, have alerted statistics educators to the fact that involving students in statistical investigations does not appear to fully develop statistical thinking. Gal et al. (1995, p. 25) believe the reason for this is "both an issue of skill transfer, as well as the fact that a somewhat different set of cognitive skills and dispositions is called for." Therefore it would seem that specific instruction in the evaluation of statistically based reports is required to fully develop statistical thinking.

Probabilistic and Deterministic Thinking

Apart from Biehler's (1994) work, educationists have not paid a great deal of attention to explicating statistical thinking from a practitioner perspective. Biehler (1994) believes there are two cultures of thinking in statistics, deterministic and probabilistic. This deterministic thinking is demonstrated in the methods of exploratory data analysis (EDA), which does not try to calibrate variability in data against a formal probability model. Patterns are sought in an attempt to search for causes; but there is the awareness that people often "see" patterns in randomness, and a filter is needed for such a phenomenon. "EDA people seem to appreciate subject matter knowledge and judgment as a background for interpreting data much more than traditional statisticians seem to" (Biehler, 1994, p. 7).

Probabilistic thinking occurs when reasoning with theoretical probability models, for example, in situations where the argument is based on the data being a random sample from a particular model. Biehler (1999, p. 261) argues strongly that the modeling of a system by a probability distribution can "reveal new types of knowledge, new causes, explanations and types of factors that cannot be detected at the individual level." Systematic and random variation and their complementary roles also need to be understood (Konold et al., 1991) in these situations. Therefore Biehler suggests that statistical thinking requires both probabilistic and deterministic thinking as well as both aggregate-based and individual-based reasoning. This shift toward EDA in statistics, which was influenced by the 1962 landmark paper of Tukey (Kotz & Johnson, 1992) and further developed by him (see Tukey, 1977), has focused statistics educators' attention on the fact that statistical thinking involves a context knowledge base, a statistical knowledge base, variation as a core component, a search for causes, and reasoning with statistical and probability models.

Variation as Fundamental in Statistical Thinking

The notion that variation is fundamental in statistical thinking was not recognized by educationists until recently (Shaughnessy, 1997; Pfannkuch, 1997), although the idea was being vigorously promoted by statisticians with an interest in education (e.g., Moore, 1990). Shaughnessy (1997) believes the lack of research and mention of variation is that research largely reflects the emphasis in curricula materials. This situation is now being addressed by Shaughnessy, Watson, Moritz, & Reading (1999) who, in their research, have found a lack of clear growth in students' conceptions of variability for a particular task.

From this brief overview of research into students' thinking, we note that the fundamental elements of statistical thinking have been identified in statistics education research. The variation element has only recently been addressed. It is a powerful underlying conception that allows us to relate behavior we can actually observe to the abstract ideas of pattern, exceptions, and randomness. Statistics education research has added important insights into statistical thinking by identifying the way students think and by recognizing that statistical thinking is not an innate, nor a community way of thinking. It must be specifically learned and developed in an educational environment and in the statistics discipline. Statistics education researchers have highlighted the difficulties students have in making the transition to a statistical way of thinking. They have also promoted awareness that statistical thinking involves a different set of cognitive skills in the arena of empirical enquiry and in the arena of the evaluation of statistically based reports.

Some Contributions from Statisticians

In the last decade in the statistics literature, David Moore has been vigorously promoting the idea that the development of a statistical way of thinking must be central in the education process and that the variation-type thinking should be at the heart of statistics education. By 1996 the board of directors of the American Statistical Association (ASA) had approved recommendations that the curriculum should emphasize the elements of statistical thinking (Moore, 1997) and adopted a definition very similar to that given by Moore (1990, below).

Variation Is the Core of Statistical Thinking

Moore (1990, p. 135) summarizes statistical thinking as:

- The omnipresence of variation in processes. Individuals are variable; repeated measurements on the same individual are variable. The domain of strict determinism in nature and in human affairs is quite circumscribed.
- The need for data about processes. Statistics is steadfastly empirical rather than speculative. Looking at the data has first priority.

- The design of data production with variation in mind. Aware of sources of uncontrolled variation, we avoid self-selected samples and insist on comparison in experimental studies. And we introduce planned variation into data production by use of randomization.
- The quantification of variation. Random variation is described mathematically by probability.
- The explanation of variation. Statistical analysis seeks the systematic effects behind the random variability of individuals and measurements.

Moore (1992a, p. 426) extends this notion of the centrality of variation by stating that "pupils in the future will bring away from their schooling a structure of thought that whispers 'variation matters.'" What specifically that structure of thought is and how it would be articulated or modeled in the teaching process is a matter of conjecture. At the root of that structure appears to be ideas about determinism and indeterminism.

There is a minefield of interrelated and overlapping concepts surrounding variation, randomness, chance, and causation. Section 3 of Wild and Pfannkuch (1999) attempts to explicate the distinctions.

Arguing with a Context Knowledge Base

Cobb and Moore (1997, p. 801) also believe that context plays an important role in how to think with data: "statistics requires a different kind of thinking, because data are just not numbers, they are numbers with a context." They emphasize that the data "literature" must be known in order to make sense of data distributions. When looking for patterns, data analysts must ultimately decide "whether the patterns have meaning and whether they have any value"; this will depend on "how the threads of those patterns interweave with the complementary threads of the story line," since the "context provides meaning" (Cobb and Moore, 1997, p. 803). Hawkins (1996) concurs, stating that students are statistically illiterate if they think that the statistical distribution is the final product.

Context knowledge is also essential for judging (1) the quality of the data arising from a particular data collection design and (2) the relevance of the data to the problem. Mallows (1998, p. 2) believes that statisticians have not paid enough attention to thinking about what he calls the zeroth problem: "considering the relevance of the observed data, and other data that might be observed, to the substantive problem." He is concerned that thinking about the relevance of the data to the problem should not be neglected when statisticians attempt to capture measures from the real situation, since "statistical thinking concerns the relation of quantitative data to a real-world problem, often in the presence of variability and uncertainty. It attempts to make precise and explicit what the data has to say about the problem of interest" (Mallows, 1998, p. 3). Moore (1997) and Hoerl, Hahn, & Doganaksoy (1997) emphasize that attention should be paid to the design of the data production process since context knowledge about the design will enable the quality of the data to be assessed. Hawkins (1996) extends this notion further by suggesting students cannot acquire statistical reasoning without knowing why and how the data were collected. Scheaffer (1997, p. 156) also emphasizes the importance of knowing "how the data originated [and] what the numbers might mean." Moore (1998, p. 1263) perhaps sums up these concerns: "effective use of statistical reasoning requires considering the zeroth problem and interpretation of formal results in the context of a specific setting." The implication is that statistical thinking involves going beyond and looking behind the data, and making connections to the context from which they came.

Transnumeration

Data reduction and data representation are an essential requirement of dealing with masses of data. Moore (1998, p. 1258) considers "statistical thinking offers simple but non-intuitive tools for trimming the mass, ordering the disorder, separating sense from nonsense, selecting the relevant few from the irrelevant many." Thus thought processes must be triggered for initiating the changing of the data into a manageable form from which information can be gleaned. Hawkins (1997, p. 144) coins the term *informacy* in an attempt to describe such reasoning and thinking. To be informate means "one requires skills in summarizing and representing information, be it qualitative or quantitative, for oneself and others." We believe this transnumeration type of thinking is fundamental for data-handling processes.

The communication of messages in the data, transnumeration-type thinking, is intimately linked with inferential thinking. Apart from considering the relevance of the data to the problem, it is also important to consider the inferences that can be made from the data. W. E. Deming first raised the important distinction between enumerative and analytical studies in 1950 (for a detailed discussion, see Hahn & Meeker, 1993). The aim of an enumerative study is to describe the current situation, whereas the aim of an analytical study is to take actions on or make predictions about a future population or process. The space for reliable statistical inference is limited to the population or process actually sampled. For example, a public opinion poll to assess the *current* view of U.S. voters on who they would vote for in the next election is an enumerative study. Formal inference will provide reasonably reliable answers. If the poll was used to predict the outcome of the next election (future process), the study then becomes analytic. Many, if not most, important problems require using data from current processes or populations to make predictions about the likely behavior of future processes or populations. There are no statistically reliable ways of doing this. Our measures of uncertainty reflect uncertainty about the true characteristics of the current process, thus understating rational levels of uncertainty about the future process. The validity of extrapolation to future processes can be justified only by contextual knowledge of the situation.

Statistical Thinking as a Way of Perceiving the World

Ullman (1995) perceives the framework in which statistical thinking operates as being broadly based, to the extent that it could be used informally in everyday life. "We utilize our quantitative intelligence all the time. ... We are measuring, estimating and experimenting all without formal statistics" (p. 6). Ullman believes this quantitative intelligence is unique to statistics. Some principles he suggests as a basis for quantitative intelligence follow: "to everything there is a purpose; most things we do involve a process; measurements inform us; typical results occur; variation is ever present; evaluation is on going; decisions are necessary" (p. 5). Quantitative intelligence allows a statistical perception of reality.

Statistical Thinking Is an Independent Intellectual Method

Statistics is an epistemology in its own right; it is not a branch of mathematics (Moore, 1992b). Hawkins (1996) suggests that a mathematically educated person can be statistically illiterate. Statistical thinking, states Moore (1998, p. 1263), "is a general, fundamental and independent mode of reasoning about data, variation and chance." Ullman (1995, p. 2) concurs that statistical thinking or quantitative intelligence is an inherently different way of thinking because the reasoning involves dealing with uncertain empirical data: "I claim that statistical thinking is a fundamental intelligence."

The statistical thinking promulgated by these statisticians is encapsulated as an independent intellectual method. Its domain is the empirical enquiry cycle, but the domain should also be extended to a way of thinking about and perceiving the world. Statistical thinking goes beyond the domain of mathematics, which statisticians use simply as a means to help them achieve their own ends. The nature of statistical thinking is explained by these statisticians as noticing, understanding, using, quantifying, explaining, and evaluating variation; thinking about the data "literature"; capturing relevant data and measurements; summarizing and representing the data; and taking account of uncertainty and data variability in decision making.

DISCUSSION AND SUMMARY

Statistical Thinking and Empirical Enquiry

The Wild & Pfannkuch (1999) four-dimensional model (Figure 1) was an attempt to characterize the way experienced statistical practitioners think when conducting empirical enquiries. As such it represents a goal for education programs to strive for. The model was developed as a result of interviewing statisticians and tertiary students about statistical projects they had been involved in; interviewing tertiary students as they performed statistical tasks; and analyzing the literature described earlier. The research focused on statistical thinking at the broad level of the statistical enquiry cycle, ranging from problem formulation to the communication of conclusions. Our four-dimensional framework (Figure 1) for statistical thinking in empirical enquiry describes a nonhierarchical, nonlinear, dynamic way of thinking that encompasses an investigative cycle, an interrogative cycle, types of thinking, and dispositions, all of which are brought to bear in the solving of a statistically based problem. The thinker operates in all four dimensions at once. For example, the thinker could be categorized as currently being in the planning stage of the investigative cycle (Dimension 1), dealing with some aspect of variation in Dimension 2 (types of thinking) by criticizing a tentative plan in Dimension 3 (interrogative cycle) driven by skepticism in Dimension 4 (dispositions).

The investigative cycle (Figure 1a) describes the procedures a statistician works through and what the statistician thinks about in order to learn more in the context sphere. The dispositions (Figure 1d) affect or even initiate entry of the thinker into the other dimensions. The interrogative cycle (Figure 1c) is a generic thinking process that is in constant use by statisticians as they carry out a constant dialogue with the problem, the data, and themselves. It is an interrogative and evaluative process that requires effort to make sense of the problem and the data with the aim of eventually coming to some resolutions about the problem and data during that dialogue. The types of thinking (Figure 1b) are divided into generic types of thinking, which are common to all problem solving, and fundamental statistical types of thinking, which we believe are inherently statistical (see the section titled "Model for Interpretation of Literature"). These types of thinking reflect that thinking, when applied in a statistical context, will enable the statistician to abstract a statistical question from the real situation; capture cogent elements of that reality in measurements and statistical models; work within models using statistical methods to draw out inferences from the data; and communicate what has been learned from the data about the real situation.

This framework was an attempt to make explicit what has previously been largely implicit—the thinking processes used by practitioners during data-based enquiry. According to Resnick (1987, p. 35), "each discipline has [its own] characteristic ways of reasoning," and such thinking processes should be embedded into the teaching and learning of that discipline. Statistical problem solving requires holistic thinking informed by statistical elements. These peculiarly statistical elements appear as the "Types Fundamental to *Statistical* Thinking" in Dimension 2 (Figure 1b).

From a survey of history, literature, and our own exploratory studies, we believe our four-dimensional framework is one way of incorporating this knowledge into a current explication of what we understand to be statistical thinking in the domain of problem solving. This framework does not, however, address statistical thinking in the arenas of evaluating enquiries and in everyday life, but it can shed light on them.

We want students to learn to interact with accounts of statistical investigations performed by others—in "the information-using domain" (Barabba, 1991; Gal, 2000). Statistically based information will be used by students to obtain information about societal issues; to make decisions about their own lives in areas such as medicine, gambling and insurance; and to make decisions in their occupations such as marketing, manufacturing, and law. Major sources include technical reports written by investigators and media reports, which are typically at least third-hand summaries. Two main processes need to be invoked. One addresses the question, "To what extent do I trust this information?" and the other extracts meaning from the information. Critical appraisal of information in a report largely consists of appraising the way in which the investigators have proceeded through the steps of PPDAC (Problem, Plan, Data, Analysis, Conclusions) in Dimension 1. We often find fatal flaws through inappropriate choices of the measures used, the study design, and the analysis used; and have learned to beware, at the conclusions stage, of extrapolations beyond the sampled inference space. Extracting meaning tends to be given less emphasis in teaching than more peripheral issues such as misleading graphics. (With a little knowledge, we can often extract correct information from a "misleading" graph.) We argue that, apart from the use of reading strategies, the extracting of meaning that goes on in the interpretation of reports is a subset of the extracting of meaning that is required during investigation. Since knowledge about investigations precedes the ability to criticize, this implies that statistical thinking in empirical enquiry is an extremely basic form of statistical thinking. Even though the evaluation of enquiries is based on knowledge of the investigation process, it still requires specific instruction to enhance the links and connections.

In addition, there is statistical thinking that affects our interpretation of the phenomena and happenstance information we come across in daily life; such thinking skills can be valuable even in the absence of data. In particular, many everyday lessons flow from an appreciation of variation, as described by Tversky and Kahneman (1982), Snee (1999), and Britz et al. (1997). We know that our statistical learning can sensitize us to such issues as bias, small sample size, and variation in the "data" that we gain through our own experience, and it can alter the way we think about risk and making decisions. It seems to us that there is potentially valuable work to be done in assembling these ideas and giving them some coherence (see Gigerenzer, Todd,& ABC Research Group, 1999; Gigerenzer, 2002). It also occurs to us that coherence might not even be possible, since people experience reality in their own unique ways. We might be dealing with inherently fragmentary side benefits of an appreciation of investigation. But someone needs to make the attempt. Unless the link is directly made, in the teaching process, to the "data" gained through people's own experience, statistical education will not help develop the way people think in everyday life.

Statistical thinking is thought processes that are triggered (1) during data-based enquiry to solve a practical problem, (2) during interaction with a data-based argument, and (3) during interaction with data-based phenomena within one's operational environment. This "art" of thinking is new and is increasingly becoming an integral part of many areas of human thought. Its importance should not be underestimated. The development of statistical thinking should be seen by educators as crucial for understanding and operating in today's environment and for perceiving a world reality. The challenge is to find ways to incorporate its explication into pedagogical practice.

Implications for Teaching and Assessing Students

The development of students' statistical thinking presents four major challenges in teaching. The first challenge for educators is to raise awareness about the characteristics of statistical thinking, to reach a common consensus on their understanding of it, and to develop a common language to describe and communicate it. The second challenge is to recognize statistical thinking in a variety of contexts and situations and be able to explain and justify how and why that type of communication constitutes statistical thinking (e.g., Chance, 2002). When educators themselves are sufficiently attuned to recognition of statistical thinking, then the third challenge is to develop teaching strategies that will promote and enhance students' statistical thinking. It will also require mapping out a developmental pathway for statistical thinking across the curriculum and learning about and recognizing the intuitive statistical thinking that is already present in students (e.g., Pfannkuch & Rubick, 2002). The final challenge is to implement teaching and assessment strategies that focus on developing students' statistical thinking. This should include acculturating students to how statisticians reason and work within the statistics discipline and developing new ways for them to view the world.

REFERENCES

Bailar, B. (1988). Statistical practice and research: The essential interactions. *Journal of the American Statistical Association, 83*(401), 1–8.

Barabba, V. (1991). Through a glass lens darkly. *Journal of the American Statistical Association, 86*(413), 1–8.

Bartholomew, D. (1995). What is statistics? *Journal of the Royal Statistical Society A, 158* (Part 1), 1–20.

Ben-Zvi, D., & Friedlander, A. (1997). Statistical thinking in a technological environment. In J. Garfield & G. Burrill (Eds.), *Research on the role of technology in teaching and learning statistics* (pp. 45–55). Voorburg, The Netherlands: International Statistical Institute.

Biehler, R. (1994). Probabilistic thinking, statistical reasoning and the search for causes: Do we need a probabilistic revolution after we have taught data analysis? In J. Garfield (Ed.), *Research Papers from The Fourth International Conference on Teaching Statistics*, Marrakech, 1994. Minneapolis, MN: University of Minnesota.

Biehler, R. (1999). Discussion: Learning to think statistically and to cope with variation. *International Statistical Review, 67*(3), 259–262.

Biehler, R., & Steinbring, H. (1991). Entdeckende Statistik, Stengel-und-Blatter, Boxplots: Konzepte, *Begrundungen und Erfahrungen eines Unterrichtsversuches. Der Mathematikunterricht, 37*(6), 5–32.

Box, J. F. (1978). *R. A. Fisher, The life of a scientist*. New York: Wiley.

Box, J. F. (1997). Fisher, Ronald Aylmer. In N. Johnson & S. Kotz (Eds.), *Leading personalities in statistical sciences: From the 17th century to the present*. New York: Wiley.

Britz, G., Emerling, D., Hare, L., Hoerl, R., & Shade, J. (1997). *How to teach others to apply statistical thinking*. Quality Progress, June 1997, 67–79.

Chance, B. (2002). Components of statistical thinking and implications for instruction and assessment. *Journal of Statistics Education, 10*(3). Retrieved February 10, 2003 from http://www.amstat.org/publications/jse/v10n3/chance.html

Cline Cohen, P. (1982). *A calculating people: The spread of numeracy in early America*. Chicago: University of Chicago Press.

Cobb, G., & Moore, D. (1997). Mathematics, statistics and teaching. *American Mathematical Monthly, 104*(9), 801–823.

Cohen, I. B. (1984). Florence Nightingale. *Scientific American, 250*(3), 98–107.

David, F. (1962). *Games, gods and gambling*. London: Charles Griffen.

Davis, P., & Hersh, R. (1986). *Descartes' dream*. Orlando, FL: Harcourt Brace Jovanovich.

Falk, R., & Konold, C. (1992). The psychology of learning probability. In F. & S. Gordon (Eds.), *Statistics for the twenty-first century*. MAA Notes, no. 29 (pp. 151–164). Washington, DC: Mathematical Association of America.

Fienberg, S., & Tanur, J. (1996). Reconsidering the fundamental contributions of Fisher and Neyman on experimentation and sampling. *International Statistical Review, 64*(3), 237–253.

Fischbein, E. (1975). *The intuitive sources of probabilistic thinking in children*. Dordrecht, The Netherlands: Reidel.

Gail, M. (1996). Statistics in action. *Journal of the American Statistical Association, 91*(433), 1–13.

Gal, I. (1997). Assessing students' interpretation of data. In B. Phillips (Ed.), *IASE papers on statistical education ICME-8*, Spain, 1996 (pp. 49–57). Hawthorn, Australia: Swinburne Press.

Gal, I. (2000). Statistical literacy: Conceptual and instructional issues. In D. Coben, J. O'Donoghue, & G. FitzSimons (Eds.), *Perspectives on adults learning mathematics* (pp. 135–150). Dordrecht, The Netherlands: Kluwer Academic Publishers.

Gal, I., Ahlgren, C., Burrill, G., Landwehr, J., Rich, W., & Begg, A. (1995). Working group: Assessment of interpretive skills. In Writing group draft summaries, *Conference on Assessment Issues in Statistics Education* (pp. 23–25). Philadelphia: University of Pennsylvania.

Gigerenzer, G. (2002). *Calculated risks: How to know when numbers deceive you*. New York: Simon & Schuster.

Gigerenzer, G., Todd, P.M., & ABC Research Group (1999). *Simple heuristics that make us smart*. New York: Oxford University Press.

Greenwood, M. (1970). Medical statistics from Graunt to Farr. In E. S. Pearson & M. G. Kendall (Eds.), *Studies in the history of statistics and probability* (pp. 47–126). London: Charles Griffen.

Hacking, I. (1975). *The emergence of probability: A philosophical study of early ideas about probability, induction and statistical inference*. Cambridge, England: Cambridge University Press.

Hahn, G., & Meeker, W. (1993). Assumptions for statistical inference. *American Statistician, 47*(1), 1–11.

Hancock, C., Kaput, J., & Goldsmith, L. (1992). Authentic enquiry with data: Critical barriers to classroom implementation. *Educational Psychologist, 27*(3), 337–364.

Hare, L., Hoerl, R., Hromi, J., & Snee, R. (1995, February). The role of statistical thinking in management. *Quality Progress, 28*(2), 53–60.

Hawkins, A. (1996). Can a mathematically-educated person be statistically illiterate? *Mathematics for the new Millennium—What needs to be changed and why?* Nuffield Foundation: pre-conference paper (pp. 107–117).

Hawkins, A. (1997). Discussion—New pedagogy and new content: The case of statistics. *International Statistical Review, 65*(2), 141–146.

Hill, A. B. (1965). The environment and disease: Association or causation. *Proceedings of the Royal Society of Medicine, 58*, 295–300.

Hoerl, R., Hahn, G., & Doganaksoy, N. (1997). Discussion—New pedagogy and new content: The case of statistics. *International Statistical Review, 65*(2), 147–153.

Joiner, B. (1994). *Fourth generation management*. New York: McGraw-Hill.

Joiner, B., & Gaudard, M. (1990, December). Variation, management, and W. Edwards Deming. *Quality Progress, 23*(12), 29–37.

Kendall, M. G. (1970). Where shall the history of statistics begin? In E. S. Pearson & M. G. Kendall (Eds.), *Studies in the history of statistics and probability* (pp. 45–46). London: Charles Griffen.

Kettenring, J. (1997). Discussion—New pedagogy and new content: The case of statistics. *International Statistical Review, 65*(2), 153.

Knight, G., Arnold, G., Carter, M., Kelly, P., & Thornley, G. (1993). *The mathematical needs of New Zealand school leavers*. Palmerston North, New Zealand: Massey University.

Konold, C., Lohmeier, J., Pollatsek, A., Well, A., Falk, R., & Lipson, A. (1991). Novice views on randomness. In *Proceedings of the Thirteenth Annual Meeting of the International Group for the Psychology of Mathematics Education—North American Chapter* (pp. 167–173). Blacksburg, VA: Virginia Polytechnic Institute and State University.

Konold, C., Pollatsek, A., Well, A., & Gagnon, A. (1997). Students analyzing data: Research of critical barriers. In J. Garfield & G. Burrill (Eds.), *Research on the role of technology in teaching and learning statistics* (Proceedings of the 1996 International Association of Statistics Education round table conference, pp. 151–167). Voorburg, The Netherlands: International Statistical Institute.

Kotz, S., & Johnson, N. (1992). *Breakthroughs in statistics*, Volumes I–III. New York: Springer-Verlag.

Krishnan, T. (1997). Fisher's contributions to statistics. *Resonance Journal of Science Education, 2*(9), 32–37.

Lightner, J. (1991). A brief look at the history of probability and statistics. *Mathematics Teacher, 84*(8), 623–630.

Mallows, C. (1998). 1997 Fisher Memorial Lecture: The zeroth problem. *American Statistician, 52*(1), 1–9.

McKean, K. (1985, June). Decisions, decisions. *Discover, 6*, 22–33.

Ministry of Education (1992). *Mathematics in New Zealand Curriculum*. Wellington, New Zealand: Learning Media.

Moore, D. (1990). Uncertainty. In L. Steen (Ed.), *On the shoulders of giants: New approaches to numeracy* (pp. 95–137). Washington, DC: National Academy Press.

Moore, D. (1992a). Statistics for all: Why? What and how? In D. Vere-Jones (Ed.), *Proceedings of the Third International Conference on Teaching Statistics*, Vol. 1 (pp. 423–428). Voorburg, The Netherlands: International Statistical Institute.

Moore, D. (1992b). Teaching statistics as a respectable subject. In F. & S. Gordon (Eds.), *Statistics for the twenty-first century*. MAA Notes, no. 26 (pp. 14–25). Washington, DC: Mathematical Association of America.

Moore, D. (1997). New pedagogy and new content: The case of statistics. *International Statistical Review, 65*(2), 123–165.

Moore, D. (1998). Statistics among the liberal arts. *Journal of the American Statistical Association, 93*(444), 1253–1259.

Pfannkuch, M. (1997). Statistical thinking: One statistician's perspective. In F. Biddulph & K. Carr (Eds.), *People in mathematics education* (Proceedings of the 20th annual conference of the Mathematics Education Research Group of Australasia, pp. 406–413). Rotorua, New Zealand: MERGA.

Pfannkuch, M., & Rubick, A. (2002). An exploration of students' statistical thinking with given data. *Statistics Education Research Journal, 1*(2), 4-21. Retrieved December 19, 2002 from http://fehps.une.edu.au/serj/

Porter, T. M. (1986). *The rise of statistical thinking 1820–1900*. Princeton, NJ: Princeton University Press.

Porter, T. M. (1995). *Trust in numbers: The pursuit of objectivity in science and public life*. Princeton, NJ: Princeton University Press.

Provost, L., & Norman, C. (1990, December). Variation through the ages. *Quality Progress, 23*(12), 39–44.

Pyzdek, T. (1990). There's no such thing as a common cause. *Proceedings of American Society for Quality Control 44th Annual Quality Congress Transactions—San Francisco* (pp. 102–108). Milwaukee, WI: ASQC.

Resnick, L. (1987). *Education and learning to think*. Washington, DC: National Academy Press.

Salsburg, D. (2001). *The lady tasting tea: How statistics revolutionized science in the twentieth century*. New York: Freeman.

Scheaffer, R. (1997). Discussion—New pedagogy and new content: The case of statistics. *International Statistical Review, 65*(2), 156–158.

Scheaffer, R. (2001). Statistics education: Perusing the past, embracing the present, and charting the future. *Newsletter of the Section on Statistical Education of the American Statistical Association, 7*(1), Winter 2001. (Reprinted in *Statistics Education Research Newsletter, 2*(2), May 2001. Retrieved May 18, 2001 from http://www.ugr.es/local/batanero/sergroup.htm)

Shaughnessy, J. M. (1992). Research in probability and statistics: Reflections and directions. In D. Grouws (Ed.), *Handbook of research on mathematics teaching and learning* (pp. 465–494). New York: Macmillan.

Shaughnessy, J. M. (1997). Missed opportunities in research on the teaching and learning of data and chance. In F. Biddulph & K. Carr (Eds.), *People in mathematics education* (Proceedings of the 20th annual conference of the Mathematics Education Research Group of Australasia, pp. 6–22). Rotorua, New Zealand: MERGA.

Shaughnessy, J. M., Garfield, J., & Greer, B. (1996). Data handling. In A. Bishop, K. Clements, C. Keitel, J. Kilpatrick, & C. Laborde (Eds.), *International handbook of mathematics education* (pp. 205–238). Dordrecht, The Netherlands: Kluwer Academic Publishers.

Shaughnessy, J. M., Watson, J., Moritz, J., & Reading, C. (1999). School mathematics students' acknowledgement of statistical variation. Paper presented at *the research pre-sessions of 77th annual meeting of the National Council of Teachers of Mathematics*, San Francisco, 1999.

Shewhart, W., & Deming, W. E. (Ed.). (1986). *Statistical method from the viewpoint of quality control.* New York: Dover Publications. (Original work published 1939)

Snee, R. (1990). Statistical thinking and its contribution to quality. *American Statistician, 44*(2), 116–121.

Snee, R. (1993). What's missing in statistical education? *American Statistician, 47*(2), 149–154.

Snee, R. (1999). Discussion: Development and use of statistical thinking: A new era. *International Statistical Review, 67*(3), 255–258.

Stigler, S. (1986). *The history of statistics—The measurement of uncertainty before 1900.* Cambridge, MA: Belknap Press of Harvard University Press.

Tukey, J. (1977). *Exploratory data analysis.* Reading, MA: Addison-Wesley.

Tufte, E. (1983). *The visual display of quantitative information.* Cheshire, CT: Graphics Press.

Tversky, A., & Kahneman, D. (1982). Judgment under uncertainty: Heuristics and biases. In D. Kahneman, P. Slovic, & A. Tversky (Eds.), *Judgment under uncertainty: Heuristics and biases* (pp. 3–20). New York: Press Syndicate of the University of Cambridge. (Originally published in *Science, 185* (1974), 1124–1131.)

Ullman, N. (1995). Statistical or quantitative thinking as a fundamental intelligence. Unpublished paper, County College of Morris, Randolph, NJ.

Watson, J. (1997). Assessing statistical thinking using the media. In I. Gal & J. Garfield (Eds.), *The assessment challenge in statistics education* (pp. 107–121). Amsterdam: IOS Press.

Watson, J., Collis, K., Callingham, R., & Moritz, J. (1995). A model for assessing higher order thinking in statistics. *Educational Research and Evaluation, 1*, 247–275.

Wild, C., & Pfannkuch, M. (1999). Statistical thinking in empirical enquiry (with discussion). *International Statistical Review, 67*(3), 223–265.

Chapter 3

STATISTICAL LITERACY[1]

Meanings, Components, Responsibilities

Iddo Gal
University of Haifa, Israel

INTRODUCTION AND NEED

Many curriculum frameworks and national and international educational initiatives, including but not limited to those focusing on the mathematical sciences, underscore the importance of enabling all people to function effectively in an information-laden society (e.g., United Nations Educational, Scientific and Cultural Organization [UNESCO], 1990; Australian Education Council, 1991; American Association for the Advancement of Science (AAAS), 1995; European Commission, 1996; National Council of Teachers of Mathematics [NCTM], 2000). The present paper focuses on *statistical literacy*, one critical but often neglected skill area that needs to be addressed if adults (or future adults) are to become more informed citizens and employees.

Statements regarding the importance of statistical reasoning or statistical knowledge in society have been eloquently made in the past. For example, Moore (1998), in his presidential address to the American Statistical Association (ASA), claimed that it is difficult to think of policy questions that have no statistical component, and argued that statistics is a general and fundamental method because data, variation and chance are omnipresent in modern life. Wallman (1993), in a 1992 ASA presidential address, emphasized the importance of strengthening understanding of statistics and statistical thinking among all sectors of the population, in part due to the various misunderstandings, misperceptions, mistrust, and misgivings that people have toward the value of statistics in public and private choices. Researchers interested in cognitive processes have emphasized the contribution of proper judgmental processes and probabilistic reasoning to people's

[1] This chapter is a reprint of "Adults' statistical literacy: Meaning, components, responsibilities," from the *International Statistical Review, 70,* pages 1–52, copyright 2002, and is reproduced here with the permission of the International Statistical Institute.

D. Ben-Zvi and J. Garfield (eds.),
The Challenge of Developing Statistical Literacy, Reasoning and Thinking, 47–78.

ability to make effective decisions (Kahneman, Slovic, & Tversky, 1982) and showed that training in statistics can aid in solving certain types of everyday problems (Kosonen & Winne, 1995). Industry trainers and education planners have pointed to the important role of statistical understanding and mathematical competencies as a component of the skills needed by workers in diverse industries (e.g., Carnevale, Gainer, & Meltzer, 1990; Packer, 1997).

While these and other sources have helped to highlight the centrality of statistical literacy in various life contexts, few attempts to describe the nature of adults' overall statistical literacy have been published to date. It is necessary to first grapple with definitional issues. In public discourse "literacy" is sometimes combined with terms denoting specific knowledge domains (e.g., "computer literacy"). In such cases the usage of "literacy" may conjure up an image of the *minimal* subset of "basic skills" expected of *all* citizens, as opposed to a more advanced set of skills and knowledge that only some people may achieve. Along these lines, statistical literacy may be understood by some to denote a minimal (perhaps formal) knowledge of basic statistical concepts and procedures. Yet increasingly the term literacy, when used as part of the description of people's capacity for goal-oriented behavior in a specific domain, suggests a broad cluster not only of factual knowledge and certain formal and informal skills, but also of desired beliefs, habits of mind, or attitudes, as well as general awareness and a critical perspective.

In line with the expanding conception of the term *literacy,* Wallman (1993) argued that statistical literacy is the ability to understand and critically evaluate statistical results that permeate daily life, coupled with the ability to appreciate the contributions that statistical thinking can make in public and private, professional and personal decisions. Watson (1997) presented a framework of statistical literacy comprised of three tiers with increasing sophistication: a basic understanding of probabilistic and statistical terminology; an understanding of statistical language and concepts when they are embedded in the context of wider social discussion; and a questioning attitude one can assume when applying concepts to contradict claims made without proper statistical foundation.

The complex and expanding meaning of domain-specific literacy can also be illustrated by examining extant conceptions of "scientific literacy." Shamos (1995) reviews prior works on scientific literacy that suggest common building blocks: basic vocabulary, understanding of science process, and understanding of the impact of science and technology on society. Jenkins (1996) suggests that scientific literacy can be characterized as scientific knowledge and attitudes, coupled with some understanding of scientific methodology.

Shamos (1995) argues that it would be a simplification to assume that somebody is either literate or illiterate in science, and suggests a continuum along which scientific literacy can be described, comprised of three overlapping levels that build upon each other in sophistication. The most basic one, "cultural" scientific literacy, refers to a grasp of basic terms commonly used in the media to communicate about science matters. Next, "functional" scientific literacy adds some substance by requiring that "the individual not only have command of a science lexicon but also be able to converse, read and write coherently, using such science terms in perhaps a non-technical but nevertheless meaningful context" (p. 88). This level also requires that the person has access to simple everyday facts of nature, such as some knowledge of the solar system (e.g., that the Earth revolves around the Sun, how eclipses occur). Finally, "true" scientific literacy requires some understanding of the overall scientific enterprise (e.g., basic knowledge of key conceptual schemes or theories that form the foundation of science and how they were arrived at), coupled with understanding of scientific and investigative processes. Examples are (see also Rutherford, 1997): appreciation of the relativity of "fact" and "theory," awareness of how knowledge accumulates and is verified, the role of experiments and mathematics in science, the ability to make sense of public communications about scientific matters, and the ability to understand and discuss how science and technology impinge on public life.

With the above broad usage of "literacy" and "statistical literacy" in mind, this paper develops a conception of statistical literacy that pertains to what is expected of adults (as opposed to students actively learning statistics), particularly those living in industrialized societies. It is proposed here that in this context, the term *statistical literacy* refers broadly to two interrelated components, primarily (a) people's ability to *interpret and critically evaluate* statistical information, data-related arguments, or stochastic phenomena, which they may encounter in diverse contexts, and when relevant (b) their ability to *discuss or communicate* their reactions to such statistical information, such as their understanding of the meaning of the information, their opinions about the implications of this information, or their concerns regarding the acceptability of given conclusions. These capabilities and behaviors do not stand on their own but are founded on several interrelated knowledge bases and dispositions which are discussed in this paper.

Statistical literacy can serve individuals and their communities in many ways. It is needed if adults are to be fully aware of trends and phenomena of social and personal importance: crime rates, population growth, spread of diseases, industrial production, educational achievement, or employment trends. It can contribute to people's ability to make choices when confronted with chance-based situations (e.g., buying lottery tickets or insurance policies, and comprehending medical advice). It can support informed participation in public debate or community action. The need for statistical literacy also arises in many workplaces, given growing demands that workers understand statistical information about quality of processes (Packer, 1997), and the contention that workers' understanding of data about the status of their organization can support employee empowerment (Bowen & Lawler, 1992).

The many examples of contexts where statistical literacy may be activated indicate that most adults are consumers (rather than producers) of statistical information. Yet, despite the centrality of statistical literacy in various life contexts, the nature of the skills and dispositions that comprise adults' statistical literacy have not received detailed discussion in the literature (Gal, 1994; Watson, 1997), and are thus the focus of this paper. Clarity on the characteristics of the building blocks of statistical literacy is needed before other questions can be addressed in earnest regarding assessment and instruction focused on statistical literacy.

A MODEL

This paper concerns itself with people's ability to act as effective "data consumers" in diverse life contexts that for brevity are termed here *reading contexts*. These contexts emerge, for example, when people are at home and watch TV or read a newspaper, when they look at advertisements while shopping, when they visit Internet sites, when they participate in community activities or attend a civic or political event, or when they read workplace materials or listen to reports at work. They include but are not limited to exposure to print and visual media, and represent the junctures where people encounter the much-heralded "information-laden" environments (European Commission, 1996). In such contexts, statistical information may be represented in three ways—through text (written or oral), numbers and symbols, and graphical or tabular displays, often in some combination. To simplify the presentation in this paper, the term *readers* will be used throughout to refer to people when they participate in reading contexts as actors, speakers, writers, readers, listeners, or viewers, in either passive or active roles.

Reading contexts should be distinguished from *enquiry contexts,* where people (e.g., students, statisticians) engage in empirical investigation of actual data (Wild and Pfannkuch, 1999). In enquiry contexts individuals serve as "data producers" or "data analyzers" and usually have to interpret their own data and results and report their findings and conclusions. Reading contexts may differ from enquiry contexts in important ways that have not been sufficiently acknowledged in the literature on statistical reasoning and are examined later.

This paper proposes a model, summarized in Table 1, of the knowledge bases and other enabling processes that should be available to adults, and by implication to learners graduating from schools or colleges, so that they can comprehend, interpret, critically evaluate, and react to statistical messages encountered in reading contexts. Based on earlier work such as cited above on statistical literacy and scientific literacy, the model assumes that people's statistical literacy involves both a *knowledge* component (comprised of five cognitive elements: literacy skills, statistical knowledge, mathematical knowledge, context knowledge, and critical questions) and a *dispositional* component (comprised of two elements: critical stance, and beliefs and attitudes).

Table 1. A model of statistical literacy

Knowledge elements	**Dispositional elements**
Literacy skills Statistical knowledge Mathematical knowledge Context knowledge Critical Questions	Beliefs and Attitudes Critical stance
Statistical Literacy	

As with people's overall numeracy (Gal, 2000), the components and elements in the proposed model should not be viewed as fixed and separate entities but as a context-dependent, dynamic set of knowledge and dispositions that together enable statistically literate behavior. *Understanding and interpretation* of statistical information requires not only statistical knowledge per se but also the availability of other knowledge bases: literacy skills, mathematical knowledge, and context knowledge. However, *critical evaluation* of statistical information (after it has been understood and interpreted) depends on additional elements as well: the ability to access critical questions and to activate a critical stance, which in turn is supported by certain beliefs and attitudes.

The model's elements are described in subsequent sections, although some overlap with each other and do not stand in isolation. The final section of the paper discusses resulting educational and policy challenges and implications for needed research. The expected contribution of this paper is to facilitate further dialogue and action by educators, practicing statisticians, policy makers, and other professionals who are interested in how citizens can be empowered to make sense of real-world messages containing statistical elements or arguments.

KNOWLEDGE ELEMENTS OF STATISTICAL LITERACY

This section reviews the five elements listed in Table 1 as comprising the knowledge component of statistical literacy. It is proposed that these elements jointly contribute to people's ability to comprehend, interpret, critically evaluate, and if needed react to statistical messages.

To provide a context for some of the ideas presented below, Figures 1, 2, 3, and 4 illustrate key modes through which statistical concepts and statistics-related information or arguments are communicated to adults in the printed media, a prime reading context. Figure 1 contains six excerpts illustrating statistical messages in daily newspapers and magazines from different countries. Figure 2 presents a statistics-related table from an American newspaper. Figure 3 presents a bar graph that appeared in a widely circulated Israeli newspaper. Figure 4 includes a pie chart used in the International Adult Literacy Survey (IALS; Statistics Canada and Organization for Economic Co-operation and Development [OECD], 1996) to simulate a newspaper graph.

Literacy Skills

A discussion of literacy skills opens the review of the knowledge bases needed for statistical literacy, given that virtually all statistical messages are conveyed through written or oral text, or require that readers navigate through tabular or graphical information displays that require the activation of specific literacy skills (Mosenthal & Kirsch, 1998).

The understanding of statistical messages requires the activation of various text-processing skills in order to derive meaning from the stimulus presented to readers. The written portion of a message may be quite long (as in some of the excerpts in Figure 1) and demand complex text comprehension skills, or may sometimes involve a graph with only a few words (Figures 3 or 4). Readers also have to comprehend surrounding text (i.e., within which the statistical portion is embedded or which explains a graph or chart presented) to place the statistical part in the proper context. Depending on the circumstances, readers may have to communicate clear opinions, orally or in writing, in which case their response should contain enough information about the logic or evidence on which it is based to enable another listener or reader to judge its reasonableness. Thus, statistical literacy and general literacy are intertwined.

In the real world, readers have to be able to make sense of a wide range of messages, formulated at different levels of complexity and in different writing or speaking styles (Wanta, 1997). Messages may be created by journalists, officials, politicians, advertisers, or others with diverse linguistic and numeracy skills. Message originators may have diverse aims in terms of the presumed facts, images, or conclusions they aim to create or instill in the mind of the reader. Some messages may be created to convince the reader or listener to adopt a specific point of view or reject another, and hence may use one-sided arguments or present selective information (Clemen & Gregory, 2000), or may use modifiers (e.g., "a startling 5% gain …") to shape a desired impression.

As several authors have pointed out (Laborde, 1990; Gal, 1999), coping with mathematical or statistical messages presents various demands on readers' literacy skills. For instance, readers have to be aware that the meanings of certain statistical terms used in the media (e.g., random, representative, percentage, average, reliable) may be different than their colloquial or everyday meaning. Messages may use technical terms in a professionally appropriate way but may also contain statistical jargon that is ambiguous or erroneous. Some newspapers and other media channels tend to employ conventions in reporting statistical findings, such as referring to "sampling error" (or "margin of error") when discussing results from polls, but without explaining the meaning of terms used.

Space and time limitations or editorial decisions may force writers (or professionals who speak on TV) to present messages that are terse, choppy, or lack essential details. Readers may need to make various assumptions and inferences, given the absence of details or the inability in many cases to interrogate the creators of messages encountered. Overall, these factors can make comprehension more challenging, complicate the interpretation task, and could place heavy demands on readers' literacy skills. This is true for adults from all walks of life, but especially of adults who are bilingual or otherwise have a weak mastery of the national/dominant language (Cocking, & Mestre, 1988). However, results from the International Adult Literacy Survey (IALS; Statistics Canada and OECD, 1996) suggest that in most of the countries surveyed, a large proportion of adults have only basic comprehension skills and are unable to cope effectively with a range of everyday literacy and computation tasks. Hence, people's literacy skills may be a bottleneck affecting their statistical literacy skills.

#1: "The study found that women of average weight in the U.S. had a 50 per cent higher chance of heart attack than did women weighing 15 per cent below average." (Watson, 1997, p. 109; from Hobart Mercury, Tasmania, February 10, 1995).

#2: "JUDGES COUNT OUT CENSUS SAMPLING: . . . at issue is far more than the accuracy of sampling in the Census held every 10 years: Billions of dollars in federal funds are allocated on the basis of how many people live in each state and city, and shifts in population can lead to the redrawing of House districts. A boost in the count of minorities would normally help Democrats." (Philadelphia Inquirer, August 25, 1998).

#3: "POLL BACKS LIMITS ON DRINKING BY TEENS: The survey of more than 7000 adults . . . which has a margin of error of 2 percentage points, found that... more than half favored restrictions on alcohol advertising . . . more than 60% would ban TV ads for beer and wine." (USA Today, October 5, 1998)

#4: "The human race held this year many more sexual intercourses than last year; the world average was 112 per person this year, compared to 109 last year. This, according to a comprehensive survey initiated and funded, for the second year, by Durex, a manufacturer of prophylactics. The survey was held in 14 countries that according to experts represent all the world citizens . . ." (Yediot Aharonot, Israel, October 28, 1997).

#5: "The Department of Education is investigating whether state scores on a national reading test were inflated by decisions states made on which students to exclude from the test . . . in both 1994 and 1998 . . . the overall exclusion rate was the same, about 6% . . . Kentucky, Connecticut and Louisiana were among states with increases in students left out of their 1998 testing sample—primarily those with learning disabilities or limited knowledge of English." (USA Today, April 13, 1999).

#6: If you care about breast cancer, [a] new risk assessment test . . . will give you a number that estimates your chances of developing breast cancer over the next 5 years. A score of 1.7 or above is considered high risk. Most likely you wont be at high risk, but you owe it to yourself to find out. The proof? In a landmark study of women 35 years or older and at high risk of breast cancer, women who took Nolvadex had fewer breast cancers than women taking sugar pills. Nolvadex decreases but does not eliminate the risk of breast cancer, and did not show an increase in survival... In the study, women taking Nolvadex were 2 to 3 times more likely to develop uterine cancer or blood clots in the lung and legs, although each of these occurred in less than 1% of women... You and your doctor must . . . discuss whether the potential benefit of Nolvadex will outweigh these potential side affects. (Excerpt from a full-page commercial advertisement in People magazine, August 30, 1999).

Figure 1. Illustrations of statistical texts in daily newspapers and magazines.

'Matrix' a virtual lock at No. 1

The Keanu Reeves sci-fi thriller *The Matrix* remained the box office champ for the second consecutive week. Newcomers had mixed results: The romantic comedy *Never Been Kissed* opened fairly strong at No. 2, ... The top 10:

	Box office (millions)		Avg.	Pct.	Weeks
Film	**Wkd.**	**Total**	**Per site**	**Chg.**	**Out**
1 *The Matrix*	$22.6	$73.3	$7,772	-19%	2
2 *Never Been Kissed*	$11.8	New	$4,821		1
3 10 Things I Hate About You	$5.05	$20.4	$2,218	-39%	2
4 *The out-of-Towners*	$5.01	$16.2	$2,380	-39%	2
5 *Analyze This*	$5.0	$85.8	$2,125	-21%	6

* Re-creation of a selected portion of a table from *USA Today* (April 13, 1999). Some details omitted to conserve space.

Figure 2. Illustration of a tabular display in a newspaper.

Graph in *Yediot Aharonot*, the daily newspaper with the largest circulation in Israel, July 11, 2000. The title says: "Women in Israel are more educated". The subtitle says: "Israel holds the world record in the percentage of women among students for Master and Doctoral degrees". The bars represent percentages for (from top to bottom): Israel (55.4%), United States, Australia, Denmark, Great Britain, Finland, Sweden, Switzerland, and Japan (21.5%). (Reprinted with permission).

Figure 3. Women's education in different countries.

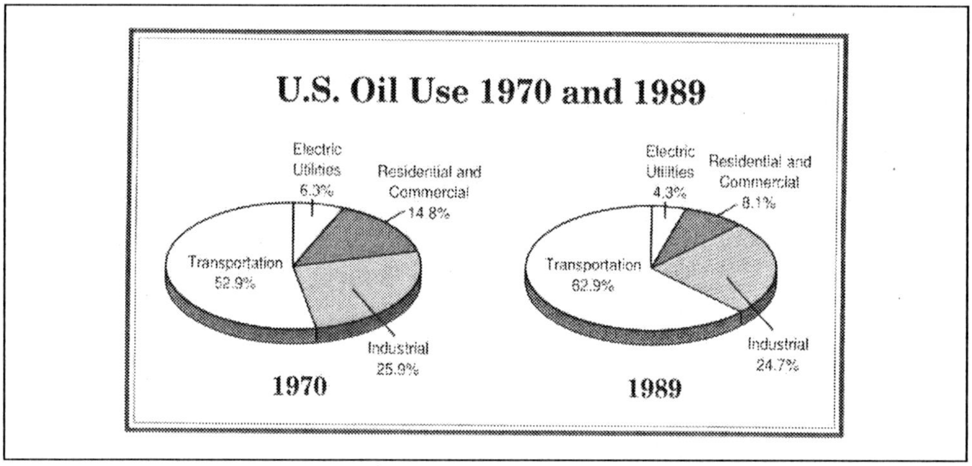

Figure 4. Oil use in two years. Stimulus from an IALS item. (Reprinted with permission).

Document Literacy

The literacy skills needed for statistical literacy are not limited to those involving processing of prose text. This subsection extends the preceding discussion by examining Document Literacy skills, which pertain to reading various nonprose texts, including graphs, charts, and tables. The growing literature on graph comprehension examines various processes involved in making sense of graphs, from simple graph-reading to making inferences based on graphs (Bright & Friel, 1998), but has seldom viewed graphs as a subtype of documents in general.

The notion of Document Literacy comes out of the influential work of Kirsch and Mosenthal (Kirsch, Jungeblut, & Mosenthal, 1998), who view literacy as comprised of three interrelated components: Prose Literacy, Document Literacy, and Quantitative Literacy. This conceptualization of literacy served as a basis for several large-scale studies, most recently the International Adult Literacy Survey (IALS; Statistics Canada and OECD, 1996; OECD & Human Resources Development Canada, 1997), and prior national studies of the literacy of adults and young adults, mainly in the United States and Canada (e.g., Kirsch, Jungeblut, Jenkins, & Kolstad, 1993), but also in Australia.

Kirsch and Mosenthal (1990) claim that documents tend to be the predominant form of literacy in nonschool settings, and serve as an important source of information and a basis for enabling actions and decisions. Document Literacy tasks require people to identify, interpret, and use information given in lists, tables, indexes, schedules, charts, and graphical displays. The information in such displays often includes explicit quantitative information, such as numbers or percentages, in addition to the quantitative or statistical information conveyed by graphs and charts. Mosenthal & Kirsch (1998) argue that documents, which include graphs and charts, are usually arranged in arrays of varying degrees of complexity: they may include "simple lists" or "combined lists," as in a simple table or a simple bar graph or pie chart (Figures 3 and 4); or "intersecting lists" or "nested lists," as in a two-way table (Figure 2) or in a complex multielement graph.

An important aspect of the Kirsch and Mosenthal work (Kirsch, Jungeblut, & Mosenthal, 1998) is the description ("grammar") provided of the cognitive operations required to locate information in documents, and the reading strategies required to match information in a question or directive to corresponding information in arrays of varying degrees of complexity. Key processes include *locating* specific information in given texts or displays, *cycling* through various parts of diverse texts or displays, *integrating* information from several locations (e.g., across two graphs, as in Figure 4), and *generating* new information (e.g., finding the difference between percentages in different parts of a table or between bars in a graph). Further, readers have to make *inferences,* quite often in the presence of irrelevant or distracting information, and perhaps apply mathematical operations as well to information contained in graphs or tables.

As Mosenthal and Kirsch (1998) argue, many types of common statistical information can be displayed in both graphs and tables, and one form is often a mere transformation of the other (e.g., when a table with a simple list is transformed into a simple bar chart). Hence, putting aside specialized aspects of graph comprehension (Tufte, 1997), their work provides a generalized way to understand literacy aspects of interpreting multiple types of documents and displays, and enables us to embed a discussion of statistical literacy within a broader framework of general literacy.

Statistical Knowledge Base

An obvious prerequisite for comprehending and interpreting statistical messages is knowledge of basic statistical and probabilistic concepts and procedures, and related mathematical concepts and issues. However, almost all authors who are concerned about the ability of adults or of school graduates to function in a statistics-rich society do *not* discuss what knowledge is needed to be statistically literate per se, but usually focus on what needs to be taught in schools and argue that all school (or college) graduates should master a range of statistical topics, assuming this will ensure learners' statistical literacy as adults. A recent example can be found in Scheaffer, Watkins, and Landwehr (1998). Based on their extensive prior work in the area of teaching statistics and on reviewing various curriculum frameworks, these authors describe numerous areas as essential to include in a study of statistical topics in high school:

- Number sense
- Understanding variables
- Interpreting tables and graphs
- Aspects of planning a survey or experiment, such as what constitutes a good sample, or methods of data collection and questionnaire design
- Data analysis processes, such as detecting patterns in univariate or two-way frequency data, or summarizing key features with summary statistics

- Relationships between probability and statistics, such as in determining characteristics of random samples, background for significance testing
- Inferential reasoning, such as confidence intervals or testing hypotheses

It is tempting to regard this list as a possible candidate for an "ideal" set of mathematical and statistical knowledge bases that can guarantee statistical literacy. (Indeed, this author would be happy if most adults possessed such knowledge.) However, what is "basic" knowledge cannot be discussed in absolute terms, but depends on the desired level of statistical literacy expected of citizens, on the functional demands of contexts of action (e.g., work, reading a newspaper), and on the characteristics of the larger societal context of living. Hence, the above list may not be appropriate for all cultural contexts, may be an overspecification in some cases, and other elements could be added to it.

Unfortunately, no comparative analysis has so far systematically mapped the types and relative prevalence of statistical and probabilistic concepts and topics across the full range of statistically related messages or situations that adults may encounter and have to manage in any particular society. Hence, no consensus exists on a basis for determining the statistical demands of common media-based messages. To date, only a single comparative study (Joram, Resnick, & Gabriele, 1995) addressed this complex issue, by analyzing the characteristics of rational numbers (especially fractions, percentages, and averages) that appear in weekly or monthly magazines written for children, teenagers, and adults in the United States. This study was based on the assumption that it is useful to view literacy not only as a skill or ability but also as a set of cultural practices that people engage in, and hence that it is important to examine the characteristics of the texts that people may have to make sense of, and ask how these characteristics shape people's literacy practices.

Regarding adults, Joram et al. (1995) sampled seven widely circulated magazines that aim at different types of readers: *Reader's Digest, National Geographic, Better Homes and Gardens, National Enquirer, Time, Consumer Reports,* and *Sports Illustrated.* They applied a complex coding scheme to capture the number of occurrences of rational numbers, especially fractions, percentages, and averages, in the middle 20 pages of one issue. Some findings that are relevant for the present paper were:

- The mean frequencies (per 20 pages) of fractions, percentages, and averages were 4.86, 10.00, and 2.00, respectively.
- Regarding percentages found in these magazines, about half expressed part/whole relations ("The nation's 113 nuclear reactors already generate 20 percent of our electricity"), and one-third referred to increase/decrease ("If ... electricity consumption increases by 2.5 percent a year, we could be headed for real problems").
- Only 14% of statements regarding rational numbers in adult magazines were modified by a part of speech such as an adjective ("An astonishing 35 percent of all ..."). This finding suggested to Joram et al. that authors in

adult magazines do not provide a great deal of interpretation of numbers in their immediate context and hence numbers are usually allowed to speak for themselves.

- Four of the seven adult magazines contained within the pages sampled at least one table or graph. Overall, the seven magazines included four tables, four bar graphs, and one pyramid graph (used to show quantities).

These and other findings reported by Joram et al. suggest that percentages are the most common rational number in magazines used to convey statistical information (see also Parker & Leinhardt, 1995), and that numerical or statistical information may appear in tables and not only in graphs. In order to make full sense of statistical information appearing in magazines, adults should be able to understand plain passages that provide the context for the rational numbers or graphs shown, and relate different elements in given passages or displays to each other. These conclusions agree with and complement the earlier discussion of literacy skills needed for interpreting statistical messages.

Beyond the data provided by Joram et al. (1995), there is no comprehensive research base from which to establish the statistical literacy requirements in the full range of domains and environments where adults function. Five key parts of the statistical knowledge base required for statistical literacy are proposed in this subsection and summarized in Table 2. These building blocks were identified on the basis of reviewing writing by mathematics and statistics educators (such as Shaughnessy, 1992, Moore, 1990, 1997b; chapters in Steen, 1997; chapters in Gal & Garfield, 1997; chapters in Lajoie, 1998; NCTM, 2000), sources on scientific literacy (e.g., Shamos, 1995; AAAS, 1995), and on mathematics and statistics in the news (e.g., Huff, 1954; Hooke, 1983; Crossen, 1994; Paulos, 1995; Kolata, 1997).

Table 2. Five parts of the statistical knowledge base

1.	Knowing why data are needed and how data can be produced
2.	Familiarity with basic terms and ideas related to descriptive statistics
3.	Familiarity with basic terms and ideas related to graphical and tabular displays
4.	Understanding basic notions of probability
5.	Knowing how statistical conclusions or inferences are reached

Knowing Why Data Are Needed and How Data Can Be Produced

Overall, adults should possess some understanding of the origins of the data on which reported findings or displays are based, understand the need to know how data were produced, and be aware of the contribution of a good design for data production to the possibility of answering specific questions (Cobb & Moore, 1997).

Adults should also be aware that public officials, organizations, employers, advertisers, and other players in the public arena need to base claims or conclusions on credible empirical evidence, and that properly produced data can inform public debate and serve as a basis for decisions and allocation of resources, much better than anecdotal evidence (Moore, 1998).

To enable critical understanding of reported findings or data-based claims, adults should possess some knowledge, at least informal, of key "big ideas" that underlie statistical investigations (Garfield & Gal, 1999). First on the list of most statisticians is the existence of variation (Moore, 1998). The need to reduce data in order to identify key features and trends despite noise and variation should be understood by adults as it provides the basis for accepting the use of statistical summaries (e.g., means, graphs) as tools for conveying information from data producers to data consumers (Wild & Pfannkuch, 1999).

Further, adults should possess some understanding of the logic behind key research designs commonly mentioned in the media, primarily experiments and the reason for using experimental and control groups to determine causal influences (see excerpt #6 in Figure 1); census (excerpt #2); polls/surveys (excerpts #3 and #4); and perhaps the role and limitations of a pilot study. Given the prevalence of polls and surveys, adults should also understand, at least intuitively, the logic of sampling, the need to infer from samples to populations, and the notions of representativeness and especially bias in this regard (Cobb & Moore, 1997; Wild & Pfannkuch, 1999). Some specific ideas to be known in this regard are the advantages of probability sampling, the dangers of convenience sampling, or the influence of the sampling process, sample size, and sample composition on researchers' ability to generalize safely and infer about a population from sample data.

Familiarity with Basic Terms and Ideas Related to Descriptive Statistics

Assuming adults understand why and how data are produced, they need to be familiar with basic concepts and data displays that are commonly used to convey findings to target audiences. Two key types of concepts whose centrality is noted by many sources are percentages (Parker & Leinhardt, 1995) and measures of central tendency, mainly the arithmetic mean (often termed "average" in newspapers) but also the median. Gal (1995) argues that it is desirable for consumers of statistical reports to know that means and medians are simple ways to summarize a set of data and show its "center"; that means are affected by extreme values, more so than medians; and that measures of center can mislead when the distribution or shape of the data on which they are based is very uneven or bimodal, or when the data or sample from which they are calculated is not representative of the whole population under study (see excerpt #5 in Figure 1). More broadly, it is useful for adults to be aware that different types of seemingly simple summary indices (i.e., percentage, mean, median) may yield different, and at times conflicting, views of the same phenomena.

Familiarity with Graphical and Tabular Displays and Their Interpretation

Adults should know that data can be displayed or reported in both graphical and tabular displays, which serve to organize multiple pieces of information and enable the detection or comparison of trends in data (Tufte, 1997). In this regard, one hopes that adults can first of all perform literal reading of data in tables or graphs, be familiar with standard conventions in creating graphs and charts, and be attentive to simple violations of such conventions (Bright & Friel, 1998) such as those in the graph in Figure 3: The relative length of the bars is not proportional to the actual percentages, and neither is the positioning of the boxes with percentages inside each bar; the decision of the graphical artist to add a female figure on the left (probably for decoration or to gain attention) masks the length of some bars and renders the visual appearance misleading. In this case, one hopes that readers realize the need to examine the actual percentages.

It is also expected that adults can do, on some level, what Curcio (1987) and Wainer (1992) call "reading between the data" and "reading beyond the data," such as understand that projections can be made from given data, and that one should look at overall patterns and not only specific points in a graph or a table (Gal, 1998). Adults should also realize that different graphs and tables may yield different (and possibly conflicting) views of the phenomena under investigation. Finally, adults should be aware that graphs can be intentionally created to mislead or highlight/hide a specific trend or difference. Various examples in this regard have been presented by Huff (1954). (See also Orcutt & Turner's [1993] analysis, discussed later, of how *Newsweek* magazine manipulated survey data on drug use to advance a specific point of view).

Understanding Basic Notions of Probability

Ideas regarding chance and random events are explicit or implicit in many types of messages adults encounter. Many statistical reports make probabilistic statements in the context of presenting findings from surveys or experiments, such as the likelihood of obtaining certain results (see excerpts #1 and #6 in Figure 1). Messages can also include probabilistic estimates made by various professionals (weather forecasters, genetic counselors, physicians, admissions administrators in colleges) regarding the likelihood of various events or the degree of confidence in their occurrence (rain, risks, side effects, or acceptance, respectively). Some of these claims may not be based on statistical studies, and could be couched in subjective estimates of individuals.

It is safe to expect that at a minimum, adults should be sensitive to the problem of interpreting correctly the "language of chance" (Wallsten, Fillenbaum, & Cox, 1986). Adults should have a sense for the many ways in which estimates of probability or risk are communicated by various sources, such as by percentages, odds, ratios, or verbal estimates. (Excerpt #6 illustrates how these combine in complex ways within a single article.)

Next, there is a need for adults to be familiar with the notion of randomness, understand that events vary in their degree of predictability or independence, yet also that some events are unpredictable (and hence that co-occurrence of certain events does not mean that they are necessarily related or cause each other). Unfortunately, while possible, it is difficult to present more advanced or explicit expectations for adults in terms of understanding random processes without appearing simplistic or naive. People from all walks of life have been shown to hold many misconceptions and discontinuities in understanding and reasoning about stochastic phenomena (Konold, 1989; Gal & Baron, 1996; Shaughnessy, Garfield, & Greer, 1997). Further, understanding of random phenomena also takes part in cognitive processes of judgment, decision making, and rationality, in which various deficiencies have been documented as well (Baron, 1988; Mellers, Schwartz, & Cooke, 1998).

Nonetheless, if adults are to understand and critically evaluate probabilistic claims, they should at least recognize the importance of ascertaining the *source* for probability estimates. Adults should realize that estimates of chance and risk may originate from diverse sources, both formal (e.g., frequency data, modeling, experimentation) and subjective or anecdotal, and that estimates may have different degrees of credibility or accuracy. Thus, they should expect that the evidence or information basis for statements of chance can be specified by those who make claims, and that judgments of chance may fluctuate and forecasts may change when additional data become available (Clemen & Gregory, 2000).

A final and more advanced expectation is that adults understand, at least intuitively, the idea of a chance variability in (random) phenomena. As Cobb and Moore (1997) explain, "When a chance mechanism is explicitly used to produce data, probability ... describes the variation we expect to see in repeated samples from the same population" (p. 813). Some understanding of probability is thus also a gateway to making sense of statements about the significance of differences between groups or likelihood of obtaining certain results, since standard statistical inference is based on probability (Cobb & Moore, 1997).

Knowing how statistical conclusions or inferences are reached.

Whereas most adults are data consumers and not producers, they do need to have a grasp on some typical ways to summarize data, such as by using means or medians, percentages, or graphs. However, given that there are different designs for collecting data, and that sampling processes or random processes may be involved, adults also need to possess some sense of how data are analyzed and conclusions reached, and be aware of relevant problems in this regard.

First, adults need to be sensitive to the possibility of different *errors* or *biases* (in sampling, in measurement, in inference) and possess a healthy concern regarding the stability and generality of findings. Second, it is useful to realize that errors may be controlled through proper design of studies, and can be estimated and described (e.g., by means of probability statements). One concept mentioned in the media in this regard is "margin of error" (see excerpt #3 in Figure 1, and the implicit mentioning of inflated scores in excerpt #5). Third, it is useful to know that there are ways to determine the significance or "trueness" of a difference between groups, but that this requires attention to the size of the groups studied, to the quality of the sampling process and the possibility that a sample is biased (understanding of these notions is needed if one is to think critically of the claims in excerpts #1 and #6). Finally, it is important to be aware that observed differences or trends may exist but may not necessarily be large or stable enough to be important, or can be caused by chance processes (as is the case with the reported increase in sexual intercourse in excerpt #4).

Mathematical Knowledge Base

A determination of the types of mathematical knowledge expected of adults to support statistical literacy should be made with caution. On the one hand, adults clearly need to be aware of some of the mathematical procedures underlying the production of common statistical indicators, such as percent or mean. At the same time, expectations regarding the amount and level of formal mathematics needed to comprehend basic statistical ideas taught at the introductory college level (or in high schools) have been changing in recent years (Moore, 1998). A brief detour to describe leading ideas in this regard is offered below to help frame later statements about the mathematical knowledge base needed for statistical literacy.

Statisticians have gradually clarified over the last few years the nature of some fundamental differences between mathematics and statistics (Moore & Cobb, 2000), and have formulated some working assumptions about the general level of mathematics one needs to learn statistics, at least at the introductory college level. Cobb and Moore (1997) summarize recommendations of the ASA/MAA committee on statistics instruction (Cobb, 1992), and suggest that while statistics makes heavy use of mathematics, statistics instruction at the introductory college level should focus on *statistical* ideas (need for data and importance of data production, omnipresence of variability, need to explain and describe variability).

Understanding the mathematical derivations that underlie key ideas presented in introductory statistics is of some importance but should be kept limited, since computers now automate many computations. While there is no intention of leading students to accept statistical derivations as magic (i.e., without knowing any of the underlying mathematics), too much emphasis on mathematical theory is not expected early on; it may disrupt the development of the necessary intuitive understanding of key statistical ideas and concepts that often do not have mathematical representations and are unique to the discipline of statistics (Moore, 1997a; Wild & Pfannkuch, 1999). Cobb and Moore (1997) further claim that probability is conceptually the hardest subject in elementary mathematics, and remind that psychological studies have documented confusion about probability even among those who master the computational side of probability theorems and can solve textbook exercises. Hence, even for understanding of the formal aspects of inference or of probability, only a limited amount of mathematical knowledge is expected.

The above logic can help in determining the mathematical knowledge that adults need to support statistical literacy. Given that most adults in any country do not study statistics at the college level (Moore & Cobb, 2000; UNESCO, 2000), the amount and level of formal knowledge of mathematics needed to support adult statistical literacy can be restricted.

Perhaps the simplest knowledge expected of adults is the realization that any attempt to summarize a large number of observations by a concise quantitative statement (percentage, mean, probability, etc.) requires some application of mathematical tools and procedures. Adults need to have numeracy skills at a sufficient level to enable correct interpretation of numbers used in statistical reports. "Number sense" is increasingly being touted as an essential skill for proper understanding of diverse types of numbers (Paulos, 1995; Curry, Schmitt, & Waldron, 1996; Scheaffer et al., 1998; NCTM, 2000), such as large numbers (e.g., trends in GNP) and small numbers, including fractions, decimals, and percents (e.g., estimates of risk or side effects).

Understanding of basic statistical findings pertaining to percentages or "averages" requires familiarity, intuitive and to some extent formal, with underlying mathematical procedures or computations used to generate these statistics (Garfield & Gal, 1999). Citizens should know *how* an arithmetic mean is computed in order to fully appreciate the meaning of the claim that an arithmetic mean can be influenced by extreme values in a data set and hence may not represent the "middle" of a set of values if the data are skewed. Excerpt #5 shows a variant on this demand, that is, understanding of the impact of *excluding* a certain proportion of extreme observations (6% in the example given) on the central tendency.

Many types of statistical information reported in the media are described in terms of percentages (Joram et al., 1995) and are sometimes included in graphs. Numerous examples can be found in Figures 1 and 2. Percentage is a seemingly simple mathematical concept, commonly perceived as expressing a proportion or ratio; it is presumably mastered in the middle grades, and hence it could be expected that the vast majority of schooled adults will understand it. Yet, its understanding is far from being simple. Parker and Leinhardt (1995) address the prevalence and complexity of percentages, and also point to specific types of percentages that normally are not encountered in routine classroom teaching but may appear in newspaper statements, such as percentages larger than 100% or percentage of percent. These authors argue that generations of students, including at the college level, have failed to fully master percentage, in part because it is a multifaceted concept that has multiple mathematical meanings and also statistical uses (e.g., a number, an expression of a relationship, a statistic, a function, an expression of likelihood). Understanding the mathematical and statistical meaning of a reported percentage can be difficult. Readers may have to make inferences and assumptions, for example, when a message does not specify the base for calculating a percentage. Percentages may represent complex relationships (e.g., conditional probabilities) and, as illustrated in Figure 1, may be linked to concepts that themselves have multiple meanings (such as "15 percent below average," "2% margin of error").

The examples pertaining to percentages and computations of means and medians imply that interpretation of even seemingly simple statistics reported in the media requires some familiarity with their derivation (though not always formal training in this regard). It follows that adults should understand, at least informally, some of the mathematics involved in generating certain statistical indicators, as well as the mathematical connection between summary statistics, graphs, or charts, and the raw data on which they are based.

Questions about the amount of mathematics one needs to know to understand more sophisticated concepts are more difficult to answer and have been the source of some debate among statistics and mathematics educators (Moore, 1997a). Terms or phrases that appear in the media, such as "margin of error" or "statistically significant difference" *can* be understood intuitively in a way that can help adults without formal statistical training make a superficial sense of news items. After all, such ideas are being successfully taught at an introductory level to children in elementary or middle schools (Friel, Russell, & Mokros, 1990). However, deeper understanding of the above or related concepts, and proper interpretation of their *exact* meaning, requires more solid understanding of underlying statistical ideas (quantification of variance, repeated sampling, sampling distributions, curves, logic of statistical inference, etc). These ideas are hard to grasp for college-bound students (Cobb & Moore, 1997; Watson & Moritz, 2000) even without the added complication of the need to understand their mathematical underpinnings.

Context/World Knowledge Base

Proper interpretation of statistical messages by adults depends on their ability to place messages in a context, and to access their world knowledge. World knowledge also supports general literacy processes and is critical to enable "sense-making" of any message. Moore (1990) has argued that in statistics, the context motivates procedures; data should be viewed as numbers with a context, and hence the context is the source of meaning and basis for interpretation of obtained results. In reading contexts, however, people do *not* engage in generating any data or in carrying any computations or analysis. Their familiarity with the data-generation process (e.g., study design, sampling plan, questionnaires used), or with the procedures employed by the researchers or statisticians to analyze the data, depends on the details and clarity of the information given in the messages presented to them. As passive receivers of messages, they are at the mercy of message creators.

It follows that adults' ability to make sense of statistical claims or displays will depend on whatever information they can glean from the message about the background of the study or data being discussed. Context knowledge is the main determinant of the reader's familiarity with *sources for variation and error.* If a listener or reader is not familiar with a context in which data were gathered, it becomes more difficult to imagine why a difference between groups can occur, what alternative interpretations may exist for reported findings about an association detected between certain variables, or how a study could go wrong.

The ways in which a study is reported in the media can easily mask or distort the information available to the reader about the source of the evidence presented. An example is when a reporter uses the term *experiment* in a way that enhances the face validity of a study that is nonexperimental in nature. Thus world knowledge, combined with some literacy skills, is prerequisite for enabling critical reflection about statistical messages and for understanding the implications of the findings or numbers reported. Adults can be helped by having a sense for, and expectations about, elements of good journalistic writing, such as for objective writing, presentation of two-sided arguments, accuracy in reporting, or provision of background information to orient readers to the context of a story.

Critical Skills

Messages aimed at citizens in general may be shaped by political, commercial, or other agendas which may be absent in statistics classrooms or in empirical enquiry contexts. Fred Mosteller said, "Policy implies politics, and politics implies controversy, and the same data that some people use to support a policy are used by others to oppose it" (cited in Moore, 1998, p. 1255). Not surprisingly, the need for critical evaluation of messages to the public has been a recurring theme in writings of educators interested in adults' literacy and numeracy (Freire, 1972; Frankenstein, 1989).

As noted in discussing literacy skills, messages in the general media are produced by very diverse sources, such as journalists, politicians, manufacturers, or advertisers. Depending on their needs and goals, such sources may not necessarily be interested in presenting a balanced and objective report of findings or implications. A potent example is Orcutt and Turner's (1993) analysis of how the print media, especially *Newsweek* magazine, selectively analyzed and intentionally manipulated trend data collected by the Institute for Social Research (ISR) regarding drug use among American high-school students between 1975 and 1985. According to Orcutt & Turner, the media attempted to created for the public an image of a "drug plague," by selecting at its convenience only some of the data collected as part of a multiyear survey project, using graphical methods to augment small percentage differences (after truncating and censorizing), to appear visually large.

Orcutt and Turner (1993) add that later in 1992, *Newsweek* attempted again to create a sense of national danger by reporting that the use of LSD is "rising alarmingly" and that for the first time since 1976, more high-school seniors used LSD than cocaine. However, analysis of the ISR data on which *Newsweek* based this argument showed that this argument had no empirical basis. Cocaine use decreased from 6.5% in 1989 to 5.3% in 1990, a statistically significant change (given sample size used), whereas LSD use increased from 4.9% to only 5.4%, which was within the range of sampling error. The contrast between these figures, which were available to *Newsweek,* and the narrative and graphs used in the articles published, suggest an intentional misuse of data and highlights the media's tendency for sensational reporting practices.

Excerpts #4 and #6 in Figure 1 further illustrate how data can be tailored to serve the needs of specific organizations (e.g., states and manufacturers), and how reports about data are shaped to influence the opinions of the listener or reader in a specific direction. Paulos (1995, p. 79) notes that originators of messages regarding diseases, accidents, or other misfortunes that afflict humans, depending on their interest, can make them appear more salient and frightening by choosing to report absolute numbers (e.g., 2,500 people nationwide suffer from X), or in contrast can downplay them by using incidence rate (e.g., 1 in every 100,000 people suffer from X). Many examples are also presented by Huff (1954) and Crossen (1994).

In light of such examples, and the possibility for biased reporting (Wanta, 1997), adults have to worry about and examine the reasonableness of claims presented in the media. They have to be concerned about the validity of messages, the nature and credibility of the evidence underlying the information or conclusions presented, and reflect upon possible alternative interpretations of conclusions conveyed to them. It follows that adults should maintain in their minds a list of "worry questions" regarding statistical information being communicated or displayed (Gal, 1994; Moore, 1997b; Garfield & Gal, 1999). Ten such questions are listed in Table 3. When faced with an interpretive statistical task, people can be imagined running through this list and asking for each question, "Is this question relevant for the situation/message/task I face right now?"

The answers people generate to these and related questions can support the process of critical evaluation of statistical messages and lead to the creation of more informed interpretations and judgments. This list can of course be modified, and some of its elements regrouped, depending on the life contexts and functional needs of different adults. It can expand beyond basic statistical issues to cover broader issues of probability and risk, or job-specific statistical topics such as those related to statistical process control or quality assurance.

Table 3. Sample "worry questions" about statistical messages

1. Where did the data (on which this statement is based) come from? What kind of study was it? Is this kind of study reasonable in this context?
2. Was a sample used? How was it sampled? How many people did actually participate? Is the sample large enough? Did the sample include people/units which are representative of the population? Is the sample biased in some way? Overall, could this sample reasonably lead to valid inferences about the target population?
3. How reliable or accurate were the instruments or measures (tests, questionnaires, interviews) used to generate the reported data?
4. What is the shape of the underlying distribution of raw data (on which this summary statistic is based)? Does it matter how it is shaped?
5. Are the reported statistics appropriate for this kind of data? E.g., was an average used to summarize ordinal data; is a mode a reasonable summary? Could outliers cause a summary statistic to misrepresent the true picture?
6. Is a given graph drawn appropriately, or does it distort trends in the data?
7. How was this probabilistic statement derived? Are there enough credible data to justify the estimate of likelihood given?
8. Overall, are the claims made here sensible and supported by the data? E.g., is correlation confused with causation, or a small difference made to loom large?
9. Should additional information or procedures be made available to enable me to evaluate the sensibility of these arguments? Is something missing? E.g., did the writer "conveniently forget" to specify the base of a reported percent-of-change, or the actual sample size?
10. Are there alternative interpretations for the meaning of the findings or different explanations for what caused them, e.g., an intervening or a moderator variable affected the results? Are there additional or different implications that are not mentioned?

Interaction of Knowledge Bases

Five knowledge bases were described above separately for ease of presentation, but they overlap and do not operate independently from each other. For example, familiarity with possible language ambiguities and reporting conventions comprises part of the literacy skills required of adults, yet they are also part of general world knowledge, and related to the need for knowledge about intentional (and possibly biased) reporting practices listed as part of critical skills. Some aspects of the statistical knowledge base overlap with mathematical knowledge, for example regarding the difference in the computational procedures used to find medians and means and their implication for interpretation of such statistics under different conditions.

The characteristics of certain real-world messages require that adults jointly activate all the knowledge based described in order to manage tasks at hand (Gal,

1997). Figure 2 exemplifies the complex task that may face readers of print media with regard to interpreting information of a statistical nature, and illustrates the interconnected nature of the knowledge bases that underlie people's statistical literacy.

Figure 2 recreates a portion of a table that appeared in *USA Today* (a nationally circulated daily newspaper) in 1999. This table combines an offbeat opening passage with a tabular display of several simple lists, each containing information of a different nature: absolute numbers, averages, percentages. Interpretation of the table requires not only basic familiarity with averages and percentages, but also literacy skills and access to different kinds of background knowledge. Some details needed to make complete sense of the mathematical information are not fully stated, forcing the reader to perform inferences, based on his or her general world knowledge: averages are denoted as "avg." and percentages as "pct. chg," both nonstandard abbreviations; the averages are "per site," but it is not explained what is a "site" nor if the average is calculated for a whole week or a weekend only; percentages describe *change* in negative numbers, yet the base is not given, only implied.

DISPOSITIONAL ASPECTS OF STATISTICAL LITERACY

The notion of "critical evaluation," highlighted in several of the conceptions of statistical literacy cited earlier (e.g., Wallman, 1993), implies a form of action, not just passive interpretation or understanding of the statistical or probabilistic information available in a situation. It is hard to describe a person as fully statistically literate if this person does not show the *inclination to activate* the five knowledge bases described earlier or share with others his or her opinions, judgments, or alternative interpretations.

Statistically literate action can take many forms, both overt and hidden. It can be an internal mental process, such as thinking about the meaning of a passage one read, or raising in one's mind some critical questions and reflecting about them. It can be extended to more external forms, such as rereading a passage, scanning a graph one encountered in the newspaper, stopping a game of chance after one remembers reading an article about the Gambler's Fallacy, or discussing findings of a survey one heard about on TV with family members at the dinner table or with co-workers. However, for any form of action to occur and be sustained, certain dispositions need to exist and be activated.

The term *dispositions* is used here as a convenient aggregate label for three related but distinct concepts—critical stance, beliefs, and attitudes—which are all essential for statistical literacy. These concepts are interconnected (McLeod, 1992), and hence are harder to describe in a compartmentalized way, unlike the description of the five knowledge bases above. This section first describes critical stance, and then examines beliefs and attitudes that underlie a critical stance.

Critical Stance

A first expectation is that adults hold a propensity to adopt, without external cues, a questioning attitude toward quantitative messages that may be misleading, one-sided, biased, or incomplete in some way, whether intentionally or unintentionally (Frankenstein, 1989). They should be able and willing to spontaneously invoke their personal list of worry questions (see Table 3) when faced with arguments that purport to be based on data or with reports of results or conclusions from surveys or other empirical research (Gal, 1994).

It is important to keep in mind that willingness to invoke action by adults when they encounter statistical information or messages may sometimes be required under conditions of uncertainty. Examples are lack of familiarity with the background of the issues discussed or estimates conveyed, partial knowledge of concepts and their meanings, or the need to cope with technical terms that "fly above the head" of the Reader. This may be the case for many adults without much formal education or effective literacy skills, who constitute a sizable percentage of the population in many countries (Statistics Canada and OECD, 1996; UNESCO, 2000). Action or reaction in such situations may involve taking some personal risks, i.e., exposing to others that one is naive about, or unfamiliar with, certain statistical issues, and possibly suffering some embarrassment or the need to argue with others.

Beliefs and Attitudes

Certain beliefs and attitudes underlie people's critical stance and willingness to invest mental effort or occasionally take risks as part of acts of statistical literacy. There is a definitional challenge in discussing "beliefs" and "attitudes," as the distinction between them is somewhat murky. (Researchers, for example, often implicitly defined statistics attitudes or beliefs as whatever their favorite assessment instrument measures in the context of a specific target population, such as school students, college students, or adults at large).

Based on McLeod's (1992) work on affective aspects of mathematics education, a distinction should be made between emotions, attitudes, and beliefs (see also Edwards, 1990; Green, 1993). Emotions are transient positive and negative responses triggered by one's immediate experiences (e.g., while studying mathematics or statistics, or while facing a certain probabilistic situation, such as receiving medical information about the chances of side effects of a proposed treatment). Attitudes are relatively stable, intense *feelings* that develop through gradual internalization of repeated positive or negative emotional responses over time. Attitudes are expressed along a positive–negative continuum (like–dislike, pleasant–unpleasant), and may represent, for example, feelings toward objects, actions, or topics ("I don't like polls and pollsters, they always confuse me with numbers"). Beliefs are individually held *ideas* or opinions, such as about a domain ("government statistics are always accurate"), about oneself ("I am really naive about statistical information," "I am not a numbers person"), or about a social context ("The government should not waste money on big surveys"; see Wallman,

1993). Beliefs take time to develop and cultural factors play an important part in their development. They have a larger cognitive component and less emotional intensity than attitudes, and are stable and quite resistant to change compared to attitudes.

Adults should develop a positive view of themselves as individuals capable of statistical and probabilistic reasoning as well as a willingness and interest to "think statistically" in relevant situations. This assumes that adults hold some *appreciation for the power of statistical processes,* and accept that properly planned studies have the potential to lead to better or more valid conclusions than those obtained by relying on anecdotal data or personal experiences (Moore, 1998). Broader metacognitive capacities that are considered part of people's general intellectual functioning can further support statistically literate behavior, such as having a propensity for logical reasoning, curiosity, and open-minded thinking (Baron, 1988).

Gal, Ginsburg, and Schau (1997) examined the role of attitudes and beliefs in statistics education, and argued that to enable productive problem solving, learners need to feel safe to explore, conjecture, and feel comfortable with temporary confusion or a state of uncertainty. It was argued earlier that reading contexts, where people are data consumers, differ in several ways from those encountered in inquiry contexts such as those addressed by Gal et al. (1997). Yet, some commonality between these two contexts does exist regarding the required beliefs that support action. Even in reading contexts adults have to feel safe to explore and hypothesize, feel comfortable being in the role of a critical reader or listener, and believe in their ability to make sense of messages (Gal, 1994), as a condition for developing and sustaining their motivation for critical action.

Finally, we come to a point where "critical stance" and "beliefs and attitudes" mesh together. For a critical stance to be maintained, adults should develop a *belief in the legitimacy of critical action.* Readers should uphold the idea that it is legitimate to be critical about statistical messages or arguments, whether they come from official or other sources, respectable as they may be. Adults should agree that it is legitimate to have concerns about any aspect of a reported study or a proposed interpretation of its results, and to raise pertinent "worry questions," even if they have not learned much formal statistics or mathematics, or do not have access to all needed background details.

DISCUSSION AND IMPLICATIONS

This paper's main goal was to propose a conceptualization of statistical literacy and describe its key components. Given the patchy literature on statistical literacy, the availability of such a model was seen as a necessary prefatory step before further scholarly discussion can ensue regarding the issues involved in developing or studying adult statistical literacy. Statistical literacy was portrayed in this paper as the ability to interpret, critically evaluate, and if needed communicate about statistical information, arguments, and messages. It was proposed that statistically literate behavior requires the joint activation of five interrelated knowledge bases (literacy, statistical, mathematical, context/world, and critical), yet that such behavior is predicated on the presence of a critical stance and supporting beliefs and attitudes.

The proposed conceptualization highlights the key role that *nonstatistical* factors and components play in statistical literacy, and reflects the broad and often multifaceted nature of the situations in which statistical literacy may be activated. That said, several observations should be made. First, the five knowledge bases discussed in this paper were sketched in broad strokes to clarify the key *categories* of knowledge to be considered when thinking of what adults need to know to be statistically literate. Each could be modified or elaborated, depending on the cultural context of interest, and on the sophistication of statistical literacy expected of citizens or workers in a given country or community. As with conceptions of other functional skills, the particulars viewed as essential for statistical literacy in a specific country will be dynamic and may have to change along with technological and societal progress.

Secondly, although five knowledge bases and a cluster of beliefs, attitudes, and a critical stance were proposed as jointly essential for statistical literacy, it does not necessarily follow that a person should fully possess all of them to be able to effectively cope with interpretive tasks in all reading and listening contexts. Following current conceptions of adult literacy (Wagner et al., 1999) and numeracy (Gal, 2000), statistical literacy should be regarded as a set of capacities that can exist to different degrees within the same individual, depending on the contexts where it is invoked or applied. Descriptions of what constitutes statistical literacy may differ in work contexts, in personal/home contexts, in public discourse contexts, and in formal learning contexts.

In light of the centrality of statistical literacy in various life contexts, yet also its complex nature, educators, statisticians, and professionals interested in how well citizens can interpret and communicate about statistical messages face numerous challenges and responsibilities. Below is a preliminary discussion regarding two key areas, education for statistical literacy, and suggested research in this area.

Educational Challenges

Several countries and organizations have introduced programs to improve school-level education on data analysis and probability, sometimes called data handling, stochastics, or chance (Australian Education Council, 1991; NCTM, 2000). Yet, at the school level, where most individuals will receive their only formal exposure to statistics (Moore, 1998), these topics overall receive relatively little curricular attention compared to other topics in the mathematical sciences. The most credible information in this regard comes from the curriculum analysis component of TIMSS, the Third International Mathematics and Science Study (Schmidt, McKnight, Valverde, Houang, & Wiley, 1997), which examined curriculum documents and textbooks and consulted with expert panels from over 40 countries. TIMSS data also pointed to an enormous diversity in curricular frameworks. Various gaps have been documented by TIMSS between the intended and implemented curriculum, (i.e., between curriculum plans and what actually appears in mainstream textbooks, which tend to be conservative).

TIMSS tests included few statistics items; hence, it was not possible to create a separate scale describing student performance in statistics. However, achievement on individual statistical tasks was problematic. For example, Mullis, Martin, Beaton, Gonzalez, Kelly, & Smith (1998) reported performance levels of students in their *final* year of schooling (usually grade 12) on a task directly related to statistical literacy: Explain whether a reporter's statement about a "huge increase" was a reasonable interpretation of a bar graph showing the number of robberies in two years that was manipulated to create a specific impression. The graph included a bar for each year but a truncated scale, causing a small difference between years to appear large. Performance levels varied across countries; on average, *less than half* of all *graduating* students appeared to be able to cope (at least partially) with this task that exemplifies one of the most basic skills educators usually use as an example for a statistical literacy skill expected of all citizens—the ability to detect a discrepancy between displayed data and a given interpretation of these data. Keeping in mind that in many countries a sizable proportion of students drop out or leave *before* the final year of high school, the overall percentage of all school leavers who can cope with such tasks is bound to be even lower.

Efforts to improve statistics education at the secondary or postsecondary levels examine needed changes in a range of areas, including in content and methods, teacher preparation and training, assessments, and the use of technology (e.g., Cobb, 1992; Pereira-Mendoza, 1993; Gal & Garfield, 1997; Lajoie, 1998). Yet a crucial question is, to what extent can such efforts develop students' interpretive and statistical literacy skills? To appreciate the complexity of the issues implicated by this question, consider the situation in the related area of scientific literacy. Eisenhart, Finkel, & Marion (1996) have argued that the broad, progressive, and inclusive vision of scientific literacy in reform proposals is being implemented in narrow and conventional ways; hence reform efforts may not lead to significant changes in national scientific literacy. To help define educational goals, it may be possible to identify levels of statistical literacy (Watson, 1997; Watson & Moritz, 2000) in a similar fashion to the continuum proposed to describe levels of scientific literacy (Shamos, 1995).

This paper argues that statistical literacy depends on possession of elements from *all* five different knowledge bases; and that literacy skills, contextual knowledge, critical skills, and needed dispositions play a significant role in this regard. It is not at all clear that learning statistical facts, rules, and procedures, or gaining personal statistical experience through a data-analysis project in a formal classroom enquiry context can in itself lead to an adequate level of statistical literacy.

Calls to change traditional approaches to teaching statistics have been repeatedly made in recent years, and met with some success (Moore & Cobb, 2000). Yet, educators have to distinguish between teaching more statistics (or teaching it better) and teaching statistics *for a different (or additional) purpose.* Literacy demands facing students who are learning statistics are more constrained than those described in the section on "Literacy skills" as characterizing reading contexts. When students who learn statistics read or listen to project reports created by their fellow students (Starkings, 1997), or when they read academic research papers, findings and conclusions are likely to be shared through language that is less varied than what appears in real-world sources. This may happen because academic conventions inhibit or channel the type of expressions and styles that authors, students, and teachers are expected to use, or due to logistical limitations in large introductory statistics courses that restrict the richness and scope of classroom discourse that teachers can afford to conduct (Wild, Triggs, & Pfannkuch, 1997). Unlike consumers of the media, when students encounter an unfamiliar or ambiguous term, they can clarify its interpretation by talking with their teacher. The upshot is that the literacy demands in statistics classes do not necessarily represent the heterogeneous communicative environment within which adults in general have to cope with statistical messages.

To develop statistical literacy, it may be necessary to work with learners, both younger students and adults, in ways that are different from, or go beyond, instructional methods currently in use. To better cover all knowledge bases supporting statistical literacy, topics and skills that are normally not stressed in regular statistics modules or introductory courses, for lack of time or teacher preparation, may have to be addressed. Some examples are

- Understanding results from polls, samples, and experiments (Landwehr, Swift, & Watkins, 1987; MacCoun, 1998) *as reported in newspapers or other media channels*
- Understanding probabilistic aspects of statements about risk and side effects (Clemen & Gregory, 2000) *as reported in newspapers or other media channels*
- Learning about styles, conventions, and biases in journalistic reporting or advertisements
- Gaining familiarity with "worry questions" (Table 3), *coupled* with experience in applying them to real examples (such as one-sided messages, misleading graphs), or seeing someone else (e.g., a teacher) model their application
- Developing a critical stance and supporting beliefs, including positive beliefs and attitudes about the domain (usefulness of statistical investigations) and oneself

TIMSS reports on curriculum planning and other school-related variables imply that young people who will be leaving schools in coming years may continue to have insufficient preparation in data analysis and probability. An important and presently much larger population is that of adults in general. The majority of the current adult population in any country has not had much if any formal exposure to the statistical or mathematical knowledge bases described earlier, given known education levels across the world (Statistics Canada & OECD, 1996; UNESCO, 2000). As IALS (OECD & Human Resources Development Canada, 1997) and other studies have shown, even in industrialized countries, literacy levels of many adults are low. This paper argues that literacy skills, including document literacy skills, are an important component of the knowledge base needed for statistical literacy. It follows that achieving the vision of "statistical literacy for all" will require a concerted effort by various educational and other systems, both formal and informal.

Large numbers of adult learners receive important educational services from adult basic education centers, adult literacy programs, workplace learning and union-based programs, and continuing education or tertiary institutions. These services have an important role in promoting statistical literacy of adults, and some have began to formally recognize the need to attend to statistical issues and to critical evaluation of messages as part of designing curricula for adult learners (European Commission, 1996; Curry et al., 1996; Stein, 2000). Yet, media organizations and media professionals (Orcutt & Turner, 1993), public and private agencies and institutes that communicate with the public on statistical matters, such as national statistical offices (Moore, 1997b), and even marketers and advertisers (Crossen, 1994), all have some responsibility in this regard. All the above stakeholders will have to devise innovative and perhaps unorthodox ways in order to jointly reach and increase statistical literacy in the general population.

Research and Assessment Challenges

As pointed out earlier, the current knowledge base about statistical literacy of school or university students and of adults in general is patchy. In the absence of solid empirical information, the speculative ideas raised in this paper may not translate into action by decision makers who are in a position to allocate resources to educational initiatives. Three related areas where further research is needed are as follows.

Research on Students' and Adults' Statistical Literacy Skills

Studies such as TIMSS (aimed at school students) and IALS (aimed at adults) provided useful but only preliminary data on restricted aspects of people's statistical literacy, mainly because their main thrust was planned to address other mathematical topics. Many knowledge elements basic to statistical literacy were left out of these assessments (e.g., understanding of averages and medians, knowledge about sampling or experimental designs, or understanding of chance-related statements). New international large-scale assessments, such as OECD's Program for International Student Achievement (http://www.pisa.oecd.org), or the Adult Literacy and Lifeskills survey (http://nces.ed.gov) will include broader coverage of statistical matters, in line with expanded notions of mathematical literacy and numeracy developed for these projects. However, given the restrictions on testing time in large-scale studies and the number of domains competing for item coverage, focused studies are needed that can provide more comprehensive information on statistical literacy skills and related attitudes, and on gaps in this regard. Qualitative studies should further enable in-depth examination of thinking processes, comprehension, and effects of instruction in this regard.

Research on Statistical Literacy Demands of Various Functional Environments

The Joram et al. (1995) findings reported earlier shed some light on the range of ways in which selected statistical and numerical information can be conveyed to readers of magazines, and point to the strong linkage between literacy and statistical elements in print media. Yet, little is known about the demands facing consumers of other media channels, such as daily newspapers, workplace materials, or TV broadcasts, and with regard to a range of statistical and probabilistic topics beyond rational numbers. The absence of credible data from which to establish the statistical literacy requirements in the full range of domains where adults have to function is alarming. Research in this area, taking into account variation both within and between countries, is a prerequisite for designing effective and efficient instruction that aims at different levels of statistical literacy.

Research on Dispositional Variables

This paper argued that a view of statistical literacy as an action-oriented set of interrelated knowledge bases and skills, one which people will actually use in everyday contexts, must consider people's inclination to apply a critical stance and the motivations, beliefs, and attitudes that affect or support statistically literate behavior. However, the conceptualization and assessment of these variables present many challenges (Gal et al., 1997). Development of research methods in this regard is essential for understanding the forces that shape statistically literate behavior in different contexts. Changes in dispositions should be measured as part of evaluating the impact of educational interventions aimed at improving statistical literacy of people in all walks of life.

REFERENCES

American Association for the Advancement of Science (AAAS) (1995). *Benchmarks for science literacy.* Washington, DC: Author.

Australian Education Council (1991). *A national statement on mathematics for Australian schools.* Carlton, Victoria: Curriculum Corporation.

Baron. J. (1988). *Thinking and deciding*. New York: Cambridge University Press.

Bowen, D. E., & Lawler, E. E. (1992, Spring). The empowerment of service workers: What, why, how, and when. *Sloan Management Review*, 31–39.

Bright, G. W., & Friel, S. N. (1998). Graphical representations: Helping students interpret data. In S. P. Lajoie (Ed.), *Reflections on statistics: Learning, teaching, and assessment in grades K–12* (pp. 63–88). Mahwah, NJ: Erlbaum.

Carnevale, A. P., Gainer, L. J., & Meltzer, A. S. (1990). *Workplace basics: The essential skills employers want*. San Francisco: Jossey-Bass.

Cobb, G. W. (1992). Teaching statistics. In L. A. Steen (Ed.), *Heeding the call for change: Suggestions for curricular action* (pp. 3–43). Washington, DC: Mathematical Association of America.

Cobb, G. W., & Moore, D. S. (1997). Mathematics, statistics, and teaching. *American Mathematical Monthly, 104*, 801–823.

Cocking, R. R., & Mestre, J. P. (Eds.). (1988). *Linguistic and cultural influences on learning mathematics*. Hillsdale, NJ: Erlbaum.

Clemen, R., & Gregory, R. (2000). Preparing adult students to be better decision makers. In I. Gal (Ed.), *Adult Numeracy Development: Theory, Research, Practice*. Cresskill, NJ: Hampton Press.
Crossen, C. (1994). *Tainted truth: The manipulation of fact in America*. New York: Simon & Schuster.
Curcio, F. R. (1987). Comprehension of mathematical relationships expressed in graphs. *Journal for Research in Mathematics Education, 18*, 382–393.
Curry, D., Schmitt, M. J., & Waldron, W. (1996). *A Framework for adult numeracy standards: The mathematical skills and abilities adults need to be equipped for the future*. Final report from the System Reform Planning Project of the Adult Numeracy Network. Washington, DC: National Institute for Literacy. Available online at http://www.std.com/anpn/
Edwards, K. (1990). The interplay of affect and cognition in attitude formation and change. *Journal of Personality and Social Psychology, 59*, 202–216.
Eisenhart, M., Finkel, E., & Marion, S. F. (1996). Creating the conditions for scientific literacy: A re-examination. *American Educational Research Journal, 33*(2), 261–295.
European Commission. (1996). *White paper on education and training: Teaching and learning—towards the learning society*. Luxembourg: Office for official publications of the European Commission.
Frankenstein, M. (1989). *Relearning mathematics: A different "R"—radical mathematics*. London: Free Association Books.
Freire, P. (1972). *Pedagogy of the oppressed*. New York: Penguin.
Friel, S. N., Russell, S., & Mokros, J. R. (1990). *Used numbers: Statistics: middles, means, and in-betweens*. Palo Alto, CA: Dale Seymour Publications.
Gal, I. (1994, September). *Assessment of interpretive skills*. Summary of working group, Conference on Assessment Issues in Statistics Education. Philadelphia, PA.
Gal, I. (1995). Statistical tools and statistical literacy: The case of the average. *Teaching Statistics, 17*(3), 97–99.
Gal, I., & Baron, J. (1996). Understanding repeated simple choices. *Thinking and Reasoning, 2*(1), 1–18.
Gal, I. (1997). Numeracy: Reflections on imperatives of a forgotten goal. In L. A. Steen (Ed.), *Quantitative literacy* (pp. 36–44). Washington, DC: College Board.
Gal, I. (1998). Assessing statistical knowledge as it relates to students' interpretation of data. In S. Lajoie (Ed.), *Reflections on statistics: Learning, teaching, and assessment in grades K–12* (pp. 275–295). Mahwah, NJ: Erlbaum.
Gal, I. (1999). Links between literacy and numeracy. In D. A. Wagner, R. L. Venezky, and B. Street (Eds.), *Literacy: An international handbook* (pp. 227–231). Boulder, CO: Westview Press.
Gal, I. (2000). The numeracy challenge. In I. Gal (Ed.), *Adult numeracy development: Theory, research, practice* (pp. 1–25). Cresskill, NJ: Hampton Press.
Gal, I., & Garfield, J. (Eds.). (1997). *The assessment challenge in statistics education*. Amsterdam, Netherlands: International Statistical Institute/IOS Press.
Gal, I., Ginsburg, L., & Schau, C. (1997). Monitoring attitudes and beliefs in statistics education. In I. Gal & J. B. Garfield (Eds.), *The assessment challenge in statistics education* (pp. 37–54). Amsterdam, Netherlands: International Statistical Institute/IOS Press.
Garfield, J. B., & Gal, I. (1999). Assessment and statistics education: Current challenges and directions. *International Statistical Review, 67*(1), 1–12.
Green, K. E. (1993, April). *Affective, evaluative, and behavioral components of attitudes toward statistics*. Paper presented at the annual meeting of the American Educational Research Association, Atlanta, GA.
Hooke, R. (1983). *How to tell the liars from the statisticians*. New York: Marcel Dekker.
Huff, D. (1954). *How to lie with statistics*. New York: Norton.
Jenkins, E. W. (1996). Scientific literacy: A functional construct. In D. Baker, J. Clay, & C. Fox (Eds.), *Challenging ways of knowing in English, maths, and science* (pp. 43–51). London: Falmer Press.
Joram, E., Resnick, L., & Gabriele, A. J. (1995). Numeracy as a cultural practice: An examination of numbers in magazines for children, teenagers, and adults. *Journal for Research in Mathematics Education, 26*(4), 346–361.
Kahneman, D., Slovic, P., & Tversky, A. (Eds.) (1982). *Judgment under uncertainty: Heuristics and biases*. New York: Cambridge University Press.
Kolata, G. (1997). Understanding the news. In L. A. Steen (Ed.), *Why numbers count: Quantitative literacy for tomorrow's America* (pp. 23–29). New York: The College Board.
Kirsch, I., & Mosenthal, P. (1990). Understanding the news. *Reading Research Quarterly, 22*(2), 83–99.

Kirsch, I. S., Jungeblut, A., Jenkins, L., & Kolstad, A. (1993). *Adult literacy in America: A first look at the results of the National Adult Literacy Survey*. Washington, DC: National Center for Education Statistics.

Kirsch, I. S., Jungeblut, A., & Mosenthal, P. B (1998). The measurement of adult Literacy. In S. T. Murray, I. S. Kirsch, & L. B. Jenkins (Eds.), *Adult literacy in OECD countries: Technical report on the first International Adult Literacy Survey* (pp. 105-134). Washington, DC: National Center for Education Statistics, U.S. Department of Education.

Konold, C. E. (1989a). Informal conceptions of probability. *Cognition and Instruction, 6*, 59–98.

Kosonen, P. & Winne, P. H. (1995). Effects of teaching statistical laws on reasoning about everyday problems. *Journal of education psychology, 87*(1), 33-46.

Laborde, C. (1990). Language and mathematics. In P. Nesher & J. Kilpatrick (Eds.), *Mathematics and cognition* (pp. 53–69). New York: Cambridge University Press.

Lajoie, S. P. (Ed.). (1998). *Reflections on statistics: Learning, teaching, and assessment in grades K–12*. Mahwah, NJ: Erlbaum.

Landwehr, J. M., Swift, J., & Watkins, A. E. (1987). *Exploring surveys and information from samples*. (Quantitative literacy series). Palo Alto, CA: Dale Seymour Publications.

MacCoun, R. J. (1998). Biases in the interpretation and use of research results. *Annual Review of Psychology 49*, 259–287.

McLeod, D. B. (1992). Research on affect in mathematics education: A reconceptualization. In D. A. Grouws (Ed.), *Handbook of research on mathematics teaching and learning* (pp. 575–596). New York: Macmillan.

Mellers, B. A., Schwartz, A., & Cooke, D. J. (1998). Judgment and decision making. *Annual Review of Psychology, 49*, 447–477.

Moore, D. S. (1990). Uncertainty. In L. A. Steen (Ed.), *On the shoulders of giants: New approaches to numeracy* (pp. 95–137). Washington, DC: National Academy Press.

Moore, D. S. (1997a). New pedagogy and new content: The case of statistics. *International Statistical Review, 65*(2), pp. 123–165.

Moore, D. S. (1997b). *Statistics: Concepts and Controversies*. San Francisco: Freeman.

Moore, D. S. (1998). Statistics among the liberal arts. *Journal of the American Statistical Association, 93*(444), 1253–1259.

Moore, D. S., & Cobb, G. W. (2000). Statistics and mathematics: Tension and cooperation. *American Mathematical Monthly, 107*(7), 615–630

Mosenthal, P. B., & Kirsch, I. S. (1998). A new measure for assessing document complexity: The PMOSE/IKIRSCH document readability formula. *Journal of Adolescent and Adult Literacy, 41*(8), 638–657.

Mullis, I. V. S., Martin, M. O., Beaton, A. E., Gonzalez, E. J., Kelly, D. L., & Smith, T. A. (1998). *Mathematics and science achievement in the final year of secondary school: IEA's Third International Mathematics and Science Study (TIMSS)*. Boston: Center for the Study of Testing, Evaluation, and Educational Policy.

National Council of Teachers of Mathematics (NCTM). (2000). *Principles and standards for school mathematics*. Reston, VA: Author.

Orcutt, J. D., & Turner, J. B. (1993). Shocking numbers and graphic accounts: Quantified images of drug problems in the print media. *Social Problems, 40*(2), 190–206.

Organization for Economic Co-operation and Development (OECD) and Human Resources Development Canada (1997). *Literacy for the knowledge society: Further results from the International Adult Literacy Survey*. Paris and Ottawa: OECD and Statistics Canada.

Packer, A. (1997). Mathematical competencies that employers expect. In L. A. Steen (Ed.), *Why numbers count: Quantitative literacy for tomorrow's America* (pp. 137–154). New York: The College Board.

Parker, M., & Leinhardt, G. (1995). Percent: A privileged proportion. *Review of Educational Research, 65*(4), 421–481.

Paulos, J. A. (1995). *A mathematician reads the newspaper*. New York: Anchor Books/Doubleday.

Pereira-Mendoza, L. (Ed.). (1993). *Introducing data-analysis in the schools: Who should teach it and how?* Voorburg, Holland: International Statistical Institute.

Rutherford, J. F. (1997). Thinking quantitatively about science. In L. A. Steen (Ed.), *Why numbers count: Quantitative literacy for tomorrow's America* (pp. 69–74). New York: The College Board.

Shamos, M. H. (1995). *The myth of scientific literacy*. New Brunswick, NJ: Rutgers University Press.

Scheaffer, R. L., Watkins, A. E., & Landwehr, J. M. (1998). What every high-school graduate should know about statistics. In S. P. Lajoie (Ed.), *Reflections on statistics: Learning, teaching and assessment in grades K–12* (pp. 3–31). Mahwah, NJ: Lawrence Erlbaum.

Schmidt, W. H., McKnight, C. C., Valverde, G. A., Houang, R. T., & Wiley, D. E. (1997). *Many visions, many aims (Vol. 1): A cross-national investigation of curricular intentions in school mathematics.* Dordrecht, The Netherlands: Kluwer.

Shaughnessy, J. M. (1992). Research in probability and statistics: Reflections and directions. In D. A. Grouws, (Ed.), *Handbook of research on mathematics teaching and learning* (pp. 465–494). New York: Macmillan.

Shaughnessy, J. M., Garfield, J. B., Greer, B. (1997). Data handling. In A. Bishop (Ed.), *International handbook on mathematics education* (pp. 205–237). Dordrecht, The Netherlands: Kluwer.

Starkings, S. (1997). Assessing student projects. (1997). In I. Gal & J. Garfield, (Eds.), *The assessment challenge in statistics education* (pp. 139–151). Voorburg, The Netherlands: International Statistical Institute and IOS Press.

Statistics Canada and Organization for Economic Co-operation and Development (OECD). (1996). *Literacy, economy, and society: First results from the International Adult Literacy Survey.* Ottawa, Ontario: Author.

Steen, L. A. (Ed.). (1997). *Why numbers count: Quantitative literacy for tomorrow's America.* New York: The College Board.

Stein, S (2000). *Equipped for the future content standards: What adults need to know and be able to do in the 21st century.* Washington, DC: National Institute for Literacy. Retrieved January 1, 2001, from http://www.nifl.gov/lincs/collections/eff/eff_publications.html

Tufte, E. R. (1997). *Visual explanations: Images and quantities, evidence and narrative.* Cheshire, CT: Graphics Press.

UNESCO. (1990). *Final Report on the World Conference on Education for All* (Jomtien, Thailand). Paris: Author.

UNESCO. (2000). *World Education Report: The right to education—Towards education for all throughout life.* Paris: Author.

Wagner, D. A. (1991). *Literacy: Developing the future.* (International Yearbook of Education, vol. XLIII). Paris: UNESCO.

Wagner, D. A., Venezky, R. L., & Street, B. V. (Eds.). (1999). *Literacy: An International Handbook.* Boulder, CO: Westview Press.

Wainer, H. (1992). Understanding graphs and tables. *Educational Researcher, 21*(1), 14–23.

Wallman, K. K. (1993). Enhancing statistical literacy: Enriching our society. *Journal of the American Statistical Association, 88*, 1–8.

Wallsten, T. S., Fillenbaum, S., & Cox, J. A. (1986). Base rate effects on the interpretations of probability and frequency expressions. *Journal of Memory and Language, 25*, 571–587.

Wanta, W. (1997). *The public and the national agenda: How people learn about important issues.* Mahwah, NJ: Lawrence Erlbaum.

Watson, J. (1997). Assessing statistical literacy through the use of media surveys. In I. Gal & J. Garfield, (Eds.), *The assessment challenge in statistics education* (pp. 107–121). Amsterdam, The Netherlands: International Statistical Institute/IOS Press.

Watson, J. M., & Moritz, J. B. (2000). Development of understanding of sampling for statistical literacy. *Journal of Mathematical Behavior, 19*, 109–136.

Wild, C., Triggs, C., & Pfannkuch, M. (1997). Assessment on a budget: Using traditional methods imaginatively. In I. Gal & J. B. Garfield (Eds.), *The assessment challenge in statistics education* (pp. 205–220). Amsterdam, The Netherlands: International Statistical Institute/IOS Press.

Wild, C. J., & Pfannkuch, M. (1999). Statistical thinking in empirical enquiry. *International Statistical Review, 67*(3), 223–256.

Chapter 4

A COMPARISON OF MATHEMATICAL AND STATISTICAL REASONING

Robert C. delMas
University of Minnesota, USA

INTRODUCTION

The focus of this chapter is on the nature of mathematical and statistical reasoning. The chapter begins with a description of the general nature of human reasoning. This is followed by a description of mathematical reasoning as described by mathematicians along with recommendations by mathematics educators regarding educational experiences to improve mathematical reasoning. The literature on statistical reasoning is reviewed and findings from the general literature on reasoning are used to identify areas of statistical reasoning that students find most challenging. Statistical reasoning and mathematical reasoning are compared and contrasted, and implications for instruction and research are suggested.

THE NATURE OF HUMAN REASONING

While human beings are very intelligent and have produced notable advances of mind over the millennia, people are still prone to systematic errors of judgment. Wason and Johnson-Laird (1972) reported on a variety of studies that systematically explored conditions under which people make reasoning errors. One of the difficulties faced by researchers of human reasoning is a lack of agreement in the definition of the phenomenon. Wason and Johnson-Laird (1972) state, "There is, of course, no clear boundary surrounding this topic. ... In our view, it is fruitless to argue about definitions of terms, and we shall be concerned with how humans draw explicit conclusions from evidence" (p. 1). In a review of the literature on reasoning research, Galotti (1989) argues that this lack of agreement on what constitutes reasoning produces some problems for the interpretation of results. Galotti points out that "reasoning" is often used interchangeably with terms such as thinking, problem solving, decision making, critical thinking, and brain storming. The confusion is compounded in that these different types of thinking are considered to involve common processes and mental activity, such as the transformation of given

D. Ben-Zvi and J. Garfield (eds.),
The Challenge of Developing Statistical Literacy, Reasoning and Thinking, 79–95.

information on the basis of stored knowledge in order to draw an inference or a conclusion.

Galotti (1989) offers a definition of reasoning that attempts to distinguish it from other forms of thinking. According to Galotti, reasoning involves mental activity that transforms given information, is focused on at least one goal (typically to make an inference or draw a conclusion), is consistent with initial premises (modified or unmodified), and is consistent with systems of logic when all premises are specified. She also adds some caveats: The mental activity does not have to be self-contained (i.e., the premises may be modified by the reasoner) and the conclusions do not have to be deductively valid. Therefore, when conducting research on reasoning, it is important to determine whether or not a person has modified the premises and to judge the quality of the reasoning accordingly.

Errors in Human Reasoning

Despite the potential for disagreement on the phenomenon being investigated, there has been a long history of research on the degree to which humans are naturally rational thinkers. Most of the studies have looked at performance on abstract, formal reasoning tasks (e.g., syllogisms; tasks solved by propositional or predicate calculus) where all necessary information is provided (Evans, 1989; Evans, Newstead, & Byrne, 1993; Oaksford & Chater, 1998; Wason & Johnson-Laird, 1972). Some studies have looked at practical and informal reasoning where the purpose is more functional and situation specific (Evans et al., 1993; Galotti, 1989). Some general findings that can be summarized from reviews of the literature (e.g., Evans, 1989; Evans et al., 1993; Galotti, 1989; Gilovich, Griffin, & Kahneman, 2002; Wason & Johnson-Laird, 1972) are as follows:

- People have difficulty with drawing a valid conclusion by denying a negatively stated assumption. In general, people find it hard to track the effect of double negation in an argument (Evans, 1989; Evans, Newstead, & Byrne, 1993; Galotti, 1989).
- People often change the nature or meaning of premises, even when explicitly trained in the interpretation of premises (Galotti, Baron, & Sabini, 1986).
- When presented with a conditional statement, people act as if a causal relationship is implied between the antecedent ("If the skies are clear tonight") and consequent ("it will be cold tomorrow morning"). Therefore, they incorrectly believe the antecedent is true if the consequent is affirmed: "It is very cold this morning, therefore, the skies must have been clear last night" (Wason & Johnson-Laird, 1972).
- Human reasoning is deductive, but it tends to be of a practical nature. People, in general, do not reason well with purely abstract information. People show impressive reasoning abilities with complex tasks, but primarily when they are highly familiar with the materials and situation (Evans, Newstead, & Byrne, 1993; Kahneman & Tversky, 1982; Wason & Johnson-Laird, 1972).

- When given an abstract task, people inadvertently modify the given information or premises by including personal knowledge that may or may not be relevant (Wason & Johnson-Laird, 1972).
- While human reasoning is deductive in nature and quite powerful, it does not seem to act in full accordance with the truth-functional relations of the propositional calculus in formal logic (Galotti, 1989).
- People do not tend to consider all possible interpretations of a premise (Erickson, 1978; Johnson-Laird, 1983) or multiple ways of combining premises (Johnson-Laird, 1983). This leads to a consideration of information and implications that are not exhaustive, which in turn may lead to erroneous conclusions (Baron, 1985). One particular example of this is confirmation bias (Evans, 1989; Nisbett & Ross, 1980; Ross & Anderson, 1982; Wason, 1977), which is the tendency to look only for confirmatory evidence and not to consider evidence that could potentially discredit an argument. People readily accept conclusions they believe to be true and have difficulty accepting conclusions they believe to be false.
- There is evidence that some biases in reasoning can be overcome if feedback produces a conceptual inconsistency for an individual (Nisbett & Ross, 1980). People tend to adjust their reasoning when they encounter contradictory evidence, although not all the time.
- Reasoning is easily affected by factors that, from a logical standpoint, should not have an effect. For example, people provide higher frequency estimates if asked to recall only a few instances (e.g., 3) of an event and lower estimates when asked to recall many instances (e.g., 9 to 12) relative to just being asked for a frequency estimate. These effects can be further mediated by having people consider their level of expertise in an area or by manipulations that increase or decrease motivation (Schwarz & Vaughn, 2002).
- Possibly as a result of confirmation bias and recall effects, people tend to be overconfident in the validity of their reasoning (Fischhoff, 1982; Lichtenstein, Fischhoff, & Phillips, 1982).

These general observations have implications for the nature of mathematical and statistical reasoning. Of most importance are the observations that people have difficulty with abstract reasoning, that people can reason well in highly familiar situations, that personal knowledge often intrudes when reasoning, and that people often fail to consider all possibilities. One possible explanation for biases in human reasoning is offered by two-system theories of reasoning (Evans, 1995; Evans & Over, 1996; Sloman, 2002). These theories propose that two separate but interactive systems of reasoning are employed for most reasoning tasks. One system is associative in nature and uses regularities in perceived characteristics and temporal structures to produce automatic responses. The other system, being rule-based in nature, is more deliberate and systematic (Evans & Over, 1996), which allows it to override some output from the associative system (Stanovich & West, 2002). When a person is faced with a problem, both systems may be activated and arrive at separate responses. While the two responses may be the same (or at least supportive) in most cases, it is possible for the responses to conflict. In addition, the more automatic associative system may finish first, producing a response before the rule-based system has a chance to check its validity. Even when both systems run to conclusion, the associative response can interfere with the output of the rule-based system (Sloman, 2002). In this way, first impressions can govern a decision before more rule-based, logical operations are brought into play, producing invalid or irrelevant conclusions under certain conditions. Evidence for systematic, nonnormative biases in reasoning that are consistent with predictions from a two-system reasoning process is found even when factors such as cognitive ability are accounted for (Stanovich & West, 2002).

THE NATURE OF MATHEMATICAL REASONING

It has been argued that mathematical ideas are essentially metaphorical in nature; therefore mathematics should not be taught only as methods of formal proof or a set of calculation techniques. According to Lakoff and Nunez (1997), mathematics "is all about ideas, and it should be taught as being about ideas" (p. 85). They argue that the metaphorical nature of mathematics must be taught if instruction is to affect students' mathematical reasoning. Lakoff and Nunez believe that an emphasis on metaphorical thinking can counter the idea that mathematics exists independent of human minds (because reasoning by metaphor is a characteristic of human intelligence). However, equating mathematical reasoning solely with metaphorical reasoning can be taken as evidence that mathematics is a product of mind, a product that does not necessarily have to correspond to objects or events in the objective world.

Mathematics, Symbols, and Language

As a discipline, mathematics can be viewed as the study of patterns; therefore, mathematical reasoning involves reasoning about patterns. Devlin (1998) notes that mathematics deals with abstract patterns that are distilled from the real world or "from the inner workings of the human mind" (p. 3). Adding to the level of abstraction is a reliance of modern mathematics on the use of abstract notation (e.g., algebraic expressions). Mathematical fields develop abstract notation systems in order to work with patterns in efficient and facile ways, but at the cost of added complexity and a high degree of remoteness from everyday experience and knowledge. Modern computers can help students visualize some of the notational representation, but only to a certain extent since a relatively small portion of modern mathematics lends itself to computer simulation.

Symbolic notation, in and of itself, is not mathematics. To have meaning, the symbols require mental models of real mathematical entities to serve as referents (Devlin, 1998). This aspect of mathematics, Devlin argues, is often overlooked because of an emphasis on procedures and computations in mathematics instruction. Nonetheless, he sees mathematics as a purely human creation built of entities that do not exist in the physical world for they are "pure abstractions that exist only in humanity's collective mind" (p. 9). Working with the highly abstract content of mathematics has proven difficult even for talented mathematicians. Devlin notes that Newton and Leibniz developed the calculus because they were able to represent processes of motion and change as functions, and then work with those functions as mathematical entities. The calculus was derived from a process of successive approximations and the idea of a limit, a concept for which they could not provide an acceptable definition. It took some 200 years of development in mathematical thinking before Weierstrauss conceived of the process of successive approximations as an entity and presented a precise definition for a limit.

Both language and mathematics can be considered abstract artifacts of human intellect and culture. Devlin (2000) argues that the mental facilities that humans use to process language are the very facilities needed to carry out abstract mathematical thought. Even if this is the case, there may be several reasons why a facility with language does not directly translate to a facility with mathematics. While language is abstract in nature (e.g., references can be made to objects in the past and future), its reference base is often concrete. Even when abstract concepts are the referents (e.g., love, happiness, despair), there are still human counterparts in emotion and experience that provide a foundation for meaning. Mathematical thought seems to require the individual to create mental referents, a process that can result in mental entities with no physical counterparts. Another factor that may add to difficulties in mathematical thinking is that it often requires the use of a mathematical proof. The study of mathematical proof has essentially produced systems of formal logic, which, as noted earlier, many people find difficult to employ.

Instruction and Mathematical Reasoning

Due to the highly abstract nature of mathematics, modern researchers in mathematics education place a strong emphasis on instructional methods that help students learn abstract mathematical concepts by relating them to familiar concepts and processes. The importance of image-based reasoning in mathematics is well documented (Devlin, 1998; English, 1997). Mathematicians often find that image or graphic representations facilitate their reasoning more than other types of symbolic representation do. However, the ultimate goal is to move the student from "actual reality" to what Sfard (2000) calls "virtual reality" discourse. Actual reality discourse can be bounded and mediated by real-world referents. For example, someone could state, "Her name is Cedar" and point to his dog. By pointing to his pet the speaker makes it clear that he is not referring to a tree that he thinks is female. The discourse object exists independent of the concept and can be used to perceptually mediate the discussion. However, Sfard (2000) argues that a statement such as $\frac{1}{4}$ is equal to $\frac{3}{12}$ is an instance of virtual reality discourse because perceptual mediation is enacted, at best, with real-world objects that substitute for, but do not fully represent the concept under discussion. Sfard sees virtual reality discourse as the primary mode of mathematical communication. As such, mathematical discourse may not carry a direct impact on human action or reality. This can create a setting of freedom and exploration for some; but it can also render mathematics as meaningless, of little importance, and of little interest to others (Sfard, 2000).

Modern mathematics curriculums recognize that human reasoning is pragmatic and incorporates real-world problems as devices for making mathematical concepts and structures meaningful. English (1997) states that "mathematical reasoning entails reasoning with structures that emerge from our bodily experiences as we interact with our environment" (p. 4). According to English, four "vehicles of thought" are used in mathematical reasoning: analogy, metaphor, metonymy, and imagery. They constitute generic mental devices that are not exclusively used in mathematical reasoning. All four of the mental devices provide a way to map concrete experience to mental models or representations of the environment. She argues that humans require experience with mapping structural information from concrete experience to a mathematically abstract mental representation (the foundation of analogy and metaphor) in order to develop mathematical reasoning. Sfard (2000) notes that both actual and virtual reality discourse are object mediated. She sees virtual reality discourse as emerging from actual reality discourse in a process that reminds one of object-oriented programming in computer science; if actual reality discourse is considered the root for all other discourse, then virtual reality discourse is seen to inherit templates and properties from real-world referents through iterative extensions of a concept to abstract contexts. This is similar to Thompson's (1985) development of instructional approaches that go beyond teaching skills and procedures and motivate students to develop abstract, figurative imagery that encapsulates the structural relationships, operations, and transformations that apply to mathematical objects. As such, mathematical discourse can be difficult because there may be no physical referent to serve as the focus of reasoning and communication. Ultimately, the purpose of mathematical inquiry is to develop an understanding of mathematical objects that is independent of real-world contexts (Cobb & Moore, 1997).

Statistical Reasoning and Thinking

In recent years, statisticians have pointed out distinctions between statistics and mathematics in order to establish statistics as a separate and unique discipline (e.g., Moore, 2000; Cobb & Moore, 1997). Statistics may be viewed as similar to disciplines such as physics that utilize mathematics, yet have developed methods and concepts that set it apart from mathematical inquiry. Unlike mathematical reasoning, statistical inquiry is dependent on data (Chance, 2002) and typically grounded within a context (Cobb & Moore, 1997; Moore, 1998; Pfannkuch & Wild, 2000; Wild & Pfannkuch, 1999). A practicing statistician may use mathematics to assist in solving a statistical problem, but only after considerable work has been done to identify the question under investigation, explore data for both patterns and exceptions, produce a suitable design for data collection, and select an appropriate model for data analysis (see Chapter 2).

Statistical thinking and statistical reasoning have often been used interchangeably to represent the same types of cognitive activity. If reasoning in general is considered a type of thinking, then how are statistical reasoning and statistical thinking related? Recent work by Wild and Pfannkuch (1999) has helped provide a model for statistical thinking that allows it to be distinguished from statistical reasoning. Lovett (2001) defines statistical reasoning as "the use of statistical tools and concepts ... to summarize, make predictions about, and draw conclusions from data" (p. 350). This definition does not distinguish statistical reasoning because it is too similar to the depiction of statistical thinking offered by Pfannkuch and Wild (see Chapter 2) and Chance (2002). Garfield (2002) offered a similar definition, but with more emphasis on the "ways" statistical knowledge is used to make sense of data. Nonetheless, Garfield found that there is very little consensus on what is involved in statistical reasoning and that research on statistical reasoning is still in a state of development.

It can be argued that both statistical thinking and reasoning are involved when working the same task, so that the two types of mental activity cannot necessarily be distinguished by the content of a problem (delMas, 2002). However, it may be possible to distinguish the two by the nature of the task. For example, a person who knows when and how to apply statistical knowledge and procedures demonstrates statistical thinking. By contrast, a person who can explain why results were produced or why a conclusion is justified demonstrates statistical reasoning. This treatment of statistical reasoning is consistent with the definition presented earlier by Galotti (1989). Examples of statistical reasoning are likely to be found at stages in people's thinking where they are asked to state implications, justify a conclusion, or make an inference. Given this perspective, statistical reasoning is demonstrated when a person can explain why a particular result is expected or has occurred, or explain why it is appropriate to select a particular model or representation. Statistical reasoning is also expressed when a selected model is tested to see if it represents a reasonable fit to a specified context. This type of explanation typically requires an understanding of processes that produce data. When students develop an understanding of processes that produce samples and, consequently, statistics derived from samples, they may be better prepared to predict the behavior of sampling distributions and understand procedures that are based on the behavior of samples and statistics (see Chapter 13).

With this type of understanding, the student can provide reasons and justification for the statistical methodology that is applicable in a context (i.e., they can think statistically). These justifications, however, are not context free, and require an interplay between the concrete and the abstract as the statistical thinker negotiates the best approach to take in solving a problem. In this way, statistics differs from mathematical reasoning in that the latter is most often context free (i.e., independent of the objective world).

DIFFICULTIES IN STATISTICAL REASONING

It seems reasonable to argue that because statistical thinking always occurs within a concrete context, students should have very little difficulty with statistical reasoning. This might be expected given the general findings from research on reasoning that people tend to draw valid conclusions when working with familiar and concrete materials even when they draw invalid conclusions for isomorphic problems rendered purely in the abstract (see Evans et al., 1993). Yet, most instructors of statistics find that students have difficulty with statistical content, let alone statistical reasoning. Why is this the case?

The Abstract Nature of Statistical Content

The answer may be that many of the concepts used in statistics are abstract in nature, let alone unfamiliar, and reasoning about abstract content is difficult for many. One source of abstraction comes from the mathematical content of statistics. For example, mathematical procedures that are used to calculate the mean for a set of data are likely to produce a value that does not exist in the data set. Many students may find it difficult to develop an understanding for something that does not necessarily exist. Just as in mathematics, statistics instruction can use analogies, metaphors, and images to represent abstract concepts and processes to help students foster meaning. A common metaphor for the mean is the process of moving a fulcrum along a beam to balance weights, where the fulcrum plays the counterpart of the mean. Just as in mathematics, developing an appropriate mental model of the statistical mean may require extensive experience with the balance beam metaphor. This type of understanding, therefore, is akin to the mathematical reasoning presented in the previous section. It should not be surprising that statistics students have as much difficulty with these aspects of their statistical education as they do with the abstract content of mathematics.

Even though statistical reasoning may involve an understanding of data and context, this does not mean that all statistical concepts are concrete and accessible. A great deal of statistical content requires the type of virtual reality thinking described by Sfard (2000). It has been suggested that statistics instruction begin with exploratory data analysis because its hands-on, concrete nature is more accessible (Cobb & Moore, 1997). Even at this elementary level, students are expected to understand and reason with numerous abstractions. Instruction in exploratory data analysis presents a variety of graphical techniques that are used to represent and explore trends and patterns in data. While many aspects of these graphical techniques are nonmathematical, using them to identify patterns may require a level of abstraction that students find just as difficult as the abstract patterns encountered in mathematics. Although graphic representations are based on real data imbedded within a context, they are nonetheless abstractions that highlight certain characteristics of the data and ignore others.

Data analysis is dependent on data that is generated by taking measurements. A measurement can be a very abstract entity (e.g., what does IQ measure?) or very unfamiliar (e.g., nitrous oxide concentrations in the blood), so it can be important to begin instruction with concrete or familiar measurements (e.g., city and highway miles per gallon [mpg] ratings of automobiles). Even when the data are familiar, a measurement is an abstraction that represents only one aspect of a complex entity. Focusing attention on only one "measurement of interest" may be difficult for some students who are familiar with a context and find it difficult not to consider aspects they see as more important or more interesting.

Students move to another level of abstraction when asked to graph the data. A stem-and-leaf plot often requires students to separate the data from the context (e.g., the car make and model are not represented in a graph of mpg), and they often lose some of the measurement detail in order to construct a visual picture of the distribution. Stems are separated from leaves, and leaves often do not represent all of the remaining information in a numerical value (e.g., the stem represents the digit in the one-hundreds place, the leaf represents the digit in the tens place, and the digit in the ones place is not used at all). Further abstraction can result if the graph is expanded or contracted in order to search for a meaningful pattern in the data. This is likely to be a very unfamiliar type of representation for many students, and the level of abstraction may compound difficulties with attempts to reason from graphic displays.

Another level of abstraction is created when students are asked to further explore a data set with a box plot. The box plot is a graphic display commonly used for the comparison of two or more data sets (see Cobb & Moore, 1997 [p. 89] for an illustrative example). Box plots remove much of the detail from a data set to make certain features stand out (e.g., central tendency, variability, positive or negative skew). Understanding how the abstract representation of a "box" can stand for an abstract aspect of a data set (a specific, localized portion of its variability) is no small task. The student must build a relationship between the signifier and the signified as described by Sfard (2000), yet both the signifier and the signified are based on abstract constructions of mind. It seems reasonable to expect that many students will find it difficult to understand graphical representations, even though the devices appear basic and elementary to the seasoned statistician.

Logic Errors and Statistical Reasoning

As noted earlier, people do not tend to generate multiple possibilities for a given situation and are prone to confirmation bias. It is reasonable to expect, therefore, that some students will find it difficult to identify exceptions to trends in order to test a model, an ability that is associated with sound statistical thinking. This same difficulty is likely to express itself when students are asked to generate alternatives during the interrogative cycle of statistical thinking as described by Wild and Pfannkuch (1999), as well as when instructors try to promote a disposition of skepticism in their students.

Cobb and Moore (1997) identify several other areas of statistics instruction that are nonmathematical and uniquely define statistics as a discipline. Experimental design is a topic found in statistics (and other disciplines) that is typically not part of the mathematics curriculum. This is an area requiring very little mathematical background, and it is highly dependent on context. Experimental design does, however, follow a particular logic. Typically, several assumptions (i.e., premises) are adhered to; for example, a control condition and a treatment condition differ on only one characteristic, with all other aspects of the two conditions being equal. If a reliable difference between two conditions is found in a controlled experiment, then the difference is attributable to the difference in the characteristic on which the conditions vary. Although the preceding is certainly an oversimplification of the details that go into the design of any experiment, it is sufficient for considering how conclusions are drawn from experimental results. If a reliable difference between conditions is found, affirmation of the antecedent occurs from which the conclusion follows that the varied characteristic was responsible. In formal logic this is known as *modus ponens* (see Evans et al., 1993). Conversely, if the characteristic that is varied is not a causal agent, then logic dictates that a reliable difference between the two conditions will not be found. This is referred to as *modus tollens.* While people appear to handle *modus ponens* reasoning naturally, many have difficulty with *modus tollens* (Evans et al., 1993). Students are likely to have similar difficulty understanding the logic of experimental design.

Formal Inference in Statistics

Formal inference is typically introduced in a first course of statistics. Formal inference involves rules for drawing conclusions about the characteristics of a population based on empirical observations of samples taken from the population. This is often taught using one (or both) of two approaches: confidence intervals or significance tests (Cobb & Moore, 1997). Either approach requires the disposition that Wild and Pfannkuch (1999) refer to as "being logical." Both approaches derive, in part, from probability theory; but they also involve a logic that is statistical in nature. Because a complete understanding of these approaches requires logical and mathematical thinking, many students will find this topic difficult to understand. The type of logical thinking involved may provide additional insight as to why formal inference is problematic for many students. As described by Cobb and Moore (1997), a significance test starts by assuming that an effect of interest is not present in a population. The reasoning goes something like this: If there is no effect in the population, then the probability for the size of the effect observed in the sample data will be high. Conversely, if the effect in the sample data is determined to be of a sufficiently low probability, this is taken as evidence that the original premise is false and that the effect does exist in the population.

Mathematics provides knowledge about the expected probability distribution of observed sample effects when there is no effect in the population. Statistics adds a probabilistic determination for the cutoff point that establishes when a probability is sufficiently low. The reasoning that follows is provided by the formal logic of predicate calculus. The logic of significance tests involves a negative statement in the premise, a situation that typically results in poorer performance on formal reasoning tasks. The logical reasoning that establishes evidence of an effect in the population follows from *modus tollens* (i.e., negation of the consequent validates negation of the antecedent). As noted earlier, people find *modus tollens* to be a difficult type of reasoning. On both accounts, students will find the logic of significance tests difficult to follow. The logic could be made easier by using an example where negation of the consequent matches commonsense understanding for a very familiar setting. However, under this condition people may draw a valid conclusion simply because they "know it is so" and not because they understand the underlying logic.

Reasoning with Confidence Intervals

A confidence interval takes a different approach to formal inference by providing an interval estimate of a population characteristic. The interval is based on data from a single sample and, therefore, is not guaranteed to capture the true value of the population characteristic due to sampling variability. Probability theory can provide an estimate of how likely (or how often) a random sample drawn from a population will capture the population value. This probability is taken as the level of confidence. Therefore, the meaning of a 95% confidence interval is based on the understanding that there is a 95% chance that a single randomly selected sample will be one of the samples that provides a confidence interval that captures the population characteristic. This understanding requires a complex mental model of several related concepts, which alone may make reasoning from confidence intervals difficult. In addition, formal inference based on confidence intervals appears to follow the same logic as significance tests. The confidence interval has a reasonably high probability of capturing the true population characteristic. Under the assumption of no effect in the population (e.g., two groups really come from the same population, so the difference between the two groups should be zero), the confidence interval is very likely to contain no effect (i.e., to capture zero). The conclusion that there is an effect in the population follows if the confidence interval does not contain zero (i.e., the consequent is negated). Once again, the situation requires reasoning based on a negated premise and *modus tollens.*

COMPARISON OF STATISTICAL REASONING AND MATHEMATICAL REASONING

It is reasonable to ask at this point how mathematical and statistical reasoning compare and contrast with each other. Mathematical and statistical reasoning should place similar demands on a student and display similar characteristics when the student is asked to reason with highly abstract concepts and relationships. When students are asked to reason primarily with abstract concepts, a great deal of concentration and persistence may be required to find relationships among the concepts. This can lead to erroneous judgments and conclusions if a student is unable to sustain the effort. Solutions may be based on the output of associative processes that fall short of the reflection and integration needed for a complete understanding.

A statistical problem can provide an illustrative example for both mathematical and statistical reasoning. A problem on a statistics exam might present a bivariate plot, summary statistics including the value of the correlation, and formulas for calculating the slope and ordinate of the y-intercept. When asked to find the slope and y-intercept, many students will not use the formulas that are provided. Instead, they may pick two plotted points that seem "typical" of the bivariate plot, derive a value for the slope using procedures learned in linear algebra, and subsequently calculate a value for the ordinate of the y-intercept. This "reasoning" may not be reasoning at all, but merely the result of well-rehearsed associative memory where "find the slope" retrieves a familiar procedure without questioning the fit of the procedure to the context. A student acting in this fashion seems to lack either a rudimentary mathematical understanding (e.g., that the model requires all points to form a straight line) or statistical understanding (e.g., that the model must take into account the inherent variability in the bivariate plot).

When students work within very familiar contexts or with well-rehearsed concepts and procedures, very few difficulties and errors are expected to occur, regardless of whether the content is statistical or mathematical. The previous example illustrates a common misunderstanding among students that, when recognized, provides an opportunity to help students develop a deeper understanding of both mathematical and statistical concepts by promoting an understanding of the contexts under which it is appropriate to apply the respective models. Once ample opportunity is provided to distinguish between the mathematical and statistical contexts, and to apply the respective procedures, errors are more likely to be mechanical than rational in nature.

While mathematical and statistical reasoning appear similar, there are some differences in the common practices of each discipline that may result in different sources of reasoning difficulty. Model abstraction is a general task that is common to both disciplines. The nature of the task, however, is somewhat different between statistics and mathematics. In mathematics, context may or may not play a large role. Initially, mathematics instruction may use familiar contexts to motivate and make accessible the underlying structure of abstract concepts. During this period of instruction, students might be misled by aspects of the context that are familiar yet irrelevant to an understanding of the underlying mathematical concept. Through guided inquiry or constructivist approaches that require the student to test models and assumptions against feedback derived from the context, students may eventually develop well-structured mental models of the mathematical object. At that point, the student may no longer require problem contextualization to reason with the mathematical concept. Further work with the concept may be conducted in a purely imaginary, figurative, and abstract way that does not require the student to relate back to any of the original contexts used to promote understanding. At this point, the student manipulates mathematical concepts and coordinates multiple relationships in a purely mental world that may have no real-world referents other than symbolic representations. This can produce significant cognitive demands that make the mathematical reasoning quite difficult.

In the practice of statistics, model abstraction always begins with a context. When this practice is taught in the statistics classroom, the student is dependent on the characteristics of the context to guide model selection and development. In some respects, this may be a more difficult task than the purely mental activity required in mathematical reasoning. During model selection and construction, the student faces some of the same cognitive demands that are required by abstract reasoning while having to check the model's validity against the context. As demonstrated in numerous studies, reasoning from a context can produce a variety of errors. Therefore, no matter how practiced and skilled the student (or statistician), she must always guard against the intrusion of everyday knowledge that is irrelevant or misleading. She must also guard against the use of heuristic, associative processes that may naturally come into play, yet lead to erroneous interpretations or the perception of relationships that do not actually exist. If the student successfully navigates these pitfalls, statistical analyses suggested by the model can be conducted. The student must then take the results and relate them back to the original context. This translation or mapping represents another potential source of error as multiple relationships must be tracked and validated, and context once again has an opportunity to influence reasoning.

In summary, it is likely that many aspects of statistical and mathematical reasoning are highly similar. The task demands of each discipline, however, may produce different sources of reasoning error. While instruction can be driven and facilitated by contextualization in both disciplines, statistical practice is highly dependent on real-world context whereas mathematical practice tends to be removed from real-world context (Cobb & Moore, 1997). The dependence on context in statistical reasoning may lead to errors in reasoning, some of which are difficult to overcome even for well-educated and experienced professionals.

IMPLICATIONS FOR STATISTICS EDUCATION AND RESEARCH

Instruction

Statistical reasoning needs to become an explicit goal of instruction if it is to be nourished and developed. Just as in mathematics instruction, experiences in the statistics classroom need to go beyond the learning of procedures to methods that require students to develop a deeper understanding of stochastic processes. Given that there is mathematical content in statistics along with the abstract nature of many statistical concepts, research on the use of analogy, metaphor, and imagery by mathematics educators should not be overlooked (e.g., English, 1997a; Thompson, 1985). Such approaches may help students map data and processes between abstract representations and context, and help them to generate and test their own representations. Both mathematics (e.g., Kelly & Lesh, 2000) and statistics educators (Cobb & Moore, 1997) recommend instruction that is grounded in concrete, physical activities to help students develop an understanding of abstract concepts and reasoning.

To promote statistical reasoning, students must experience firsthand the process of data collection and explore the behavior of data, experiences that everyday events do not readily provide (Moore, 1998). This should help students gain familiarity and understanding with concepts that are difficult to experience in everyday life (e.g., the sampling distribution of a statistic). These experiences should include the opportunity to ask why and how data is produced, why and how statistics behave, and why and how conclusions can be drawn and supported (delMas, 2002). Students will more than likely need extensive experience with recognizing implications and drawing conclusions in order to develop a disposition for "being logical." Methods for presenting statistical content in ways that match natural ways of thinking and learning should be sought. One promising approach involves instruction that is based on frequency representations of situations (e.g., Sedlmeier, 1999), which can be seen as a natural extension of incorporating data and data production into instruction. Another promising approach is the use of predict-and-test activities (e.g., delMas, Garfield, & Chance, 1999), which provide students the opportunity to confront and correct misunderstandings about stochastic processes.

Statistics Education Research

The past decade witnessed the initiation of a reform movement in statistics education that focuses on statistical thinking, conceptual understanding, use of technology, authentic assessment, and active learning (e.g., Cobb, 1992). Much of this movement has been motivated by research in mathematics education, education, and psychology (e.g., Garfield, 1995), and there appears to have been significant impact on teaching practices from these recommendations (Garfield, Hogg, Schau, & Whittinghill, 2002). Statistics is being taught to increasing numbers of students at all ages as quantitative reasoning is seen as essential for effective citizenship (e.g., National Council of Teachers of Mathematics [NCTM] Standards, 2000). The content, pedagogy, and use of technology in introductory statistics courses have been modernized to focus on concepts, real data, effective use of technology, and statistical thinking (e.g., Cobb, 1992; Moore, 1997). New resources are now available to enable instructors to implement these changes (e.g., Moore, 2001).

However, while statistics instruction has seen dramatic growth and attention, research devoted exclusively to issues in statistics education has not. One of the most neglected areas is research devoted to understanding students' statistical reasoning. For example, a great deal is known about the errors and misconceptions that students make when reasoning about problems in probability (e.g., Gilovich, Griffin, & Kahneman, 2002; Kahneman, Slovic, & Tversky, 1982; Sedlmeier, 1999; Shaughnessy, 1992). Most of these studies use forced-choice items in comparative studies as measures of students' thinking. Very few studies use clinical methods to document and model students' thought processes as they reason (Shaughnessy, 1992), although there are certainly some exceptions (e.g., Konold, 1989; Mokros & Russell, 1995).

The research programs presented at the Statistical Reasoning, Thinking, and Literacy forums (SRTL-1 and SRTL-2) indicate that classroom research and clinical interview methodologies are starting to be utilized in the study of students' statistical thinking. These methodologies have developed to a point where they can provide considerable insight into students' reasoning (e.g., Kelly & Lesh, 2000). Future research needs to go beyond the documentation of errors and misunderstandings to probing for an understanding of the processes and mental structures that support both erroneous and correct statistical reasoning. The previous section discussed areas of statistics instruction where students are likely to encounter difficulty in understanding the expected statistical reasoning. While it may make sense to expect such difficulties, empirical evidence is needed to establish if difficulties exist and to explicate their nature. A deeper understanding of students' mental models and processes will improve the design of educational approaches for developing students' statistical reasoning. More detailed descriptions of the cognitive processes and mental structures that students develop during instruction should provide a richer foundation from which to interpret the effects of instructional interventions.

REFERENCES

Baron, J. (1985). *Rationality and intelligence*. Cambridge, UK: Cambridge University Press.

Chance, B. L. (2002). Components of statistical thinking and implications for instruction and assessment. *Journal of Statistics Education, 10*(3). Retrieved April 7, 2003, from http://www.amstat.org/publications/jse/

Cobb, G. (1992). Teaching statistics. In L. A. Steen (Ed.), *Heeding the call for change: Suggestions for curricular action* (Notes: vol. 22, 3–43). Washington, DC: Mathematical Association of America.

Cobb, G. W., & Moore, D. (1997). Mathematics, statistics, and teaching. *American Mathematical Monthly, 104*, 801–823.

delMas, R. (2002). Statistical literacy, reasoning, and thinking: A commentary. *Journal of Statistics Education, 10*(3). Retrieved April 7, 2003 from http://www.amstat.org/publications/jse/

delMas, R., Garfield, J., & Chance, B. (1999). A model of classroom research in action: Developing simulation activities to improve students' statistical reasoning. *Journal of Statistics Education, 7*(3). Retrieved April 7, 2003 from http://www.amstat.org/publications/jse/

Devlin, K. (1998). *The language of mathematics: Making the invisible visible*. New York: Freeman.

Devlin, K. (2000). *The math gene: Why everyone has it, but most people don't use it*. London: Weidenfeld & Nicolson.

English, L. D. (1997). Analogies, metaphors, and images: Vehicles for mathematical reasoning. In L. D. English (Ed.), *Mathematical reasoning: Analogies, metaphors, and images*. Hove, UK: Erlbaum, 3–18.

Erickson, J. R. (1978). Research on syllogistic reasoning. In R. Revlin & R. E. Mayer (Eds.), *Human reasoning*. Washington, DC: Winston.

Evans, J. St. B. T. (1989). *Bias in human reasoning: Causes and consequences*. Hillsdale, NJ: Erlbaum.

Evans, J. St. B. T. (1995). Relevance and reasoning. In S. E. Newstead & J. St. B. T. Evans (Eds.), *Perspectives on thinking and reasoning: Essays in honour of Peter Wason*. Hillsdale, NJ: Erlbaum, 147–171.

Evans, J. St. B. T., Newstead, S. E., & Byrne, R. M. J. (1993). *Human reasoning: The psychology of deduction*. Hillsdale, NJ: Erlbaum.

Evans, J. St. B. T., & Over, D. E. (1996). *Rationality and reasoning*. Hove, England: Psychology Press.

Fischhoff, B. (1982). For those condemned to study the past: Heuristics and biases in hindsight. In D. Kahneman, P. Slovic, & A. Tversky (Eds.), *Judgment under uncertainty: Heuristics and biases*. New York: Cambridge University Press, 335–351.

Galotti, K. M. (1989). Approaches to studying formal and everyday reasoning. *Psychological Bulletin, 105*(3), 331–351.

Galotti, K. M., Baron, J., & Sabini, J. P. (1986). Individual differences in syllogistic reasoning: Deduction rules or mental models? *Journal of Experimental Psychology: General, 115*, 16–25.

Garfield, J. (1995). How students learn statistics. *International Statistical Review, 63*, 25–34.

Garfield, J. B. (2002). The challenge of developing statistical reasoning. *Journal of Statistics Education, 10*(3), Retrieved April 7, 2003 from http://www.amstat.org/publications/jse/.

Garfield, J., Hogg, B., Schau, C., and Whittinghill, D. (2002). First courses in statistical science: The status of educational reform efforts. *Journal of Statistics Education, 10*(2). Retrieved April 7, 2003 from http://www.amstat.org/publications/jse/

Gilovich, T., Griffin, D., & Kahneman, D. (Eds.) (2002). *Heuristics and biases: The psychology of intuitive judgment*. New York: Cambridge University Press.

Johnson-Laird, P. N. (1983). *Mental models*. Cambridge, MA: Harvard University Press.

Kahneman, D., & Tversky, A. (1982). Judgment under uncertainty: Heuristics and biases. In D. Kahneman, P. Slovic, & A. Tversky (Eds.), *Judgment under uncertainty: Heuristics and biases*. New York: Cambridge University Press, 3–20.

Kahneman, D., Slovic, P., & Tversky, A. (eds.) (1982). *Judgment under uncertainty: Heuristics and biases*. New York: Cambridge University Press.

Kelly, A. E., & Lesh, R. A. (Eds.). (2000). *Handbook of research design in mathematics and science education*. Mahwah, NJ: Erlbaum.

Konold, C. (1989). Informal conceptions of probability. *Cognition and Instruction, 6*, 59–98.

Lakoff, G., & Nunez, R. E. (1997). The metaphorical structure of mathematics: Sketching out cognitive foundations for a mind-based mathematics. In L. D. English (Ed.), *Mathematical reasoning: Analogies, metaphors, and images*. Hove, UK: Erlbaum, 21–89.

Lichtenstein, S., Fischhoff, B., & Phillips, L. D. (1982). Calibration of probabilities: The state of the art to 1980. In D. Kahneman, P. Slovic, & A. Tversky (Eds.), *Judgment under uncertainty: Heuristics and biases*. New York: Cambridge University Press, 306–334.

Lovett, M. (2001). A collaborative convergence on studying reasoning processes: A case study in statistics. In S. M. Carver & D. Klahr (Eds.), *Cognition and instruction: Twenty-five years of progress*. Hillsdale, NJ: Erlbaum, 347–384.

Mokros, J., & Russell, S. J. (1995). Children's concepts of average and representativeness. *Journal for Research in Mathematics Education, 26*(1), 20–39.

Moore, D. S. (1997). New pedagogy and new content: The case of statistics. *International Statistical Review, 65*, 123–137.

Moore, D. (1998). Statistics among the liberal arts. *Journal of the American Statistical Association*, 1253–1259.

Moore, D. (2000). Statistics and mathematics: Tension and cooperation. *American Mathematical Monthly*, 615–630.

Moore, D. (2001). Undergraduate programs and the future of academic statistics. *American Statistician, 55*(1), 1–6.

National Council of Teachers of Mathematics. (2000). *Principles and standards for school mathematics*. Reston, VA: NCTM.

Nisbett, R., & Ross, L. (1980). *Human inference: Strategies and shortcomings of social judgment*. Englewood Cliffs, NJ: Prentice-Hall.

Oaksford, M., & Chater, N. (1998). *Rational models of cognition*. New York: Oxford University Press.

Pfannkuch, M., & Wild, C. J. (2000). Statistical thinking and statistical practice: Themes gleaned from professional statisticians. *Statistical Science, 15*(2), 132–152.

Ross, L., & Anderson, C. A. (1982). Shortcomings in the attribution process: On the origins and maintenance of erroneous social assessments. In D. Kahneman, P. Slovic, & A. Tversky (Eds.), *Judgment under uncertainty: Heuristics and biases*. New York: Cambridge University Press, 129–152.

Schwarz, N., & Vaughn, L. A. (2002). The availability heuristic revisited: Ease of recall and content of recall as distinct sources of information. In T. Gilovich, D. Griffin, & D. Kahneman (Eds.), *Heuristics and biases: The psychology of intuitive judgment*. New York: Cambridge University Press, 103–119.

Sedlmeier, P. (1999). *Improving statistical reasoning: Theoretical models and practical implications*. Hillsdale, NJ: Erlbaum.

Sfard, A. (2000). Symbolizing mathematical reality into being—or how mathematical discourse and mathematical objects create each other. In P. Cobb, E. Yackel, & K. McClain (Eds.), *Symbolizing and communicating in mathematics classrooms: Perspectives on discourse, tools, and instructional design*. Mahwah, NJ: Erlbaum, 37–98.

Shaughnessy, M. (1992). Research in probability and statistics: Reflections and directions. In A. Grouws (Ed.), *Handbook of research on mathematics teaching and learning*. New York: Macmillan, 465–494.

Sloman, S. A. (2002). Two systems of reasoning. In T. Gilovich, D. Griffin, & D. Kahneman (Eds.), *Heuristics and biases: The psychology of intuitive judgment*. New York: Cambridge University Press, 379–396.

Stanovich, K. E., & West, R. F. (2002). Individual differences in reasoning. In T. Gilovich, D. Griffin, & D. Kahneman (Eds.), *Heuristics and biases: The psychology of intuitive judgment*. New York: Cambridge University Press, 421–440.

Thompson, P. W. (1985). Experience, problem solving, and learning mathematics: Considerations in developing mathematics curricula. In E. A. Silver (Ed.), *Teaching and learning mathematical problem solving: Multiple research perspectives*. Hillsdale, NJ: Erlbaum, 189–243.

Wason, P. C. (1977). On the failure to eliminate hypotheses—a second look. In P. N. Johnson-Laird & P. C. Wason (eds.), *Thinking*. Cambridge, UK: Cambridge University Press, 89–97.

Wason, P. C., & Johnson-Laird, P. N. (1972). *Psychology of reasoning: Structure and content*. Cambridge, MA: Harvard University Press.

Wild, C. J., & Pfannkuch, M. (1999). Statistical thinking in empirical enquiry. *International Statistical Review, 67*(3), 223–265.

Chapter 5

MODELS OF DEVELOPMENT IN STATISTICAL REASONING

Graham A. Jones[1], Cynthia W. Langrall[2], Edward S. Mooney[2], and Carol A. Thornton[2]
Griffith University, Gold Coast Campus, Australia[1], and Illinois State University, USA [2]

OVERVIEW

In recent years, key reform groups in school mathematics (e.g., Australian Education Council [AEC], 1994; National Council of Teachers of Mathematics [NCTM], 1989, 2000; Department of Education and Science and the Welsh Office [DFE], 1995) have focused on the importance of students' thinking and reasoning in all areas of the mathematics curriculum including statistics. Consistent with this perspective, our chapter examines cognitive models of development in statistical reasoning and the role they can play in statistical education. The cognitive models we will describe and analyze examine statistical reasoning processes like decision making, prediction, inference, and explication as they are applied to the exploration of both univariate and multivariate data.

As a preface to our analysis of models of development in statistical reasoning we consider models of development from a psychological perspective and then look at how models of statistical reasoning have evolved historically from models of development in probability. Our survey of the research literature begins with comprehensive models of cognitive development that deal with multiple processes in statistical reasoning and suggest that school students' statistical reasoning passes through a number of hierarchical levels and cycles. Subsequently, the chapter focuses on models of cognitive development that characterize students' statistical reasoning as they deal with specific areas of statistics and data exploration: data modeling, measures of center and variation, group differences, bivariate relationships, sampling, and sampling distributions.

The models of development in statistical reasoning documented in this chapter have been formulated through structured interviews, clinical studies, and teaching

D. Ben-Zvi and J. Garfield (eds.),
The Challenge of Developing Statistical Literacy, Reasoning and Thinking, 97–117.

experiments. Research studies involving teaching experiments are especially powerful because they enable researchers and teachers to trace students' individual and collective development in statistical reasoning during instruction. Because the cognitive models provide a coherent picture of students' statistical reasoning, they have implications for curriculum design, instruction, and assessment. We will discuss these implications, particularly those relating to the role that models of statistical reasoning can play in providing a knowledge base for teachers in designing and implementing instruction.

THE MEANING OF MODELS OF DEVELOPMENT IN STATISTICAL REASONING

The psychology of cognitive development has focused on understanding the structure and dynamics of change in people's understanding of mathematics and other domains since the time of Piaget (1954, 1962). This strong psychological focus on the dynamics of change in people's understanding of the world has been accompanied by controversial debate on the issue of whether children's intellectual growth passes through a sequence of stages. More specifically, there has always been tension in Piagetian theory between its constructivist framework and its structuralist stage model. On the one hand, constructivism characterizes the acquisition of knowledge as a product of the child's creative self-organizing activity in particular environments. In other words, Piaget's perspective on constructivism affords some recognition of the presence of environment and of educational intervention. On the other hand, the stage model depicts knowledge in terms of biologically driven universal structures that are independent of specific contexts or are context neutral. That is, environment and educational intervention seemingly have no role in the evolving cognitive developmental stages.

Subsequent research by neo-Piagetian cognitive development theorists (Bidell & Fischer, 1992; Biggs & Collis, 1982, 1991; Case, 1985; Case & Okamoto, 1996; Fischer, 1980) has strengthened the place of stage-theory models but has also resulted in the replacement of Piaget's universal stage model with domain-specific theories. According to domain-specific theories, knowledge is not organized in unitary structures that cut across all kinds of tasks and situations; rather, knowledge is organized within specific domains defined by particular content or tasks such as those involved in data exploration and statistical reasoning. Moreover, contemporary neo-Piagetian theories connect rather than separate organism and environment. For example, the research studies of Biggs and Collis and those of Case have examined the process of cognitive development as it occurred in everyday environments including school settings.

The discussion of cognitive models of development in this chapter recognizes that contemporary models of cognitive development deal with domain-specific knowledge such as statistical reasoning and are essentially seamless with respect to organism and environment. Hence our use of the term *cognitive models of development* will incorporate both organism and environmental effects; or as Reber (1995) states, "maturational and interactionist effects" (p. 749). For us, the term *cognitive model of development in statistical reasoning* refers to a theory suggesting different levels or patterns of growth in statistical reasoning that result from maturational or interactionist effects in both structured and unstructured learning environments.

AN INFLUENTIAL GENERAL MODEL OF COGNITIVE DEVELOPMENT

In the previous section we referred to several neo-Piagetian models that focus on the development of domain-specific knowledge, including various aspects of mathematical knowledge. For example, models like Biggs & Collis (1982, 1991), Case (1985), Case & Okamoto, (1996), and Fischer (1980) have been consistently used as the research base for studying students' mathematical thinking and reasoning in number, number operations, geometry, and probability. In this section we examine the Biggs and Collis model in more detail because it has been widely used in developing cognitive models of development in students' statistical reasoning (e.g., Chance, delMas, & Garfield, Chapter 13 in this text; see also Jones et al., 2000; Mooney, 2002; Watson, Collis, Callingham, & Moritz, 1995).

The Biggs and Collis model has been an evolutionary one beginning with the structure of observed learning outcomes (SOLO) taxonomy (Biggs & Collis, 1982). The SOLO taxonomy postulated the existence of five modes of functioning (sensorimotor—from birth, ikonic—from around 18 months, concrete-symbolic—from around 6 years, formal—from around 14 years, and postformal—from around 20 years) and five cognitive levels (prestructural, unistructural, multistructural, relational, and extended abstract) that recycle during each mode and represent shifts in complexity of students' reasoning. Later extensions to the SOLO model (Biggs & Collis, 1989, 1991; Collis & Biggs, 1991; Pegg & Davey, 1998) acknowledged the existence and importance of multimodal functioning in many types of learning. That is, rather than earlier-developed modes being subsumed by later modes, development in earlier modes actually supports development in later modes. In fact, growth in later modes is often linked with actions or thinking associated with the earlier ones. As the models of statistical reasoning discussed later in this chapter cover students from elementary through college, we will be interested in all modes of functioning and interactions between these modes.

As noted earlier, this multimodal functioning also incorporates, within each mode, a cycle of learning that has *five* hierarchical levels (Biggs & Collis, 1982, 1989, 1991; Biggs, 1992; Watson, Collis, & Callingham et al., 1995). At the prestructural (P) level, students engage a task but are distracted or misled by an irrelevant aspect belonging to an earlier mode. For the unistructural (U) level, the student focuses on the relevant domain and picks up on one aspect of the task. At the multistructural (M) level, the student picks up on several disjoint and relevant aspects of a task but does not integrate them. In the relational (R) level, the student integrates the various aspects and produces a more coherent understanding of the task. Finally, at the extended abstract (EA) level, the student generalizes the structure to take in new and more abstract features that represent thinking in a higher mode of functioning. Within any mode of operation, the middle three levels are most important because, as Biggs and Collis note, prestructural responses belong in the previous mode and extended abstract responses belong in the next.

The levels of the Biggs and Collis learning cycle have provided a powerful theoretical base for situating research on students' statistical reasoning from the elementary school years through college (Chapter 13; Jones et al., 2000; Mooney, 2002; Watson, Collis, & Callingham et al., 1995). Even though Biggs and Collis highlight the importance of the three middle levels, some researchers have developed characterizations of students' statistical reasoning that are consistent with the first four levels (Jones et al., 2000, Mooney, 2002) while others have characterized students' statistical reasoning according to all five levels (Chapter 13). These studies also reveal that statistical reasoning operates across different modes in accord with the multimodal functioning of the Biggs and Collis model; this is especially noteworthy in relation to the modal shifts associated with the ikonic and concrete-symbolic modes.

Recent studies in mathematics, science, and statistical reasoning have identified the existence of two U-M-R cycles operating within the concrete-symbolic mode (Callingham, 1994; Campbell, Watson, & Collis, 1992; Levins & Pegg, 1993; Pegg, 1992; Pegg & Davey, 1998; Watson, Collis, & Campbell, 1995; Watson, Collis, & Callingham et al., 1995). More specifically, these researchers have identified two cycles when students engage in reasoning about fractions, volume measurement, and higher order statistical thinking. The first of these cycles is associated with the *development* of a concept and the second with the *consolidation and application* of the concept (Watson, Collis, Callingham et al., p. 250).

At opportune times in later sections of this chapter, we refer to the Biggs and Collis model in considering various models of development in statistical reasoning. Other authors in this book (e.g., Reading & Shaughnessy, Chapter 9; Watson, Chapter 12) will also elaborate on how their research has been situated in the work of Biggs and Collis.

A HISTORICAL PERSPECTIVE ON MODELS OF DEVELOPMENT IN STOCHASTICS

Cognitive models of development have frequented the literature on stochastics (a term commonly used in Europe when referring to both probability and statistics [Shaughnessy, 1992]) from the time of Piaget and Inhelder's (1951/1975) seminal work on probability. As their clinical studies demonstrated, probability concepts are acquired in stages that are in accord with Piaget's more general theory of cognitive development. Since the Piaget and Inhelder studies, there has been a strong focus on cognitive models in stochastics, most of them focused on probabilistic rather than statistical reasoning (Fischbein, 1975; Fischbein & Gazit, 1984; Fischbein & Schnarch, 1997; Green, 1979, 1983; Jones, Langrall, Thornton, & Mogill, 1997; Olecka, 1983; Polaki, Lefoka, & Jones, 2000; Tarr & Jones, 1997; Watson, Collis, & Moritz, 1997, Watson & Moritz, 1998). Some of these models on probabilistic reasoning have been situated in neo-Piagetian theories such as those of Biggs and Collis (e.g., Jones, Langrall, Thornton, & Mogill; Watson, Collis, & Moritz; Watson & Moritz) and Case (e.g., Polaki, Lefoka, & Jones). Scholz (1991) presented a review of psychological research on probability that included developmental models like those of Piaget and Fischbein. He also described his own information-processing model of probabilistic thinking that was predicated on giving students time to solve and reflect on probability tasks. Scholz's emphasis on reflection rather than on intuitive probabilistic reasoning seems to have influenced research on probabilistic reasoning in the latter part of the 1990s, and it may well have influenced the research on statistical reasoning that we discuss later in this chapter.

One cognitive development model (Shaughnessy, 1992) described stochastic conceptions in a way that has relevance for both statistical and probabilistic reasoning. Shaughnessy's broad characterization identified four types of conceptions: non-statistical (responses are based on beliefs, deterministic models, or single-outcome expectations); naïve-statistical (nonnormative responses based on judgmental heuristics or experience that shows little understanding of probability); emergent-statistical (responses are based on normative mathematical models and show evidence that the respondent is able to distinguish between intuition and a model of chance); and pragmatic-statistical (responses reveal an in-depth understanding of mathematical models and an ability to compare and contrast different models of chance). Shaughnessy did not claim that these four conceptions are linearly ordered or mutually exclusive; however, he did see the third and fourth conceptions resulting from instructional invention, and he noted that few people reach the pragmatic-statistical stage.

The research on cognitive models in probabilistic reasoning was undoubtedly the forerunner to research on models of development in statistical reasoning. However, research endeavors in statistical reasoning have also been stimulated by instructional models postulating that teachers can facilitate mathematical thinking and learning by using research-based knowledge of how students think and learn mathematics (Carpenter, Fennema, Peterson, Chiang, & Loef, 1989). Such instructional models have led researchers like Cobb et al. (1991) and Resnick (1983) to advocate the need for detailed cognitive models of students' reasoning to guide the planning and development of mathematics instruction. According to Cobb and Resnick, such cognitive models should incorporate key elements of a content domain and the processes by which students grow in their understanding of the content within that domain. Hence, in the case of statistical reasoning, it appears that we should be focusing on cognitive models that incorporate processes like decision making, prediction, and inference as they occur when students collect and explore data and begin to deal with the existence of variation, data reduction through summaries and displays, population parameters by considering samples, the logic of sampling processes, estimation and control of errors, and causal factors (Gal & Garfield, 1997).

COMPREHENSIVE MODELS OF DEVELOPMENT IN STATISTICAL REASONING

Several researchers have formulated models of cognitive development that incorporate multiple statistical processes (Jones et al., 2000; Mooney, 2002, Watson, Collis, Callingham, & Moritz, 1995). Jones et al. (2000) and Mooney (2002) characterize elementary and middle school students' statistical reasoning according to four processes: describing data, organizing and reducing data, representing data, and analyzing and interpreting data. Watson, Collis, & Callingham et al. (1995) characterize middle school students' higher order statistical reasoning as they engage in a data-card task that incorporated processes like organizing data, seeking relationships and associations, and making inferences.

Jones et al. and Mooney Models

The related research programs of Jones et al. (2000, 2001) and Mooney (2002) have produced domain-specific frameworks characterizing the development of elementary and middle school students' statistical reasoning from a more comprehensive perspective. These researchers' frameworks are grounded in a twofold theoretical view. First, it is recognized that for students to exhibit statistical reasoning, they need to understand data-handling concepts that are multifaceted and develop over time. Second, in accord with the general developmental model of Biggs and Collis (1991), it is assumed that students' reasoning can be characterized as developing across levels that reflect shifts in the complexity of their reasoning. From this theoretical perspective, Jones et al. and Mooney describe students' statistical reasoning with respect to the four statistical processes listed earlier. They assert that for each of these four processes, students' reasoning can be characterized as developing across four levels of reasoning referred to as idiosyncratic, transitional, quantitative, and analytical.

The four key statistical processes described in the Jones et al. (2000, 2001) and Mooney (2002) frameworks coincide with elements of data handling identified by Shaughnessy, Garfield, and Greer (1996) and reflect critical areas of research on students' statistical reasoning. These four processes are described as follows.

Describing Data

This process involves the explicit reading of raw data or data presented in tables, charts, or graphical representations. Curcio (1987) considers "reading the data" as the initial stage of interpreting and analyzing data. The ability to read data displays becomes the basis for students to begin making predictions and discovering trends. Two subprocesses relate to describing data: (a) showing awareness of display features and (b) identifying units of data values.

Organizing Data

This process involves arranging, categorizing, or consolidating data into a summary form. As with the ability to describe data displays, the ability to organize data is vital for learning how to analyze and interpret data. Arranging data in clusters or groups can illuminate patterns or trends in the data. Measures of center and dispersion are useful in making comparisons between sets of data. Three subprocesses pertain to organizing data: (a) grouping data, (b) summarizing data in terms of center, and (c) describing the spread of data.

Representing Data

This process involves displaying data in a graphical form. Friel, Curcio, and Bright (2001) stated that the graphical sense involved in representing data "includes a consideration of what is involved in constructing graphs as tools for structuring data and, more important, what is the optimal choice for a graph in a given situation" (p. 145). Representing data, like the previous two processes, is important in analyzing and interpreting data. The type of display used and how the data are represented will determine the trends and predictions that can be made. Also, different data displays can communicate different ideas about the same data. Two subprocesses underlie representing data: (a) completing or constructing a data display for a given data set and (b) evaluating the effectiveness of data displays in representing data.

Analyzing and Interpreting Data

This process constitutes the core of statistical reasoning. It involves recognizing patterns and trends in the data and making inferences and predictions from data. It incorporates two subprocesses that Curcio (1987) refers to using the following descriptors: (a) *reading between the data* and (b) *reading beyond the data*. The former involves using mathematical operations to combine, integrate, and compare data (interpolative reasoning); the latter requires students to make inferences and predictions from the data by tapping their existing schema for information that is not explicitly stated in the data (extrapolative reasoning). Some examples of tasks that relate to reading between and beyond the data are presented in the next few pages when we examine the elementary and middle school statistical reasoning frameworks.

With regard to levels of statistical reasoning, the Jones et al. (2000, 2001) and Mooney (2002) statistical reasoning frameworks characterize students' reasoning across four levels: idiosyncratic, transitional, quantitative, analytical. At the idiosyncratic level, students' reasoning is narrowly and consistently bound to idiosyncratic or subjective reasoning that is unrelated to the given data and often focused on personal experiences or subjective beliefs. This level corresponds to the prestructural level described by Biggs and Collis (1991). Students reasoning at this level may be distracted or misled by irrelevant aspects of a problem situation. At the transitional level students begin to recognize the importance of reasoning quantitatively, but are inconsistent in their use of such reasoning. Students reasoning at this level engage a task in a relevant way but generally focus on only one aspect of the problem situation. In the Biggs and Collis model, this is the unistructural level. At the quantitative level, students' reasoning is consistently quantitative in that they can identify the mathematical ideas of the problem situation and are not distracted or misled by the irrelevant aspects. However, students who reason at this level do not necessarily integrate these relevant mathematical ideas when engaged in the task. Biggs and Collis consider this the multistructural level. At the analytical level, students' reasoning is based on making connections between the multiple aspects of a problem situation. Their reasoning at this level can integrate the relevant aspects of a task into a meaningful structure (e.g., creating multiple data displays, or making a reasonable prediction); this is what Biggs and Collis refer to as the relational level.

The Jones et al. (2000) framework characterizes the development of *elementary school* children's statistical reasoning across the four levels just described. For each of the four statistical processes, their framework provides specific descriptors of children's reasoning at each level. In Figure 1, we have shown that part of the Jones et al. framework that pertains to *analyzing and interpreting data.* There are four descriptors, relating to each of the four levels, for the two subprocesses reading between the data and reading beyond the data. For reading between the data, a relevant task is to compare the number of students who attended a butterfly garden display before 1 p.m. with those who attended after 1 p.m., when we know each student's name and the time she attended. In the case of reading beyond the data, a relevant task is to predict the number of friends who would visit a boy named Sam in a month, when the students are given data on the number of friends who visited Sam each day of *one* week.

Mooney's framework (Mooney, 2002; Mooney, Langrall, Hofbauer, & Johnson, 2001) characterizes the development of middle school students' statistical reasoning across the same four levels and processes as described in the Jones et al. framework. The part of Mooney's framework that pertains to *analyzing and interpreting data* is presented in Figure 2. There are descriptors pertaining to the two subprocesses reading between and beyond the data as well an additional subprocess involving the use of relative and proportional reasoning. For reading between the data, a relevant task is to compare the number of medals won by five countries when given data on the number of gold, silver, and bronze medals won by each country. A reading beyond the data task is to ask students to compare the concert tours of several groups when given the number of cities where they performed, number of shows performed, and total concert earnings (see Figure 3). This latter inferential task requires proportional reasoning.

Process	Level 1 Idiosyncratic	Level 2 Transitional	Level 3 Quantitative	Level 4 Analytical
Analyzing & Interpreting Data	**Reading Between the Data**			
	Gives an idiosyncratic or invalid response when asked to make comparisons.	Makes some comparisons between single data values, but does not look at global trends.	Makes local or global comparisons, but does not link comparisons.	Makes both local and global comparisons and relates comparisons to each other.
	Reading Beyond the Data			
	Gives an idiosyncratic or invalid response when asked to make predictions.	Gives vague or inconsistent predictions that are not well linked to the data.	Uses the data in a consistent way to engage in sense-making predictions.	Uses both the data and the context to make complete and consistent predictions.

Figure 1. Elementary framework descriptors for analyzing and interpreting data.

Process	Level 1 Idiosyncratic	Level 2 Transitional	Level 3 Quantitative	Level 4 Analytical
Analyzing & Interpreting Data	**Reading Between the Data**			
	Makes incorrect comparisons within and between data sets.	Makes a single correct comparison or a set of partially correct comparisons within or between data sets.	Makes local or global comparisons within and between data sets.	Makes local and global comparisons within and between data sets.
	Reading Beyond the Data			
	Makes inferences that are not based on the data or inferences based on irrelevant issues.	Makes inferences that are partially based on the data. Some inferences may be only partially reasonable.	Makes inferences primarily based on the data. Some inferences may be only partially reasonable.	Makes reasonable inferences based on data and the context.
	Using Proportional Reasoning Where Necessary			
	Does not use relative thinking.	Uses relative thinking qualitatively.	Uses relative and proportional reasoning in an incomplete or invalid manner.	Uses relative and proportional reasoning.

Figure 2. Middle school framework descriptors for analyzing and interpreting data.

To illustrate these descriptors of students' statistical reasoning and to contrast the statistical reasoning of elementary students with middle school students, we look at student responses to the *Best Concert Tour* problem—a task that required students to analyze and interpret data. The task is presented in Figure 3; typical responses for elementary and middle school students at each of the four levels of the respective frameworks are presented in Table 1.

Task: Here are three graphs showing information on concert tours for Barbra Streisand, the Rolling Stones, Boyz II Men, and the Eagles. Who had the most successful concert tour? Justify your decision.

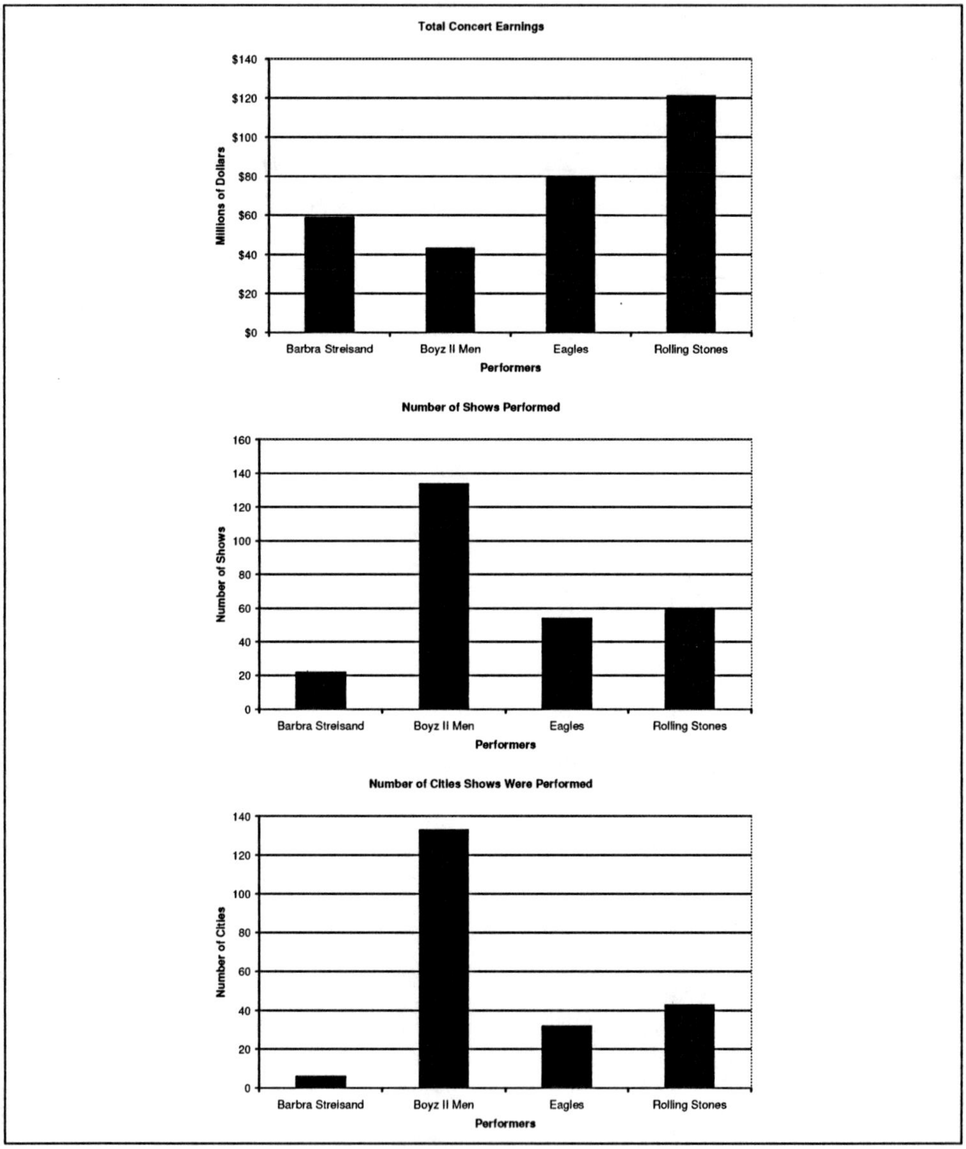

Figure 3. Best concert tour problem.

Table 1. Typical student responses at each level of reasoning on the best concert tour task

Level	Elementary Responses	Middle School Responses
Idiosyncratic	Boyz II Men, I went to one of their concerts	If you took these bars [for each performer] and put them on top of each other and you stacked them all up, Boyz II Men would be the tallest and most successful.
Transitional	Boyz II Men, the bars are tall.	The Rolling Stones performed three times as many shows as Barbara Streisand but only make twice as much money as she did. I think she did better.
Quantitative	I looked at each of the graphs and picked this one [the total concert earnings graph] and decided that the Rolling Stones are best because they got more money.	For Barbara Streisand it was 60 [total concert earnings] to 20 [number of shows] or 3 to 1. I don't need to look at Boyz II Men. The Eagles is about 2 to 1. For the Rolling Stones it is exactly 2 to 1. That makes Barbara Streisand the best.
Analytical	Boyz II Men performed a lot of shows but they didn't make much money. The Rolling Stones made a lot of money but didn't perform as many shows. I'd go with the Rolling Stones.	I calculated the earnings per show for each of the performers. Streisand is about 2.8 million dollars per show. Boyz II Men is about 0.3 million, the Eagles are about 1.45 million, and the Rolling Stones are about 2 million. I'd go with Barbara Streisand but there are some other things you would want to know, like how many people are in the band and the size of the audience.

At the idiosyncratic level, elementary students tend to base their reasoning on their own data sets (I went to one of their concerts), while middle school students often use the given data but in an inappropriate way (combine all the bars). Elementary and middle school students who exhibit transitional reasoning tend to focus on one aspect of the data, for example, the height of the bars in the case of the elementary student and ratios that are not fully connected in the case of the middle school student. The middle school student applies more sophisticated mathematical ideas than the elementary student, but neither student provides a complete justification. At the quantitative level, both elementary and middle school students make multiple quantitative comparisons but have difficulty linking their ideas. For example, the elementary student compares the data in the three graphs and then makes a local comparison within the "best" data set (total concert earnings); the middle school student makes multiple comparisons based on total earnings versus number of shows, but does not actually link the ratios to the context. The main difference between the elementary and middle school students' responses at this level is that the middle school student has access to proportional reasoning. Students who exhibit analytical reasoning use local and global comparisons of data and knowledge of the context to make valid inferences. For example, both the elementary and the middle school students recognize the need to relate money earned with number of shows performed; the main difference is that the middle school student actually determines and compares appropriate rates derived from the context. In fact, the middle school student even raises some additional factors that may act as limitations to the solution presented.

The differences between the responses of typical elementary and middle school students, at the four levels of the frameworks, can be related to the SOLO model (Biggs & Collis, 1991). These differences seem to reflect statistical reasoning that is associated with two different cycles in the concrete-symbolic mode (see Pegg & Davey, 1998; Watson, Collis, & Callingham et al., 1995). In essence, the cycle associated with the elementary students' statistical reasoning deals with the conceptual development of statistical concepts while the second cycle, demonstrated in the reasoning of the middle school students, deals with the application of statistical and mathematical concepts and procedures that have already been learned. Watson and her colleagues examine statistical reasoning associated with two developmental cycles in more detail in the next comprehensive model.

Watson et al. Model

Watson, Collis, Callingham, & Moritz (1995) used the Biggs and Collis (1991) cognitive development model to characterize middle school students' higher order statistical reasoning. More specifically, these researchers hypothesized that students' higher order statistical reasoning could be characterized according to two hierarchical unistructural-multistructural-relational [U-M-R] cycles, the first dealing with the *development* of statistical concepts and the second with the *consolidation and application* of these statistical concepts.

There were two parts to the study: clinical interviews with six 6th-grade students and one 9th-grade student and three instructional sessions with two 6th-grade classes working largely in groups. An interview protocol based on a set of 16 data cards containing information like student's name, age, favorite activity, eye color, weight, and number of fast-food meals per week was developed by the authors for use in both parts of the study.

In the clinical interview, students were asked to think of some interesting questions that could be answered using the cards; they were further prompted to imagine they were doing a school project with the cards. Following the analysis of the interview data, the researchers adapted the data-card task for use in the instructional setting. In the first class session, the students were introduced to ideas about looking for statistical associations and were then given a project that asked them to look for interesting questions and *connections* in the data. During the second session, the students were introduced to methods of displaying data (e.g., graphs) using examples that were unrelated to the data cards. The students continued working on their projects during the rest of the session and for part of the third session. They then presented their projects in the form of reports and posters.

The findings from this study demonstrated that the statistical reasoning of all 7 students in the interviews could be characterized according to the first U_1-M_1-R_1

cycle: students at the U_1 level focused on individual data with imaginative speculation on what caused certain data values; students at the M_1 level sorted the cards into different groups, focused on one variable at a time, and described that variable; students at the R_1 level sorted the cards into different groups, focused on more than one variable at a time, and appreciated the need to relate variables. Three students were classified as reasoning at U_1, 3 students at M_1, and 1 student at R_1. By contrast, during the instructional program, two U-M-R cycles were needed to characterize students' statistical reasoning. Moreover, all of the group or individual projects were classified beyond U_1. The characterizations of the second U_2-M_2-R_2 cycle moved into reasoning that involved justification and application: students at the U_2 level recognized the need to justify conjectured associations but did not proceed beyond that; students at the M_2 level used tables or graphs to support claims of association or cause, and students at the R_2 level used statistics such as the mean to support claims of association. Watson, Collis, & Callingham et al. (1995) also noted some evidence of multimodal functioning with ikonic intuitions and perceptions supporting students' reasoning and decision making in the concrete-symbolic mode. Both the learning cycle model and multimodal functioning have implications for informing instruction and enhancing teachers' knowledge of how students might respond to contextual data exploration tasks.

From a research perspective, it is interesting that Watson, Collis, & Callingham et al. (1995) uncovered two learning cycles in building models of higher order statistical reasoning; whereas Jones et al. (2000), working with elementary students, and Mooney (2002), working with middle school students, each found that one learning cycle was sufficient to characterize students' statistical reasoning. On the one hand, this difference may result from Watson and her colleagues' intimate knowledge of the Biggs and Collis model and their caveat that the additional cycles appear "when student understanding is viewed in considerable detail" (p. 250). On the other hand, it is possible that the two learning cycles identified by Jones et al. and Mooney represented two different cycles within the concrete symbolic mode—the first focusing on conceptual development of statistical concepts and the second incorporating applications of statistical concepts. Notwithstanding these possible rationalizations, there is clearly a need for researchers involved in formulating models of development in statistical reasoning to be aware of emerging research that suggests the existence of multiple learning cycles within a mode of operation like concrete-symbolic or formal (Callingham, 1994; Campbell, Watson, & Collis, 1992; Pegg & Davey, 1998; Watson, Collis, & Campbell, 1995).

COGNITIVE MODELS OF DEVELOPMENT FOR SPECIFIC STATISTICAL CONCEPTS AND PROCESSES

In this section we survey the research literature focusing on models of cognitive development that relate to specific statistical concepts. In particular, we focus on the following key concepts and processes: data modeling, measures of center and variation, group differences, covariation and association, and sampling and sampling distributions. In examining these models we do not claim to have exhausted all models of development in the field; rather, our review presages the concepts and processes that are considered in more detail in the following chapters of this book.

Data Modeling

Many researchers have examined patterns of growth in statistical reasoning when students have been engaged in data-modeling problems or model-eliciting problems that involve data (Ben-Zvi, Chapter 6; Ben-Zvi & Arcavi, 2001; Doerr, 1998; Doerr & Tripp, 1999; Lehrer & Romberg, 1996; Lehrer & Schauble, 2000; Lesh, Amit, & Schorr, 1997; Wares, 2001). Because of their inherent nature, data-modeling problems provide a distinctive context for observing students' statistical reasoning in open-ended situations. Modeling problems focus on organizing and representing data, pattern building, and seeking relationships (Lesh & Doerr, 2002), and they involve students in statistical reasoning such as decision making, inference, and prediction. Moreover, data-modeling problems often reveal students' innermost conceptual ideas about statistical reasoning—especially fundamental processes like dealing with variation, transforming data, evaluating statistical models, and integrating contextual and statistical features of the problem (Wild & Pfannkuch, 1999).

Measures of Center and Variation

Most of the research pertaining to measures of center has focused on the concepts of average, representativeness, or mean. Several studies have described students' varying conceptions of measures of center (Bright & Friel, 1998; Konold & Pollatsek, 2002; Konold & Pollatsek, Chapter 8; Morkros & Russell, 1995; Strauss & Bichler, 1988) but have not necessarily traced the development of students' understandings. Two studies that have addressed developmental aspects of students' reasoning with measures of center are the work of Reading and Pegg (1996) and Watson and Moritz (2000a). The few studies that have addressed the concept of variation or spread have examined the development of students' reasoning about variation (Shaughnessy, Watson, Moritz, & Reading, 1999; Reading & Shaughnessy, 2001; Reading & Shaughnessy, Chapter 9; Torok & Watson, 2000).

Comparing Two Data Sets

Making statistical inferences is a key aspect of statistical reasoning, and the importance of statistical inference is acknowledged in several curriculum documents (AEC, 1991; NCTM, 2000; DFE, 1995). One way that students can be introduced to statistical inference is by having them compare two or more sets of numerical data in contexts where the number in each set may be equal or unequal. Various researchers

(Cobb, 1999; McClain & Cobb, 1998; Mooney, 2002; Watson & Moritz, 1999) have produced models of development that characterize students' reasoning as they make statistical inferences involving the comparison of two data sets.

Bivariate Relationships

The study of correlation (association) and regression is important in statistics because these processes are used to identify statistical relationships between two or more variables and, where appropriate, to seek causal explanations. Accordingly, an understanding of association and regression has become important in the school mathematics curriculum (e.g., AEC, 1994; NCTM, 1989, 2000); thus, some researchers have examined the development of students' conceptions in relation to association and regression (Batanero, Estepa, Godino, & Green, 1996; Ross & Cousins, 1993; Wavering, 1989; Mevarech & Kramarsky, 1997). These studies have foreshadowed the more definitive cognitive models of Moritz and Watson (2000), Moritz (2001), and Mooney (2002).

Sampling and Sampling Distributions

The notion of sample is one of the most fundamental ideas in statistics, since samples enable us to gain information about the whole by examining the part (Moore, 1997). More specifically, sampling is used to make inferences about populations, that is, to predict population parameters from sample statistics. Processes like inference and prediction are grounded in the concept of sampling distributions, which is a complex idea for students to grasp. Research in this area has examined the development of students' statistical reasoning, not only in relation to the concept of sample, sample size, and sampling procedures (Watson, Chapter 12; Watson & Moritz, 2000b) but also in relation to more sophisticated ideas like sampling distributions (Chapter 13; Saldanha & Thompson, 2001) and the Central Limit Theorem (Chapter 13; delMas, Garfield, & Chance, 1999).

IMPLICATIONS FOR STATISTICAL EDUCATION

In statistical education, as in mathematics education, there is a continuing drive toward research that makes connections between the learning process and the teaching process. This has been brought into even sharper focus with the advent of constructivist approaches to learning and the need for pedagogies that facilitate students' mathematical constructions. The importance of this connection between teaching and learning is evident across the international scene; curriculum documents (AEC, 1991; NCTM, 1989, 2000; DFE, 1995) espouse reforms in mathematics education that encourage teachers to focus on "understanding what students know and need to know" and advocate that learners should "learn mathematics with understanding, actively building new knowledge from experience and prior knowledge" (NCTM, 2000, p. 11).

Due to this increased emphasis on teaching and learning and the need to have students actively building mathematical and statistical knowledge, powerful new instructional models have emerged during the last 15 years: Realistic Mathematics Education (RME; Gravemeijer, 1994), Cognitively Guided Instruction (CGI; Carpenter et al., 1989), and the Mathematics Teaching Cycle (MTC; Simon, 1995). Although these instructional models have many differences, they share the common perspective that students' learning is not only central to the instructional process; it must drive the instructional process. For example, RME evolved in order to create a shift from a mechanistic orientation to teaching and learning to an approach that emphasized student learning through reconstructive activity grounded in reality and sociocultural contexts (Streefland, 1991); CGI has as its major tenet the need to use research-based knowledge of students' reasoning to inform instruction; and MTC stresses "the reflexive relationship between the teacher's design of activities and consideration of the reasoning that students might engage in as they participate in those activities" (Simon, p. 133). All of these instructional theories highlight the need for teachers to understand and use the reasoning that students bring to mathematics classes.

Given these directions in teaching and learning, models of development in statistical reasoning have a key role in statistical instruction. Because these models incorporate domain-specific knowledge of students' statistical reasoning across key statistical concepts and processes, they arm teachers with the kind of knowledge that can be used in the design, implementation, and assessment of instruction in statistics and data exploration.

With respect to the *design of instruction,* cognitive models of development provide a coherent picture of the diverse range of statistical reasoning that a teacher might expect students to bring to the classroom. The use of cognitive models in designing instruction can be amplified by examining Simon's (1995) notion of hypothetical learning trajectory. By *hypothetical learning trajectory,* Simon means the formulation of learning goals, learning activities, and a conjectured learning process. In the first instance, many of the cognitive models discussed in this chapter identify key processes and concept goals, by their very nature indicating where children might be in relation to these goals. For example, the Jones et al. (2000) model identifies key processes like describing data, organizing data, representing data, and analyzing and interpreting data; it also documents, through the level descriptors, the kind of goals that might be appropriate for individual children or the class as a whole. In considering learning activities, the research on cognitive models invariably incorporates tasks and activities that have been used to engage students' statistical reasoning. For example, tasks like those incorporated in the technology simulation on sampling distributions (Chapter 13) have widespread potential in college and high school instructional settings. Finally, in relation to conjecturing the possible direction of the classroom learning process, the cognitive model provides a database for the teacher on the range of statistical reasoning that he or she might expect to find during instruction. For example, in Grade 3 instruction dealing with sampling, the Watson and Moritz (2000b) model suggests that all children would be small samplers with more than 50% of them using idiosyncratic methods of selecting samples.

With respect to the *implementation of instruction,* models of development can serve as a filter for analyzing and characterizing students' responses. In our teaching experiments (Jones et al., 2001), we have found that filtering students' responses using a model of development helps teachers to build a much richer knowledge base than they would without such a filter. In particular, a model helps teachers to frame questions and written tasks that accommodate the diversity of reasoning reflected in a group or class. Such accommodation and sensitivity by the teacher may enable children to develop more mature levels of reasoning. For example, a teacher who was aware from earlier group work that one student was reasoning about the dimensions of the sampling process in an integrated way (Level 5; see Chapter 13) might use that student's response as a focal point for a formative or summative discussion on the dimensions of the sampling. Alternatively, the teacher might use the response of a student who was using transitional reasoning (Level 2; see Chapter 13) on the dimensions of sampling as a means of focusing on the need for completeness and connections.

Finally, we believe that models of development in statistical reasoning can be helpful in *assessing* and monitoring students' performances over time, as well as in evaluating the effectiveness of classroom instruction. We are not suggesting that middle school students, for example, might move in a linear way through the four levels of Mooney's (2002) model of development in relation to analyzing and interpreting data. However, we are suggesting that teachers can observe differences in middle school students' collective and individual statistical reasoning that are recognizable based on the levels of the Mooney model. In a similar way, teachers can evaluate their instruction or instructional technology using models of development. For example, Chance et al. (Chapter 13) have used their cognitive model to evaluate and refine the simulation technology on sampling distributions. By assessing and observing changes in students' reasoning according to the model, they have identified weaknesses in the technology, have further refined and changed the technology, and have then reassessed the students' reasoning. This cycle of assessment and refinement has great potential in evaluating the pedagogical effectiveness of technology, in particular the use of microworlds.

As the research in this chapter reveals, students' statistical reasoning from elementary through college is diverse and often idiosyncratic. Moreover, students' statistical reasoning is constantly changing and hence is dynamic rather than static. Notwithstanding the diversity and dynamics of students' statistical reasoning, recurring patterns or levels of statistical reasoning are consistently observable when students are involved in key statistical processes like decision making, inferring, and predicting; and when they deal with concepts like sampling, organizing and representing data, center and variation, and analysis and interpretation. These recurring patterns of statistical reasoning, and the models of development that have evolved from them, offer a powerful resource for informing instructional programs that focus on having students learn statistical reasoning by building on or reformulating the statistical ideas they bring to the classroom.

REFERENCES

Australian Education Council. (1991). *A national statement on mathematics for Australian schools.* Carlton, VIC: Curriculum Corporation.

Australian Education Council. (1994). *Mathematics: A curriculum profile for Australian schools.* Carlton, VIC: Curriculum Corporation.

Batanero, C., Estepa, A., Godino, J. D., & Green, D. R. (1996). Intuitive strategies and preconceptions about association in contingency tables. *Journal for Research in Mathematics Education, 27,* 151–169.

Ben-Zvi, D., & Arcavi, A. (2001). Junior high school students' construction of global views of data and data representation. In J. Garfield, D. Ben-Zvi, & C. Reading (Eds.), *Background Readings of the Second International Research Forum on Statistical Reasoning, Thinking, and Literacy* (pp. 73–110). Armidale, Australia: Centre for Cognition Research in Learning and Teaching, University of New England.

Bidell, T. R., & Fischer, K. W. (1992). Cognitive development in educational contexts: Implications of skill theory. In A. Demetriou, M. Shayer, & A. Efklides (Eds.), *Neo-Piagetian theories of cognitive development: Implications and applications for education* (pp. 11–30). London: Routledge.

Biggs, J. B. (1992). Modes of learning, forms of knowing, and ways of schooling. In A. Demetriou, M. Shayer, & A. Efklides (Eds.), *Neo-Piagetian theories of cognitive development: Implication and applications for education* (pp. 31–51). London: Routledge.

Biggs, J. B., & Collis, K. F. (1982). *Evaluating the quality of learning: The SOLO taxonomy.* New York: Academic.

Biggs, J. B., & Collis, K. F. (1989). Towards a model of school-based curriculum development and assessment using the SOLO taxonomy. *Australian Journal of Education, 33,* 151–163.

Biggs, J. B., & Collis, K. F. (1991). Multimodal learning and intelligent behavior. In H. Rowe (Ed.), *Intelligence: Reconceptualization and measurement* (pp. 57–76). Hillsdale, NJ: Erlbaum.

Bright, G. W., & Friel, S. N. (1998, April). *Interpretation of data in a bar graph by students in grades 6 and 8.* Paper presented at annual meeting of the American Educational Research Association, San Diego, CA.

Callingham, R. A. (1994). *Teachers 'understanding of the arithmetic mean.* Unpublished master's thesis. University of Tasmania, Hobart, Australia.

Campbell, K. J., Watson, J. M., & Collis, K. F. (1992). Volume measurement and intellectual development. *Journal of Structural Learning and Intelligent Systems, 11,* 279–298.

Carpenter, T. P., Fennema, E., Peterson, P. L., Chiang, C. P., & Loef, M. (1989). Using knowledge of children's mathematical thinking in classroom teaching: An experimental study. *American Educational Research Journal, 264,* 499–532.

Case, R. (1985). *Intellectual development: A systematic reinterpretation.* New York: Academic Press.

Case, R., & Okamoto, Y. (1996). The role of conceptual structures in the development of children's thought. Monographs of the *Society for Research in Child Development, 61* (1–2, Serial No. 246).

Cobb, P., Wood, T., Yackel, E., Nicholls, J., Wheatley, G., Trigatti, B., & Perlwitz, M. (1991). Assessment of a problem-centered second-grade mathematics project. *Journal for Research in Mathematics Education, 22,* 3–29.

Cobb, P. (1999). Individual and collective mathematical development: The case of statistical data analysis. *Mathematical Thinking and Learning, 1,* 5–43.

Collis, K. F., & Biggs, J. B. (1991). Developmental determinants of qualitative aspects of school learning. In G. Evans (Ed.), *Learning and teaching cognitive skills* (pp. 185–207). Melbourne: Australian Council for Educational Research.

Curcio, F. R. (1987). Comprehension of mathematical relationships expressed in graphs. *Journal for Research in Mathematics Education, 18,* 382–393.

delMas, R. C., Garfield, J., & Chance, B. (1999). A model of classroom research in action: Developing simulation activities to improve students' statistical reasoning [Electronic version]. *Journal of Statistics Education, 7*(3), 1–16.

Department of Education and Science and the Welsh Office (DFE). (1995). *Mathematics in the national curriculum*. London: Author.

Doerr, H. M. (1998). A modeling approach to non-routine problem situations. In S. Berenson, K. Dawkins, M. Blanton, W. Coulombe, J. Kolb, K. Norwood, & L. Stiff (Eds.), *Proceedings of the nineteenth annual meeting, North American Chapter of the International Group for the Psychology of Mathematics Education* (Vol. 2, pp. 441–446). Columbus, OH: ERIC Clearinghouse for Science, Mathematics, and Environmental Education.

Doerr, H. M., & Tripp, J. S. (1999). Understanding how students develop mathematical models. *Mathematical Thinking and Learning, 1*, 231–254.

Fischbein, E. (1975). *The intuitive sources of probabilistic thinking in children*. Dordrecht, The Netherlands: Reidel.

Fischbein, E., & Gazit, A. (1984). Does the teaching of probability improve probabilistic intuitions? *Educational Studies in Mathematics, 15*, 1–24.

Fischbein, E., & Schnarch, D. (1997). The evolution with age of probabilistic, intuitively based misconceptions. *Journal for Research in Mathematics Education, 28*, 96–105.

Fischer, K. W. (1980). A theory of cognitive development: The control and construction of hierarchies of skill. *Psychological Review, 87*, 477–531.

Friel, S. N., Curcio, F. R., & Bright, G. W. (2001). Making sense of graphs: Critical factors influencing comprehension and instructional implications. *Journal for Research in Mathematics Education, 31*, 124–158.

Gal, I., & Garfield, J. B. (1997). Curricular goals and assessment challenges in statistics education. In I. Gal & J. B. Garfield (Eds.), *The assessment challenge in statistical education* (pp. 1–15). Amsterdam, The Netherlands: IOS Press.

Gravemeijer, K. (1994). Educational development and developmental research. *Journal for Research in Mathematics Education*, 25, 443–471.

Green, D. R. (1979). The chance and probability concepts project. *Teaching Statistics, 1*(3), 66–71.

Green, D. R. (1983). A survey of probability concepts in 3000 pupils aged 11–16. In D. R. Grey, P. Holmes, V. Barnett, & G. M. Constable (Eds.), *Proceedings of the first international conference on Teaching Statistics* (pp. 766–783). Sheffield, UK: Teaching Statistics Trust.

Jones, G. A., Langrall, C. W., Thornton, C. A., & Mogill, A. T. (1997). A framework for assessing and nurturing children's thinking in probability. *Educational Studies in Mathematics, 32*, 101–125.

Jones, G. A., Langrall, C. W., Thornton, C. A., Mooney, E. S., Wares, A., Jones, M. R., Perry, B., Putt, I. J., & Nisbet, S. (2001). Using students' statistical thinking to inform instruction. *Journal of Mathematical Behavior, 20*, 109–144.

Jones, G. A., Thornton, C. A., Langrall, C. W., Mooney, E. S., Perry, B., & Putt, I. J. (2000). A framework for characterizing students' statistical thinking. *Mathematics Thinking and Learning, 2*, 269–307.

Konold, C., & Pollatsek, A. (2002). Data analysis as the search for signals in noisy processes. Journal for Research in Mathematics Education, 33, 259–289.

Lehrer, R., & Romberg, T. (1996). Exploring children's data modeling. Cognition and Instruction, 14, 69–108.

Lehrer, R. & Schauble, L. (2000). Inventing data structures for representational purposes: Elementary grade children's classification models. Mathematical Thinking and Learning, 2, 51–74.

Lesh, R., Amit, M., & Schorr, R. Y. (1997). Using "real-life" problems to prompt students to construct statistical models for statistical reasoning. In I. Gal & J. Garfield (Eds.), *The assessment challenge in statistics education* (pp. 65–84). Amsterdam, The Netherlands: IOS Press.

Lesh, R., & Doerr, H. (2002). Foundations of a models and modeling perspective. In R. Lesh & H. Doerr (Eds.), *Beyond constructivist: A models and modeling perspective on mathematics teaching, learning and problem solving*. Mahwah, NJ: Lawrence Erlbaum Associates.

Levins, L., & Pegg, J. (1993). Students' understanding of concepts related to plant growth. *Research in Science Education, 23*, 165–173.

Mevarech, Z. A., & Kramarsky, B. (1997). From verbal descriptions to graphic representations: Stability and change in students' alternative conceptions. *Educational Studies in Mathematics, 32*, 229–263.

McClain, K., & Cobb, P. (1998). Supporting students' reasoning about data. In S. Berenson, K. Dawkins, M. Blanton, W. Coulombe, J. Kolb, K. Norwood, & L. Stiff (Eds.), *Proceedings of the nineteenth annual meeting, North American Chapter of the International Group for the Psychology of Mathematics Education* (Vol. 1, pp. 389–394). Columbus, OH: ERIC Clearinghouse for Science, Mathematics, and Environmental Education.

Mokros, J., & Russell, S. J. (1995). Children's concepts of average and representativeness. *Journal for Research in Mathematics Education*, 26, 20–39.

Mooney, E. S. (2002). A framework for characterizing middle school students' statistical thinking. *Mathematical Thinking and Learning, 4*, 23–63.

Mooney, E. S., Langrall, C. W., Hofbauer, P. S., & Johnson, Y. A. (2001). Refining a framework on middle school students' statistical thinking. In R. Speiser, C. A. Maher, & C. N. Walter (Eds.), *Proceedings of the twenty-third annual meeting of the North American Chapter of the International Group for the Psychology of Mathematics Education* (Vol.1, pp. 439–447). Columbus, OH: ERIC Clearinghouse for Science, Mathematics, and Environmental Education.

Moore, D. S. (1997). *Statistics: Concepts and controversies* (4th ed.). New York: Freeman.

Moritz, J. B. (2001). Graphical representations of statistical associations by upper primary students. In J. Garfield, D. Ben-Zvi, & C. Reading (Eds.), *Background Readings of the Second International Research Forum on Statistical Reasoning, Thinking, and Literacy* (pp. 256–264). Armidale, Australia: Centre for Cognition Research in Learning and Teaching, University of New England.

Moritz, J. B., & Watson, J. (2000). *Representing and questioning statistical associations*. Unpublished manuscript, University of Tasmania, Hobart, Australia.

National Council of Teachers of Mathematics (1989). *Curriculum and evaluation standards for school mathematics*. Reston, VA: Author.

National Council of Teachers of Mathematics (2000). *Principles and standards for school mathematics*. Reston, VA: Author.

Olecka, A. (1983). An idea of structuring probability teaching based on Dienes' six stages. In D. R. Grey, P. Holmes, V. Barnett, & G. M. Constable (Eds.), *Proceedings of the First International Conference on Teaching Statistics* (pp. 727–737). Sheffield, UK: Teaching Statistics Trust.

Pegg, J. (1992). Assessing students' understanding at the primary and secondary level in the mathematical sciences. In J. Izard & M. Stephens (Eds.), *Reshaping assessment practice: Assessment in the mathematical sciences under challenge* (pp. 368–365). Melbourne, VIC: Australian Council of Educational Research.

Pegg, J., & Davey, G. (1998). Interpreting student understanding in geometry: A synthesis of two models. In R. Lehrer & D. Chazan (Eds.), *Designing learning environments for developing understanding of geometry and space* (pp. 109–135). Mahwah, NJ: Erlbaum.

Piaget, J. (1954). *The construction of reality in the child*. New York: Basic Books.

Piaget, J. (1962). *The origins of intelligence in the child*. New York: Norton.

Piaget, J., & Inhelder, B. (1975). *The origin of the idea of chance in children* (L. Leake, P. Burrell, & H. D. Fischbein, Trans.). New York: Norton. (Original work published 1951.)

Polaki, M. V., Lefoka, P. J., & Jones, G. A. (2000). Developing a cognitive framework for describing and preparing Basotho students' probabilistic thinking. *Boleswa Educational Research Journal, 17*, 1–20.

Reading, C., & Pegg, J. (1996). Exploring understanding of data reduction. In L. Puig & A. Gutierrez (Eds.), *Proceedings of the 20th conference of the International Group for the Psychology of Mathematics Education* (Vol. 4, pp. 187–194). Valencia, Spain: University of Valencia.

Reading, C., & Shaughnessy, J. M. (2001). Student perceptions of variation in a sampling situation. In J. Garfield, D. Ben-Zvi, & C. Reading (Eds.), *Background Readings of the Second International Research Forum on Statistical Reasoning, Thinking, and Literacy*, (pp. 119–126). Armidale, Australia: Centre for Cognition Research in Learning and Teaching., University of New England

Reber, A. S. (1995). *The Penguin dictionary of psychology* (2nd ed.). London: Penguin.

Resnick, L. B. (1983). Developing mathematical knowledge. *American Psychologist, 44*, 162–169.

Ross, J. A., & Cousins, J. B. (1993). Patterns of students' growth in reasoning about correlational problems. *Journal of Educational Psychology, 85*, 49–65.

Saldanha, L. A., & Thompson, P. (2001). Students' reasoning about sampling distributions and statistical inference. In J. Garfield, D. Ben-Zvi, & C. Reading (Eds.), *Background Readings of the Second International Research Forum on Statistical Reasoning, Thinking, and Literacy* (pp. 291–296). Armidale, Australia: Centre for Cognition Research in Learning and Teaching, University of New England.

Scholz, R. W. (1991). Psychological research on the probability concept and its acquisition. In R. Kapadia & M. Borovcnik (Eds.), *Chance encounters: Probability in education* (pp. 213–254). Dordrecht, The Netherlands: Kluwer Academic.

Shaughnessy, J. M. (1992). Research in probability and statistics: Reflections and directions. In D. A. Grows (Ed.), *Handbook of research on mathematics teaching and learning* (pp. 465–494). New York: Macmillan.

Shaughnessy, J. M., Garfield, J., & Greer, B. (1996). Data handling. In A. J. Bishop, K. Clements, C. Keitel, J. Kilpatrick, & C. Laborde (Eds.), *International handbook of mathematics education* (Part 1, pp. 205–238). Dordrecht, The Netherlands: Kluwer Academic.

Shaughnessy, J. M., Watson, J. M., Moritz, J. B., & Reading, C. (1999, April). *School mathematics students' acknowledgment of statistical variation.* Paper presented at the 77th annual conference of the National Council of Teachers of Mathematics, San Francisco, CA.

Simon, M. A. (1995). Restructuring mathematics pedagogy from a constructivist perspective. *Journal for Research in Mathematics Education, 26*, 114–145.

Strauss, S., & Bichler, E. (1988). The development of children's concept of the arithmetic average. *Journal for Research in Mathematics Education, 19*, 64–80.

Streefland, L. (1991). *Fractions in Realistic Mathematics Education—A paradigm of developmental research.* Dordrecht, The Netherlands: Kluwer Academic.

Tarr, J. E., & Jones, G. A. (1997). A framework for assessing middle school students' thinking in conditional probability and independence. *Mathematics Education Research Journal, 9*, 39–59.

Torok, R., & Watson, J. (2000). Development of the concept of statistical variation: An exploratory study. *Mathematics Education Research Journal, 12*, 147–169.

Wares, A. (2001). *Middle school students' construction of mathematical models.* Unpublished doctoral dissertation, Illinois State University, Normal.

Watson, J. M., Collis, K. F., Callingham, R. A., & Moritz, J. B. (1995). A model for assessing higher order thinking in statistics. *Educational Research and Evaluation, 1*, 247–275.

Watson, J. M., Collis, K. F., & Campbell, K. J. (1995). Developmental structure in the understanding of common and decimal fractions. *Focus on Learning Problems in Mathematics, 17*(1), 1–24.

Watson, J. M., Collis, K. F., & Moritz, J. B. (1997). The development of chance measurement. *Mathematics Education Research Journal, 9*, 60–82.

Watson, J. M., & Moritz, J. B. (1998). Longitudinal development of chance measurement. *Mathematics Education Research Journal, 10*, 103–127.

Watson, J. M., & Moritz, J. B. (1999). The beginning of statistical inference: Comparing two data sets. *Educational Studies in Mathematics, 37*, 145–168.

Watson, J. M., & Moritz, J. B. (2000a). The longitudinal development of understanding of average. *Mathematical Thinking and Learning, 2*, 11–50.

Watson, J. M., & Moritz, J. B. (2000b). Developing concepts of sampling. *Journal for Research in Mathematics Education 31*, 44–70.

Wavering, J. (1989). Logical reasoning necessary to make line graphs. *Journal of Research in Science Teaching, 26*, 373–379.

Wild, C. J., & Pfannkuch, M. (1999). Statistical thinking in empirical enquiry. *International Statistical Review, 67*, 223–265.

PART II

STUDIES OF STATISTICAL REASONING

Chapter 6

REASONING ABOUT DATA ANALYSIS

Dani Ben-Zvi
University of Haifa, Israel

OVERVIEW

The purpose of this chapter is to describe and analyze the ways in which middle school students begin to reason about data and come to understand exploratory data analysis (EDA). The process of developing reasoning about data while learning skills, procedures, and concepts is described. In addition, the students are observed as they begin to adopt and exercise some of the habits and points of view that are associated with statistical thinking. The first case study focuses on the development of a global view of data and data representations. The second case study concentrates on design of a meaningful EDA learning environment that promotes statistical reasoning about data analysis. In light of the analysis, a description of what it may mean to learn to reason about data analysis is proposed and educational and curricular implications are drawn.

THE NATURE OF EXPLORATORY DATA ANALYSIS

Exploratory data analysis (EDA), developed by Tukey (1977), is the discipline of organizing, describing, representing, and analyzing data, with a heavy reliance on visual displays and, in many cases, technology. The goal of EDA is to make sense of data, analogous to an *explorer of unknown lands* (Cobb & Moore, 1997). The original ideas of EDA have since been expanded by Mosteller and Tukey (1977) and Velleman and Hoaglin (1981); they have become the accepted way of approaching the analysis of data (Biehler, 1990; Moore, 1990, 1992).

According to Graham (1987), and Kader and Perry (1994), data analysis is viewed as a four-stage process: (a) pose a question and formulate a hypothesis, (b) collect data, (c) analyze data, and (d) interpret the results and communicate conclusions. In reality however, statisticians do not proceed linearly in this process, but rather iteratively, moving forward and backward, considering and selecting possible paths (Konold & Higgins, 2003). Thus, EDA is more complex than the four-stage process: "data analysis is like a give-and-take conversation between the

D. Ben-Zvi and J. Garfield (eds.),
The Challenge of Developing Statistical Literacy, Reasoning and Thinking, 121–145.

hunches researchers have about some phenomenon and what the data have to say about those hunches. What researchers find in the data changes their initial understanding, which changes how they look at the data, which changes their understanding" (Konold & Higgins, 2003, p. 197).

EDA employs a variety of techniques, mostly graphical in nature, to maximize insight into a data set. Exploring a data set includes examining shape, center, and spread; and investigating various graphs to see if they reveal clusters of data points, gaps, or outliers. In this way, an attempt is made to uncover underlying structure and patterns, test underlying assumptions, and develop parsimonious models. Many EDA graphical techniques are quite simple, such as stem-and-leaf plots and box plots. Computers support EDA by making it possible to quickly manipulate and display data in numerous ways, using statistical software packages such as Data Desk (Velleman, 2003), Fathom (Finzer, 2003), and Tabletop (TERC, 2002).

However, the focus of EDA is not on a set of techniques, but on making sense of data, how we dissect a data set; what we look for; how we look; and how we interpret. EDA postpones the classical "statistical inference" assumptions about what kind of model the data follow with the more direct approach of "let the numbers speak for themselves" (Moore, 2000, p. 1), that is, allowing the data itself to reveal its underlying structure and model.

This complete and complex picture of data analysis should be reflected in the teaching of EDA and in the research on students' statistical reasoning. Simplistic views can lead to the use of recipe approaches to data analysis instruction and to research that does not go beyond the surface understanding of statistical techniques.

EDA in School Curriculum

EDA provides a pedagogical opportunity for open-ended data exploration by students, aided by educational technology. Allowing students to explore data is aligned with current educational paradigms, such as, teaching and learning for understanding (Perkins & Unger, 1999), inquiry-based learning (Yerushalmy, Chazan, & Gordon, 1990), and project-based learning (Evensen & Hmelo, 2000). However, the complexity of EDA raises numerous instructional challenges, for example, how to teach methods in a new and changing field, how to compensate for the lack of teachers' prior experience with statistics, and how to put together an effective K–12 curriculum in statistics that incorporates EDA.

Elements of EDA have been integrated into the school mathematics curriculum in several countries, such as Australia (Australian Education Council, 1991, 1994), England (Department for Education and Employment, 1999), New Zealand (Ministry of Education, 1992), and the United States (National Council of Teachers of Mathematics, 1989, 2000). In recently developed curricula—for example, *Chance and Data* (Lovitt & Lowe, 1993), *The Connected Mathematics Project* (Lappan, Fey, Fitzgerald, Friel, & Phillips, 1996), *Data: Kids, Cats, and Ads* (Rubin & Mokros, 1998), *Data Handling* (Greer, Yamin-Ali, Boyd, Boyle, & Fitzpatrick, 1995), *Data Visualization* (de Lange & Verhage, 1992), *Exploring Statistics* (Bereska, Bolster, Bolster, & Scheaffer, 1998, 1999), *The Quantitative Literacy*

Series (e.g., Barbella, Kepner, & Schaeffer, 1994), and *Used Numbers* (e.g., Friel, Mokros, & Russel, 1992)—there is growing emphasis on developing students' statistical reasoning about data analysis; on graphical approaches; on students gathering their own data and intelligently carrying out investigations; on the use of educational software, simulations, and Internet; on a cross-curricular approach; and on the exploration of misuses and distortions as points of departure for study.

Research on Reasoning about Data Analysis

Research on reasoning about data analysis is beginning to emerge as a unique area of inquiry. In a teaching experiment conducted with lower secondary school students by Biehler & Steinbring (1991), data analysis was introduced as "detective" work. Teachers gradually provided students with a data "tool kit" consisting of tasks, concepts, and graphical representations. The researchers concluded that all students succeeded in acquiring the beginning tools of EDA, and that both the teaching and the learning became more difficult as the process became more open. There appears to be a tension between directive and nondirective teaching methods in this study. A study by de Lange, Burrill, & Romberg (1993) reveals the crucial need for professional development of teachers in the teaching of EDA in the light of the difficulties teachers may find in changing their teaching strategy from expository authority to guide. It is also a challenge for curriculum developers to consider these pedagogical issues when creating innovative EDA materials. Recent experimental studies in teaching EDA around key concepts (distribution, covariation) in middle school classes have been conducted by Cobb (cf., 1999) with an emphasis on sociocultural perspectives of teaching and learning.

Ben-Zvi and Friedlander (1997b) described some of the characteristic reasoning processes observed in students' handling of data representations in four patterns: (a) *uncritical thinking,* in which the technological power and statistical methods are used randomly or uncritically rather than "targeted"; (b) *meaningful use of a representation,* in which students use an appropriate graphical representation or measure in order to answer their research questions and interpret their findings; (c) *meaningful handling of multiple representations,* in which students are involved in an ongoing search for meaning and interpretation to achieve sensible results as well as in monitoring their processes; and (d) *creative thinking,* in which students decide that an uncommon representation or method would best express their thoughts, and they manage to produce an innovative graphical representation, or self-invented measure, or method of analysis.

THE CURRENT STUDY

Theoretical Perspectives

Research on mathematical cognition in the last decades seems to converge on some important findings about learning, understanding, and becoming competent in mathematics. Stated in general terms, research indicates that becoming competent in a complex subject matter domain, such as mathematics or statistics, "may be as much a matter of acquiring the habits and dispositions of interpretation and sense making as of acquiring any particular set of skills, strategies, or knowledge" (Resnick, 1988, p. 58). This involves both cognitive development and "socialization processes" into the culture and values of "doing mathematics" (*enculturation*). Many researchers have been working on the design of teaching in order to "bring the practice of knowing mathematics in school closer to what it means to know mathematics within the discipline" (Lampert, 1990, p. 29). This chapter is intended as a contribution to the understanding of these processes in the area of EDA.

Enculturation Processes in Statistics Education

A core idea used in this study is that of *enculturation.* Recent learning theories in mathematics education (cf., Schoenfeld, 1992; Resnick, 1988) include the process of enculturation. Briefly stated, this process refers to entering a community or a practice and picking up their points of view. The beginning student learns to participate in a certain cognitive and cultural practice, where the teacher has the important role of a mentor and mediator, or the *enculturator.* This is especially the case with regard to statistical thinking, with its own values and belief systems and its habits of questioning, representing, concluding, and communicating. Thus, for *statistical enculturation* to occur, specific thinking tools are to be developed alongside collaborative and communicative processes taking place in the classroom.

Statistical Thinking

Bringing the practice of knowing statistics at school closer to what it means to know statistics within the discipline requires a description of the latter. Based on in-depth interviews with practicing statisticians and statistics students, Wild and Pfannkuch (1999, and Chapter 2) provide a comprehensive description of the processes involved in statistical thinking, from problem formulation to conclusions. They suggest that a statistician operates (sometimes simultaneously) along four dimensions: investigative cycles, types of thinking, interrogative cycles, and dispositions.

Based on these perspectives, the following research questions were used to structure the case studies and the analysis of data collected:

- How do junior high school students begin to reason about data and make sense of the EDA perspective in the context of open-ended problem-solving situations, supported by computerized tools?
- How do aspects of the learning environment promote students' statistical reasoning about data analysis?

METHOD

This study employs a qualitative analysis method, to examine seventh-grade students' statistical reasoning about data in the context of two classroom investigations. Descriptions of the setting, curriculum, and technology are followed by a profile of the students, and then by methods of data collection and analysis.

The Setting

The study took place in three seventh-grade classes (13-year-old girls and boys) in a progressive experimental school in Tel-Aviv. The classes were taught by skillful and experienced teachers, who were aware of the spirit and goals of the curriculum (described briefly later). They were part of the CompuMath curriculum development and research team, which included several mathematics and statistics educators and researchers from the Weizmann Institute of Science, Israel. The CompuMath Project is a large and comprehensive mathematics curriculum for grades 7–9 (Hershkowitz, Dreyfus, Ben-Zvi, Friedlander, Hadas, Resnick, Tabach, & Schwarz, 2002), which is characterized by the teaching and learning of mathematics using open-ended problem situations to be investigated by peer collaboration and classroom discussions using computerized environments.

The *Statistics Curriculum* (SC)—the data component of the CompuMath Project—was developed to introduce junior high school students (grade 7, age 13) to statistical reasoning and the "art and culture" of EDA (described in more detail in Ben-Zvi & Friedlander, 1997b). The design of the curriculum was based on the creation of small scenarios in which students can experience some of the processes involved in the experts' practice of data-based enquiry. The SC was implemented in schools and in teacher courses, and subsequently revised in several curriculum development cycles.

The SC was designed on the basis of the theoretical perspectives on learning and the expert view of statistical thinking just described. It stresses: (a) student's active participation in organization, description, interpretation, representation, and analysis of data situations (on topics close to the students' world such as sport records, lengths of people's names in different countries, labor conflicts, car brands), with a considerable use of visual displays as analytical tools (in the spirit of Garfield, 1995, and Shaughnessy, Garfield, & Greer, 1996); and (b) incorporation of technological tools for simple use of various data representations and transformations of them (as described in Biehler, 1993, 1997; Ben-Zvi, 2000). The scope of the curriculum is 30 periods spread over 2-1/2 months, and it includes student book (Ben-Zvi & Friedlander, 1997a) and teacher guide (Ben-Zvi & Ozruso, 2001).

Technology

During the experimental implementation of the curriculum a spreadsheet package (Excel) was used. Although Excel is not the ideal tool for data analysis (Ben-Zvi, 2000), the main reasons for choosing this software were:

- Spreadsheets provide direct access that allows students to view and explore data in different forms, investigate different models that may fit the data, manipulate a line to fit a scatter plot, etc.
- Spreadsheets are flexible and dynamic, allowing students to experiment with and alter representations of data. For instance, they may change, delete or add data entries in a table and consider the graphical effect of the change or manipulate directly data points on the graph and observe the effects on a line of fit. Spreadsheets are adaptable by providing control over the content and style of the output.
- Spreadsheets are common, familiar, and recognized as a fundamental part of computer literacy (Hunt, 1995). They are used in many areas of everyday life, as well as in other domains of mathematics curricula, and are available in many school computer labs. Hence, learning statistics with a spreadsheet helps to reinforce the idea that this is something connected to the real world.

Participants

This study focuses mainly on two students—*A* and *D* (in the first case), and on *A* and *D* and four of their peers (in the second case). *A* and *D* were above–average ability students, very verbal, experienced in working collaboratively in computer-assisted environments, and willing to share their thoughts, attitudes, doubts, and difficulties. They agreed to participate in this study, which took place within their regular classroom periods and included being videotaped and interviewed (after class) as well as furnishing their notebooks for analysis.

When they started to learn this curriculum, *A* and *D* had limited in-school statistical experience. However, they had some informal ideas and positive dispositions toward statistics, mostly through exposure to statistics jargon in the media. In primary school, they had learned only about the mean and the uses of some diagrams. Prior to, and in parallel with, the learning of the SC they studied beginning algebra based on the use of spreadsheets to generalize numerical linear patterns (Resnick & Tabach, 1999).

The students appeared to engage seriously with the curriculum, trying to understand and reach agreement on each task. They were quite independent in their work, and called the teacher only when technical or conceptual issues impeded their progress. The fact that they were videotaped did not intimidate them. On the contrary, they were pleased to speak out loud; address the camera explaining their actions, intentions, and misunderstandings; and share what they believed were their successes.

Data Collection

To study the effects of the new curriculum, student behavior was analyzed using video recordings, classroom observations, interviews, and the assessment of students' notebooks and research projects. The two students—*A* and *D*—were videotaped at almost all stages (20 hours of tapes), and their notebooks were also collected.

Analysis

The analysis of the videotapes was based on interpretive microanalysis (see, for example, Meira, 1991, pp. 62–63): a qualitative detailed analysis of the protocols, taking into account verbal, gestural and symbolic actions within the situations in which they occurred. The goal of such an analysis is to infer and trace the development of cognitive structures and the sociocultural processes of understanding and learning.

Two stages were used to validate the analysis, one within the CompuMath researchers' team and one with four researchers from the Weizmann Institute of Science, who had no involvement with the data or the SC (triangulation in the sense of Schoenfeld, 1994). In both stages the researchers discussed, presented, and advanced and/or rejected hypotheses, interpretations, and inferences about the students' cognitive structures. Advancing or rejecting an interpretation required: (a) providing as many pieces of evidence as possible (including past and/or future episodes, and all sources of data as described earlier); and (b) attempts to produce equally strong alternative interpretations based on the available evidence. In most cases the two analyses were in full agreement, and points of doubt or rejection were refuted or resolved by iterative analysis of the data.

Case Study 1: Constructing Global Views of Data

The first case study concentrates on the growth and change of the students' conceptions as they entered and learned the culture of EDA and started to develop their reasoning about data and data representations. This study focused on the shift between local observations and global observations. In EDA, *local understanding* of data involves focusing on an individual value (or a few of them) within a group of data (a particular entry in a table of data, a single point in a graph). *Global understanding* refers to the ability to search for, recognize, describe, and explain general patterns in a set of data (change over time, trends) by naked-eye observation of distributions and/or by means of statistical parameters or techniques. Looking globally at a graph as a way to discern patterns and generalities is fundamental to statistics, and it includes the production of explanations, comparisons, and predictions based on the variability in the data. By attending to where a collection of values is centered, how those values are distributed or how they change over time, statistics deals with features not inherent to individual elements but to the aggregate that they comprise.

Learning to look globally at data can be a complex process. Studies in mathematics education show that students with a sound local understanding of certain mathematical concepts struggle to develop global views (cf., Monk, 1988; Bakker, Chapter 7). Konold, Pollatsek, and Well (1997) observed that high school students—after a yearlong statistics course—still had a tendency to focus on properties of individual cases rather than on propensities of data sets.

The interplay between local and global views of data is reflected in the tools statistics experts use. Among such tools, which support data-based arguments, explanations, and (possibly) forecasts, are *time plots,* which highlight data features such as trends and outliers, center, rates of change, fluctuations, cycles, and gaps (Moore & McCabe, 1993). For the purpose of reflection (or even dishonest manipulation), trends can be highlighted or obscured by changing the scales. For example, in Cartesian-like graphs the vertical axis can be "stretched," so that the graph conveys the visual impression of a steep slope for segments connecting consecutive points, giving a visual impression that the rate of change is large. Experts propose standards in order to avoid such visual distortions (cf., Cleveland, 1994, pp. 66–67).

The Task

In the first activity of the SC, the *Men's 100 Meters Olympic Race,* students were asked to examine real data about the winning times in the men's 100 meters during the modern Olympic Games. Working in pairs, assisted by the spreadsheet, they were expected to analyze the data in order to find trends and interesting phenomena. This covariation problem concerned tables and graphical representations (time plots) and formulating verbal statements as well as translating among those representations. In the second part of this activity, a problem is presented to students in the following way:

> Two sports journalists argue about the record times in the 100 meters. One of them claims that there seems to be no limit to human ability to improve the record. The other argues that sometime there will be a record, which will never be broken. To support their positions, both journalists use graphs.

One task of this investigation asks students to design a representation, using a computer, to support different statements, such as: (a) The times recorded in the Olympic 100 meters improved considerably; and (b) Throughout the years, the changes in the Olympic times for the 100 meters were insignificant.

Analysis: Toward an Expert Reasoning

Students started their introduction to EDA by learning to make sense of general questions normally asked in data exploration. They often offered irrelevant answers, revealed an implicit sense of discomfort with these answers, asked for help, and used the teacher's feedback to try other answers. They worked on EDA tasks with partial understanding of the overall goal. By confronting the same issues with different sets of data and in different investigational contexts, they overcame some of their difficulties. The teacher's role included reinforcing the legitimacy of an observation as being of the right "kind" despite not being fully correct, or simply refocusing attention on the question. These initial steps in an unknown field are regarded as an aspect of the enculturation process (e.g., Schoenfeld, 1992; Resnick, 1988).

At the beginning stage, students also struggled with how to read and make sense of local (pointwise) information in tables and in graphs. This stage involved learning to see each row in a table (Table 1) with all its details as one whole case out of the many shown, and focusing their attention on the entries that were important for the curricular goal of this activity: the record time, and the year it occurred. This view of each single row, with its two most relevant pieces of information, was reinforced afterward when students displayed the data in a time plot (Figure 1), since the graph (as opposed to the table) displays just these two variables. Also, this understanding of pointwise information served later on as the basis for developing a global view, as an answer to "how do records change over time?"

Table 1. Part of the table of the men's 100 meters winning times in the 23 Olympiads from 1896 to 1996

Year	**City**	**Athlete's name**	**Country**	**Time (sec.)**
1896	Athens	Thomas Burke	USA	12.0
1900	Paris	Francis Jarvis	USA	10.8
1904	St. Louis	Archie Hahn	USA	11.0
1908	London	Reginald Walker	South Africa	10.8
1912	Stockholm	Ralph Craig	USA	10.8
1920	Antwerp	Charles Paddock	USA	10.8
1924	Paris	Harold Abrahams	UK	10.6

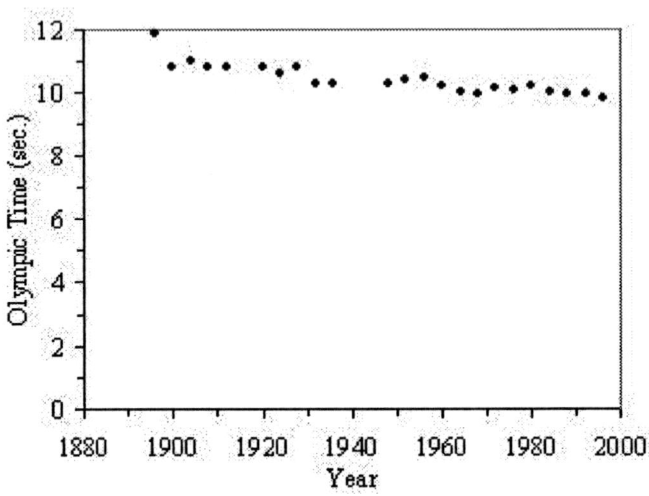

Figure 1. Time plot showing winning times for men's 100 meters.

Instead of looking at the graph as a way to discern patterns in the data, students' response focused first on the nature and language of the graph as a representation—how it displays discrete data, rather than as a tool to display a generality, a trend. When invited to use the line connecting the dots in the dot plot (Figure 1) as an artifact to support a global view, they rejected it because it lacked any meaning in light of the pointwise view they had just learned, and with which they felt comfortable.

When *A* and *D* were asked to describe what they learned from the 100 meters table (Table 1), they observed that "There isn't anything constant here." After the teacher reinforced the legitimacy of their observation, they explained more clearly what they meant by constancy in the following dialogue (the dialogues are translated from Hebrew, therefore they may not sound as authentic as in the original):

D Let's answer the first question: "What do you learn from this table?"
A There are no constant differences between …
D We learn from this table that there are no constant differences between the record times of … [looking for words]
A The results of …
D The record times of the runners in …
A There are no constant differences between the runners in the different Olympiads …

The students' attention focused on differences between adjacent pairs of data entries, and they noticed that these differences are not constant. These comparisons presumably stemmed from their previous knowledge and experiences with a spreadsheet in algebra toward finding a formula. In other words, one of the factors that moved them forward toward observation of patterns was their application of previous knowledge. Thus, the general pattern the students observed and were able to express was that the differences were not constant. Maybe they implicitly began to sense that the nature of these data in this new area of EDA, as opposed to algebra, is disorganized, and it is not possible to capture it in a single deterministic formula.

After the two students had analyzed the 100 meters data for a while, they worked on the next question: to formulate a preliminary hypothesis regarding the trends in the data. They seemed to be embarrassed by their ignorance—not knowing what trends mean, and asked for the teacher's help.

A	What are trends? What does it mean?
T	What is a trend? A trend is ... What's the meaning of the word trend?
A	Ah ... Yes, among other things, and what is the meaning in the question.
T	O.K. Let's see: We are supposed to look at what?
D	At the table.
T	At the table. More specifically—at what?
A	At the records.
T	At the records. O.K. And now, we are asked about what we see: Does it decrease all the time?
A&D	No.
T	No. Does it increase all the time?
A&D	No.
T	No. So, what does it do after all?
D	It changes.
T	It changes. Correct.
A	It generally changes from Olympiad to Olympiad. Generally, not always.
T	Sometimes it doesn't change at all. Very nice! Still, it usually changes. And, is there an overall direction?
D	No!
T	No overall direction?
A	There is no overall declining direction, namely, improvement of records. But, sometimes there is deterioration ...
T	Hold on. The overall direction is? Trend and direction are the same.
A&D	Increase, Increase!
T	The general trend is ...
D	Improvement in records.
T	What is "improvement in records"?
A	Decline in running times.
T	Yes. Decline in running times. O.K. ... But ...
A	Sometimes there are bumps, sort of steps ...
T	... But, this means that although we have deviations from the overall direction here and there, still the overall direction is this ... Fine, write it down.

The students were unfamiliar with the term *trends,* and they were vague about the question's purpose and formulation. In response, the teacher gradually tried to nudge the students' reasoning toward global views of the data. Once they understood the intention of the question, the students—who viewed the irregularity as the most salient phenomenon in the data—were somehow bound by the saliency of local values: They remained attached to local retrogressions, which they could not overlook in favor of a general sense of direction/trend.

The teacher, who did not provide a direct answer, tried to help them in many ways. First, she devolved the question (in the sense of Brousseau, 1997, pp. 33–35 and 229–235), and when this did not work, she rephrased the question in order to refocus it: "We are supposed to look at what?" and "more specifically at what?" She then hints via direct questions: "Does it increase all the time?" and "So, what does it do after all?" In addition, she appropriated (in the sense of Moschkovich, 1989) the students' answers to push the conversation forward by using their words and answers, for example: "It changes. Correct"; "increase"; "decrease." At other times she subtly transformed their language, such as changing *bumps* to *deviations;* or by providing alternative language to rephrase the original question to: "Is there an overall direction?"

After the interaction just presented, *A* and *D* wrote in their notebooks the following hypothesis: "The overall direction is increase in the records, yet there were occasionally lower (slower) results, than the ones achieved in previous Olympiads." At this stage, it seems that they understood (at least partially) the meaning of *trend,* but still stressed (less prominently than before) those local features that did not fit the pattern.

In the second part of the activity, the students were asked to delete an "outlying" point (the record of 12 sec. in the first Olympiad, 1896) from the graph (Figure 1) and describe the effect on its shape. The purpose of the curriculum was to lead students to learn how to transform the graph in order to highlight trends. It was found that by focusing on an exceptional point and the effect of its deletion directed students' attention to a general view of the graph. This finding seems consistent with Ainley (1995), who also describes how an outlier supported students' construction of global meanings for graphs.

The following transcript describes the students' comments on the effect of changing the vertical scales of the original 100 meters graph from 0–12 (Figure 2) to 0–40 (Figure 3) as requested in the second part of the activity.

A Now, the change is that the whole graph stayed the same in shape, but it went down.

D The same in shape, but much, much lower, because the column [the y-axis] went up higher. Did you understand that? [D uses both hands to signal the down and up movements of the graph and the y-axis respectively.]

A Because now the 12, which is the worst record, is lower. It used to be once the highest. Therefore, the graph started from very high. But now, it [the graph] is already very low.

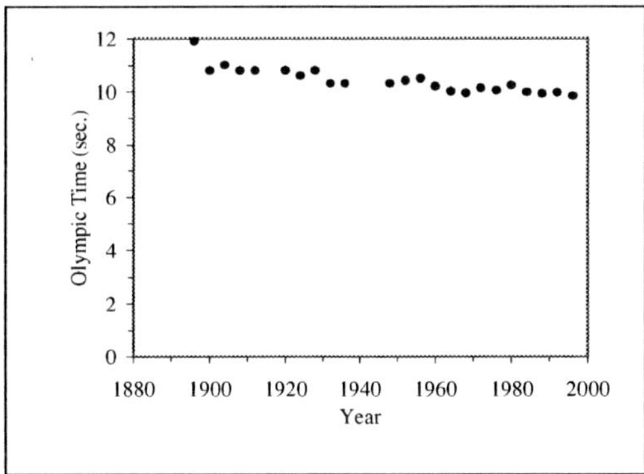

Figure 2. The original 100 meters graph.

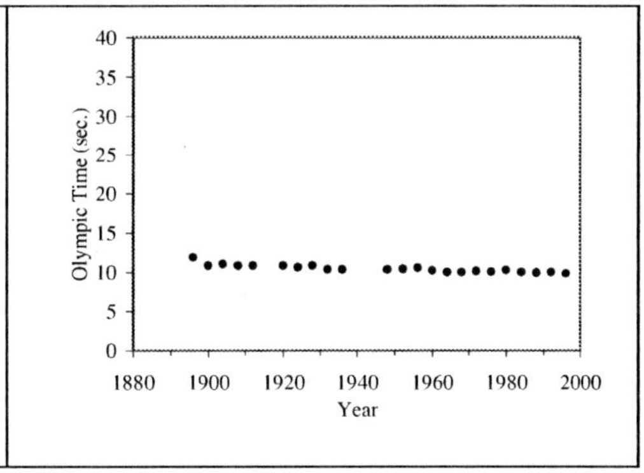

Figure 3. The 100 meters graph after the change of the y-scales.

The change of scales also focused the students' attention on the graph as a whole. They talked about the change in the overall relative position of the graph, whereas they perceived the shape itself as "the same." Their description included global features of the graph ("The whole graph ... went down"), attempts to make sense of the change via the y-axis ("Because the column went up higher"), and references to an individual salient point ("Because now the 12, which is the worst record, is lower"). Student *A* wrote the following synthesis in his notebook: "The graph remained the same in its shape, but moved downward, because before, 12—the worst record—was the highest number on the y-axis, but now it is lower."

However, the purpose of the rescaling was to enable the students to visualize the graph as a whole in a different sense. In order to take sides in the journalists' debate, the transformation was aimed at visually supporting the position that there are no significant changes in the records. Although the students' focus was global, for them the perceptually salient effect of the rescaling was on relative "location" of the whole graph rather than on its trend.

When *A* and *D* were asked to design a graph to support the (opposite) statement: "Over the years, the times recorded in the Olympic 100 meters improved considerably," they did not understand the task and requested the teacher's help:

T [Referring to the 0–40 graph displayed on the computer screen—see Figure 3.] How did you flatten the graph?
A [Visibly surprised.] How did we flatten it?
T Yes, you certainly notice that you have flattened it, don't you?
D No. The graph was like that before. It was only higher up [on the screen].

The teacher and the students seemed to be at cross purposes. The teacher assumed that the students had made sense of the task in the way she expected, and that they understood the global visual effect of the scaling on the graph's shape. When she asked, "How did you flatten the graph?" she was reacting to what she thought was their difficulty: how to perform a scale change in order to support the claim. Thus, her hint consisted of reminding them of what they had already done (scale change). However, the students neither understood her jargon ("flatten the graph") nor regarded what they had done as changing the graph's shape ("The graph was like that before"). Although this intervention is an interesting case of miscommunication, it apparently had a catalytic effect, as reflected in the dialogue that took place immediately afterward—after the teacher realized what might have been their problem.

T How would you show that there were very very big improvements?
A [Referring to the 0–40 graph; see Figure 3.] We need to decrease it [the maximum value of the y-axis]. The opposite of ... [what we have previously done].
D No. To increase it [to raise the highest graph point, i.e., 12 sec.].
A The graph will go further down.
D No. It will go further up.
A No. It will go further down.
D What you mean by increasing it, I mean—decreasing.
A Ahhh ... Well, to decrease it ... O.K., That's what I meant. Good, I understand.
D As a matter of fact, we make the graph shape look different, although it is actually the same graph. It will look as if it supports a specific claim.

When the teacher rephrased her comment ("How would you show that there were very very big improvements?") the students started to make sense of her remarks, although they were still attached to the up-down movement of the whole graph. Student *D* began to discern that a change of scale might change the perceptual impressions one may get from the graph. The teacher's first intervention ("How did you flatten the graph?"), although intended to help the students make sense of the task, can be considered unfortunate. She did not grasp the nature of their question, misjudged their position, and tried to help by reminding them of their previous actions on scale changing. The students seemed comfortable with scale changing, but their problem was that they viewed this tool as doing something different from what the curriculum intended.

The miscommunication itself, and the teacher's attempt to extricate herself from it, contributed to their progress. At first, *A* and *D* were surprised by her description of what they had done as *flattening* the graph. Then, they "appropriated" the teacher's point of view (in the sense of Moschkovich, 1989) and started directing their attention to the shape of the graph rather than to its relative position on the screen. They started to focus on scaling and rescaling in order to achieve the "most convincing" design. Briefly stated, they transferred and elaborated, in iterative steps, ideas of changing scales from one axis to the other until they finally arrived at a satisfying graph (Figure 4) with no further intervention from the teacher. (See Ben-Zvi, 1999, for a detailed description of this rescaling process.) Students *A* and *D* flexibly and interchangeably relied on pointwise observations and global considerations (both in the table and in the graph) in order to fix the optimal intervals on the axes so that the figure would look as they wished.

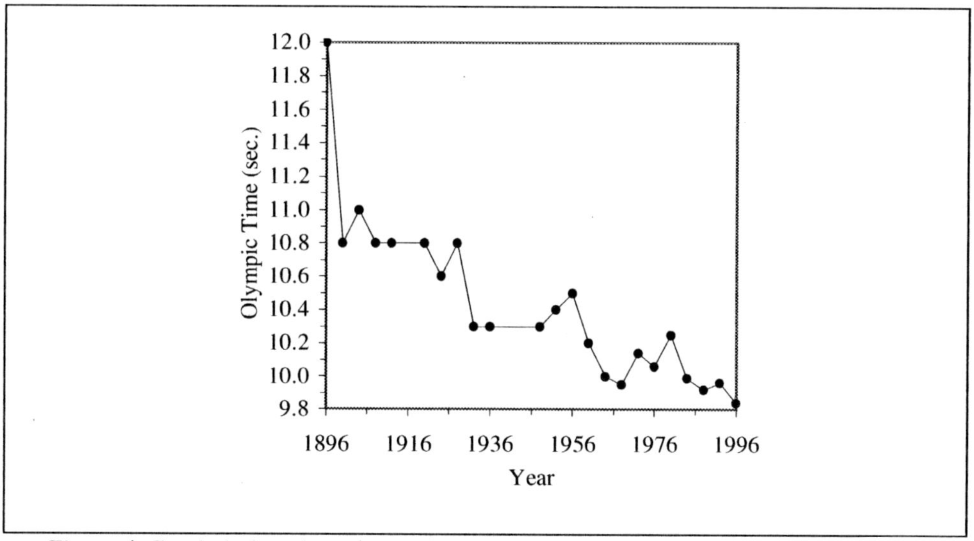

Figure 4. Graph designed to support the statement that the 100 meters times improved considerably.

In summary, at the beginning of this episode the students interpreted the effect of changing scales as a movement of the graph downward rather than as an effect on its shape. Following the teacher's intervention, they started to consider how scaling of both axes affects the shape of the graph. Moreover, they were able to develop manipulations for these changes to occur in order to achieve the desired shape. In the process, they began to move between local and global views of the data in two representations.

It is interesting to notice the students' persistent invocation of "differences" between values ("This way we actually achieved a result that appears as if there are enormous differences"). However, their focus here is on the way these differences are "blown up" by the scaling effect, rather than on them not being constant, as was the case earlier when differences were invoked. The importance of their prior knowledge appears to have been adapted to a new use and for a new purpose. The differences, which were used to drive the way the students made sense of patterns in the data, were being successfully used here as a powerful tool to evaluate their success in designing a graph to visually support a certain claim about a trend in the data.

Case Study 2: Students Taking a Stand

The second case study focused on the role of the SC learning environment in supporting students' reasoning about data analysis. The students in this study were observed as they engaged in *taking a stand* in a debate on the basis of data analysis. The purpose of the analysis was to advance the understanding of (a) how students learn in such an environment, and (b) how can we be more aware of student reasoning, in order to design "better" tasks. Better tasks are situations in which students engage seriously, work and reflect, and advance their statistical reasoning about data.

One SC activity was the *Work dispute* in a printing company. In this activity, the workers are in dispute with the management, which has agreed to an increase in the total salary amount by 10 percent. How this amount of money is to be divided among the employees is a problem—and thereby hangs the dispute. The students were given the salary list of the 100 employees, along with an instruction booklet to guide them in their work. They also received information about the national average and minimum salaries, Internet sites to look for data on salaries, and newspaper articles about work disputes and strikes. In the first part of the activity, students were required to take sides in the dispute and to clarify their arguments. Then, using the computer, they described the distribution of salaries and used statistical measures (e.g. median, mean, mode, and range) to support their position in the dispute. The students learned the effects of grouping data and the different uses of statistical measures in arguing their case. In the third part, the students suggested alterations to the salary structure without exceeding the 10 percent limit. They produced their proposal to solve the dispute, and designed representations to support their position and refute opposing arguments. Finally the class met for a general debate and voted for the winning proposal. The time spent on the full activity was about seven class periods, or a total of six hours.

This task context was familiar to students since it provided interesting, realistic, and meaningful data. The data were altered so that they were more manageable and provided points of departure for addressing some key statistical concepts. For example, the various central tendency measures were different, allowing students to choose a representative measure to argue their case. It was arranged that the mean salary (5000 IS) was above the real national averages (4350 IS—all employees, 4500 IS—printers only).

Students were expected to clarify their thoughts, learn to listen to each other, and try to make sense of each other's ideas. But, most importantly students were asked to *take sides* in the conflict situation. Their actions (e.g. handling data, choosing statistics, creating displays, and arguing) were all motivated, guided, and targeted by the stand they chose. However, their actions sometimes caused them to change their original stand.

The following transcript from a video recording of one of the experimental classes illustrates the use of concepts, arguments, and statitical reasoning that the task promoted. It is based on a group of students who chose to take the side of the workers. After clarifying their arguments, they described the distribution of the current salaries, guided by their position in the dispute. The student pairs prepared various suggested alterations to the salary structure to favor workers (as opposed to management), and then held a series of meetings with fellow student pairs (about 10 students in all), in which they discussed proposals, designed graphical representations to support their position, and prepared themselves for the general debate. This transcript is taken from the second "workers' meeting." It includes the students *A* and *D* from the previous case study along with four other students (referred to as *S*, *N*, *M*, and *H*).

D OK, we have this pie [chart] and we plan to use it [See Figure 5]. Everybody agrees?

Students Yes, yes.

D Let's see what should we say here? Actually we see that ... 60 percent of ...

A 60 percent of the workers are under the average wage [4500 IS]. Now, by adding 12 percent – there are far fewer [workers under the national average].

S OK, but I have a proposal, that brings almost everybody above the average wage. If we add 1000 shekel to the 49 workers, who are under the average ...

N It's impossible. Can't you understand that?

S This [my proposal] will leave us with 1000 shekel, that can be divided among the other workers, who are over [the average].

A Then each of them will get exactly five shekel! ...

M But we don't have any chance to win this way.

D What is the matter with you? We'll have a revolt in our own ranks. Do you want that to happen at the final debate?

S Anyway, this is my opinion! If there are no better proposals ...

D Of course there are: a rise of 12 percent on each salary [excluding the managers] ...

H OK. Show me by how much will your proposal reduce the 60 percent.

N I am printing now an amazing proposal—everybody will be above the [national] average: No worker will be under the average wage! This needs a considerable cut in the managers' salaries ...

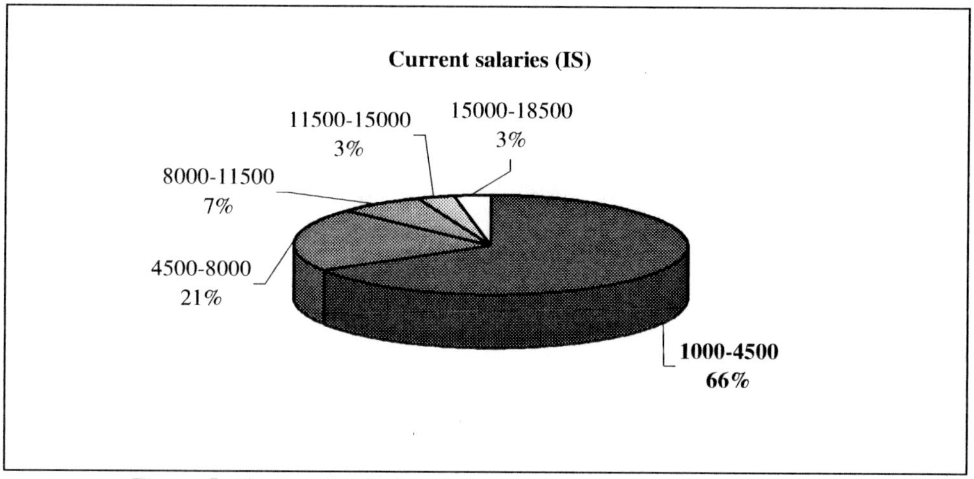

Figure 5. The "workers" description of the current salary distribution.

In this exchange, three different proposals for the alteration of the salary structure were presented. The first, offered by *A* and *D*, suggested an increase of 12 percent for all workers but the managers' salaries remained unchanged. The second proposal, originated by *S*, suggested an equal (1000 IS) increase for each of the 49 workers earning less than the national average (4350 IS), the small remainder to be divided among the other workers. Again the managers' salaries remained unchanged. The third proposal, presented by *N,* suggested a considerable cut in managers' salaries, and an increase for all workers under the national average, to bring them above the average.

Central to students' actions and motives is the stand to be taken by the workers. For example, Figure 5 is grouped to emphasize the large proportion of salaries below the printers' national average. Moreover, the workers' explanations for choosing representative measures and graphical displays emerged from their stand in the dispute. Taking a stand also made students check their methods, arguments, and conclusions with extreme care. They felt it natural to face criticism and counterarguments made by peers and teacher, and to answer them.

These observations suggest that students' reasoning about data as well as their interactions with data were strongly affected by the design of the problem situation, which includes *taking a stand*. The students were able to:

- Deal with a complex situation and the relevant statistical concepts (averages, percentages, charts, etc.).
- Select among measures of center, in relation to looking at graphs, which is an important component of EDA reasoning.
- Use critical arguments to confront conflicting alternatives.
- Use statistical procedures and concepts with a purpose and within a context, to solve problems, relying heavily on visual representations and computer.
- Demonstrate involvement, interest, enthusiasm, and motivation in their learning.
- Create their own products (proposals and their representations).

DISCUSSION

The two case studies focused on students' reasoning about data analysis as they started to develop views (and tools to support them) that are consistent with the use of EDA. Sociocultural and cognitive perspectives will now be considered in a detailed analysis of the case studies. The sociocultural perspective focuses on learning (of a complex domain, such as EDA) as the adoption of the viewpoint of a community of experts, in addition to learning skills and procedures. Thus, this study looked at learning as an enculturation process with two central components: students engaged in doing, investigating, discussing and making conclusions; and teachers engaged in providing role models by being representatives of the culture their students are entering through timely interventions. The cognitive perspective focuses on the development and change in students' conceptions and the evolution of their reasoning. Learning is perceived as a series of interrelated actions by the learner to transform information to knowledge—such as collecting, organizing, and processing information—to link it to previous knowledge and provide interpretations (Davis, Maher, & Noddings, 1990).

It is not easy to tease out the two perspectives for this analysis. Conceptions and reasoning evolve within a purposeful context in a social setting. On the other hand, developing an expert point of view, and interacting with peers or with a teacher, implies undergoing mental actions within specific tasks related to complex ideas. These actions over time are a central part of the meaningful experience within which the culture of the field is learned and the reasoning is developed. These perspectives contribute to the analysis of the data, which revealed the following factors in the process of developing students' reasoning about data in the EDA environment.

The Role of Previous Knowledge

One of the strongest visible pieces of knowledge *A* and *D* applied and repeatedly referred to was the difference between single pairs of data, which came from their practices in the algebra curriculum. This background knowledge played several roles. On the one hand, it gave these students the differences lens, which conditioned most of what they were able to conclude for quite a while. On the other hand, looking at differences helped them to refocus their attention from "pure" pointwise observations toward more global conclusions (that the differences are not constant). Also, looking at differences helped the students, in implicit and subtle ways, to start getting accustomed to a new domain in which data do not behave in the deterministic way that the students were used to in algebra, in which regularities are captured in a single exact formula.

A and *D*'s focus on the differences served more than one function in their learning. It was invoked and applied not only when they were asked to look for patterns in the data but also in a very fruitful way when they spontaneously evaluated the results of rescaling the graph. There, they used the differences in order to judge the extent to which the re-scaled graph matched their goal of designing a graph to support a certain claim about trends.

Thus *A* and *D*'s previous knowledge not only conditioned what they saw—sometimes limiting them—but also, on other occasions, empowered them. Moreover, their previous knowledge served new emerging purposes, as it evolved in the light of new contextual experiences. In conclusion, this analysis illustrates the multifaceted and sometimes unexpected roles prior knowledge may play, sometimes hindering progress and at other times advancing knowledge in interesting ways.

Moving from a Local-Pointwise View toward a Flexible Combination of Local and Global Views

In the first case study, *A* and *D* persistently emphasized local points and adjacent differences. Their views were related to their "history" (i.e., previous background knowledge about regularities with linear relationships in algebra). The absence of a precise regularity in a set of statistical data (understanding variability) was their first difficulty. When they started to adopt the notion of trend (instead of the regular algebraic pattern expected), they were still attentive to the prominence of "local deviations." These deviations kept them from dealing more freely with global views of data. Later on, it was precisely the focus on certain pointwise observations (for example, the place and deletion of one outlying point) that helped them to direct their attention to the shape of the (remaining) graph as a whole. During the scaling process, *A* and *D* looked at the graph as a whole; but rather than focusing on the trends, they discussed its relative locations under different scales. Finally, when they used the scaling and had to relate to the purpose of the question (support of claims in the journalists' debate), they seemed to begin to make better sense of trends.

It is interesting to note that the local pointwise view of data sometimes restrained the students from seeing globally, but in other occasions it served as a basis upon which the students started to see globally. In addition, in a certain context, even looking globally indicated different meanings for the students than for an expert (i.e., noting the position of the graph rather than noticing a trend).

Appropriation: A Learning Process That Promotes Understanding

The data show that most of the learning took place through dialogues between the students themselves and in conversations with the teacher. Of special interest were the teacher's interventions, at the students' request (additional examples of such interventions are described in Ben-Zvi & Arcavi, 2001). These interventions, though short and not necessarily directive, had catalytic effects. They can be characterized in general as "negotiations of meanings" (in the sense of Yackel & Cobb, 1996). More specifically, they are interesting instances of *appropriation* as a nonsymmetrical, two-way process (in the sense of Moschkovich, 1989). This process takes place, in the *zone of proximal development* (Vygotsky, 1978, p. 86), when individuals (expert and novices, or teacher and students) engage in a joint activity, each with their own understanding of the task. Students take actions that are shaped by their understanding; the teacher "appropriates" those actions—into her own framework—and provides feedback in the form of her understandings, views of relevance, and pedagogical agenda. Through the teacher's feedback, the students start to review their actions and create new understandings for what they do.

In this study, the teacher appropriated students' utterances with several objectives: to legitimize their directions, to redirect their attention, to encourage certain initiatives, and implicitly to discourage others (by not referring to certain remarks). The students appropriate from the teacher a reinterpretation of the meaning of what they do. For example, they appropriate from her answers to their inquiries (e.g., what *trend* or *interesting phenomena* may mean), from her unexpected reactions to their request for explanation (e.g., "How did you flatten the graph?"), and from inferring purpose from the teacher's answers to their questions (e.g., "We are supposed to look at what?").

Appropriation by the teacher (to support learning) or by the students (to change the sense they make of what they do) seems to be a central mechanism of enculturation. As shown in this study, this mechanism is especially salient when students learn the dispositions that accompany using the subject matter (data analysis) rather than its skills and procedures.

Curriculum Design to Support Reasoning about Data

The example described in the second case study illustrates how curriculum design can take into account new trends in subject matter (EDA)—its needs, values, and tools—as well as student reasoning. Staging and encouraging students to *take sides* pushed them toward levels of reasoning and discussion that have not been observed in the traditional statistics classroom. They were involved in selecting appropriate statistical measures, rather than just calculating them, and in choosing and designing graphs to best dispaly their views. They showed themselves able to understand and judge the complexities of the situation—engaged in preparing a proposal that in their view was acceptable, rational, and just—and were able to defend it.

Furthermore, students realized that data representations could serve *rhetorical functions,* similar to their function in the work of statisticians, who select data, procedures, tools, and representations that support their perspective. Thus, the development of students' reasoning about data is extended beyond the learning of statistical mathods and concepts, to involve students in "doing" statistics in a realistic context.

IMPLICATIONS

The learning processes described in this chapter took place in a carefully designed environment. It is recommended that similar environments be created to help students develop their reasoning about data analysis. The essential features of such learning environments include

- A curriculum built on the basis of EDA as a sequence of semi-structured (yet open) leading questions within the context of extended meaningful problem situations (Ben-Zvi & Arcavi, 1998)
- Timely and nondirective interventions by the teacher as representative of the discipline in the classroom (cf., Voigt, 1995)
- Computerized tools that enable students to handle complex actions (change of representations, scaling, deletions, restructuring of tables, etc.) without having to engage in too much technical work, leaving time and energy for conceptual discussions

In learning environments of this kind, students develop their reasoning about data by meeting and working with, from the very beginning, ideas and dispositions related to the culture of EDA. This includes making hypotheses, formulating questions, handling samples and collecting data, summarizing data, recognizing trends, identifying variability, and handling data representations. Skills, procedures and strategies (e.g., reading graphs and tables, rescaling) are learned as integrated in the context and at the service of the main ideas of EDA.

It can be expected that beginning students will have difficulties of the type described when confronting the problem situations posed by the EDA curriculum. However, what *A* and *D* experienced is an integral and inevitable component of their meaningful learning process with long-lasting effects (cf., Ben-Zvi 2002). These results suggest that students should work in environments such as the one just described, which allows for:

- Students' prior knowledge to be engaged in interesting and surprising ways—possibly hindering progress in some instances but making the basis for construction of new knowledge in others
- Many questions to be raised—some will either make little sense to them, or, alternatively, will be reinterpreted and answered in different ways than intended
- Students' work to be based on partial understandings, which will grow and evolve

This study confirmed that even if students do not make more than partial sense of the material with which they engage, appropriate teacher guidance, in-class discussions, peer work and interactions, and more importantly, ongoing cycles of experiences with realistic problem situations, will slowly support the building of meanings and the development of statistical reasoning.

Multiple challenges exist in the assessment of outcomes of students' work in such a complex learning environment: the existence of multiple goals for students, the mishmash between the contextual (real-world) and the statistical, the role of the computer-assisted environment, and the group versus the individual work (Gal & Garfield, 1997). It is recommended that extended performance tasks be used to assess students' reasoning about data, instead of traditional tests that focus on definitions and computation. Performance tasks should be similar to those given to students during the learning activities (e.g., open-ended questions, "complete" data investigations), allowing students to work in groups and use technological tools.

In EDA learning environments of the kind described in these case studies, teachers cease to be the dispensers of a daily dose of prescribed curriculum and must respond to a wide range of unpredictable events. They can play a significant role in their interactions with students by encouraging them to employ critical reasoning strategies and use data representations to search for patterns and convey ideas; expanding and enriching the scope of their proposed work; and providing reflective feedback on their performance. Thus our challenge is to assist statistics educators in their important role of mentors and mediators, or the *enculturators*.

Given that EDA is a challenging topic in statistics education and is part of the mathematics curriculum in many schools today, it is important that teaching efforts be guided not only by systematic research on understanding the core ideas in data analysis but also by how reasoning about data analysis develops. Without this research and the implementation of results, statistics classes will continue to teach graphing and data-collection skills that do not lead to the ability to reason about data analysis.

Many research questions need to be addressed, including those pertaining to the development of students' understanding and reasoning (with the assistance of technological tools), the student-teacher and student-student interactions within open-ended data investigation tasks, the role of enculturation processes in learning, and the impact of learning environments similar to those described here. The refinement of these ideas, and the accumulation of examples and studies, will contribute to the construction of an EDA learning and instruction theory.

REFERENCES

Ainley, J. (1995). Re-viewing graphing: Traditional and intuitive approaches. *For the Learning of Mathematics, 15*(2), 10–16.

Australian Education Council (1991). *A national statement on mathematics for Australian schools.* Carlton, Vic.: Author.

Australian Education Council (1994). *Mathematics—A curriculum profile for Australian schools.* Carlton, Vic.: Curriculum Corporation.

Barbella, P., Kepner, J., & Schaeffer, R. L. (1994). *Exploring Measurement.* Palo Alto, CA: Seymour Publications.

Ben-Zvi, D. (1999). Constructing an understanding of data graphs. In O. Zaslavsky (Ed.), *Proceedings of the Twenty-Third Annual Conference of the International Group for the Psychology of Mathematics Education* (Vol. 2, pp. 97–104). Haifa, Israel: Technion—Institute of Technology.

Ben-Zvi, D. (2000). Toward understanding of the role of technological tools in statistical learning. *Mathematical Thinking and Learning, 2*(1&2), 127–155.

Ben-Zvi, D. (2002). Seventh grade students' sense making of data and data representations. In B. Phillips (Ed.), *Proceedings of the Sixth International Conference on Teaching of Statistics* (on CD-ROM). Voorburg, The Netherlands: International Statistical Institute.

Ben-Zvi, D., & Arcavi, A. (2001). Junior high school students' construction of global views of data and data representations. *Educational Studies in Mathematics, 45*, 35–65.

Ben-Zvi, D., & Arcavi, A. (1998). Toward a characterization and understanding of students' learning in an interactive statistics environment. In L. Pereira-Mendoza (Ed.), *Proceedings of the Fifth International Conference on Teaching Statistics* (Vol. 2, 647–653). Voorburg, The Netherlands: International Statistical Institute.

Ben-Zvi, D., & Friedlander, A. (1997a). *Statistical investigations with spreadsheets—Student's workbook* (in Hebrew). Rehovot, Israel: Weizmann Institute of Science.

Ben-Zvi, D., & Friedlander, A. (1997b). Statistical thinking in a technological environment. In J. B. Garfield & G. Burrill (Eds.), *Research on the Role of Technology in Teaching and Learning Statistics* (pp. 45–55). Voorburg, The Netherlands: International Statistical Institute.

Ben-Zvi, D., & Ozruso, G. (2001). *Statistical investigations with spreadsheets—Teacher's guide* (in Hebrew). Rehovot, Israel: Weizmann Institute of Science.

Bereska, C., Bolster, C. H., Bolster, L. C., & Scheaffer, R. (1998). *Exploring statistics in the elementary grades*, Book 1 (Grades K–6). New York: Seymour Publications.

Bereska, C., Bolster, C. H., Bolster, L. C., & Scheaffer, R. (1999). *Exploring statistics in the elementary grades*, Book 2 (Grades 4–8). New York: Seymour Publications.

Biehler, R. (1990). Changing conceptions of statistics: A problem area for teacher education. In A. Hawkins (Ed.), *Proceedings of the International Statistical Institute Round Table Conference* (pp. 20–38). Voorburg, The Netherlands: International Statistical Institute.

Biehler, R. (1993). Software tools and mathematics education: The case of statistics. In C. Keitel & K. Ruthven (Eds.), *Learning from computers: Mathematics education and technology* (pp. 68–100). Berlin: Springer-Verlag.

Biehler, R. (1997). Software for learning and for doing statistics. *International Statistical Review 65*(2), 167–189.

Biehler, R., & Steinbring, H. (1991). *Explorations in statistics, stem-and-leaf, box plots: Concepts, justifications, and experience in a teaching experiment* (elaborated English version). Bielefeld, Germany: Author.

Brousseau, G. (1997). *Theory of didactical situations in mathematics* (Edited and translated by N. Balacheff, M. Cooper, R. Sutherland, & V. Warfield). Dordrecht, The Netherlands: Kluwer.

Cleveland, W. S. (1994). *The elements of graphing data.* Murray Hill, NJ: AT&T Bell Laboratories.

Cobb, P. (1999). Individual and collective mathematical learning: The case of statistical data analysis. *Mathematical Thinking and Learning, 1*, 5–44.

Cobb, G. W., & Moore, D. S. (1997). Mathematics, statistics, and teaching. *The American Mathematical Monthly, 104*(9), 801–823.

Davis, R. B., Maher, C. A., & Noddings, N. (Eds.) (1990). *Constructivist views on the teaching and learning of mathematics.* Reston, VA: NCTM.

de Lange, J., Burrill, G., & Romberg, T. (1993). *Learning and teaching mathematics in context—the case: Data visualization.* Madison, WI: National Center for Research in Mathematical Sciences Education.

de Lange, J., & Verhage, H. (1992). *Data visualization.* Scotts Valley, CA: Wings for Learning.

Department for Education and Employment (1999). *Mathematics: The national curriculum for England.* London: Author and Qualifications and Curriculum Authority.

Evensen, D. H, & Hmelo, C. E. (Eds.) (2000). *Problem-based learning: A research perspective on learning interactions.* Mahwah, NJ: Erlbaum.

Finzer, B. (2003). *Fathom: Dynamic Statistics Software for Deeper Understanding* (Version 1.16). Emeryville, CA: Key Curriculum Press. <http://www.keypress.com/fathom/>

Friel, S. N., Mokros, J. R., & Russell, S. (1992). *Used numbers: Middles, means, and in-betweens.* Palo Alto, CA: Dale Seymour.

Gal, I., & Garfield, J. B. (1997). Curricular goals and assessment challenges in statistics education. In I. Gal & J. B. Garfield (eds.), *The assessment challenge in statistics education* (pp. 1–13). Amsterdam, Netherlands: IOS Press.

Garfield, J. (1995). How students learn statistics. *International Statistical Review 63*(1), 25–34.

Graham, A. (1987). *Statistical investigations in the secondary school.* Cambridge, UK: Cambridge University Press.

Greer, B., Yamin-Ali, M., Boyd, C., Boyle, V., & Fitzpatrick, M. (1995). *Data handling* (six student books in the Oxford Mathematics series). Oxford, UK: Oxford University Press.

Hershkowitz, R., Dreyfus, T., Schwarz, B., Ben-Zvi, D., Friedlander, A., Hadas, N., Resnick, T., & Tabach, M. (2002). Mathematics curriculum development for computerized environments: A designer-researcher-teacher-learner activity. In L. D. English (Ed.), *Handbook of international research in mathematics education* (pp. 657–694). London: Erlbaum.

Hunt, D. N. (1995). Teaching Statistical Concepts Using Spreadsheets. In the *Proceedings of the 1995 Conference of the Association of Statistics Lecturers in Universities.* UK: The Teaching Statistics Trust.

Kader, G. D., & Perry, M. (1994). Learning statistics. *Mathematics Teaching in the Middle School 1*(2), 130–136.

Konold, C., & Higgins, T. L. (2003). Reasoning about data. In J. Kilpatrick, G. Martin, & D. Schifter (Eds.), *A research companion to principles and standards for school mathematics* (pp. 193–215). Reston, VA: NCTM.

Konold, C., Pollatsek, A., & Well, A. (1997). Students analyzing data: Research of critical barriers. In J. Garfield & G. Burrill (Eds.), *Research on the role of technology in teaching and learning statistics* (pp. 151–168). Voorburg, The Netherlands: International Statistical Institute.

Lampert. M. (1990). When the problem is not the question and the solution is not the answer: Mathematical knowing and teaching. *American Educational Research Journal, 27*, 29–63.

Lappan, G., Fey, J. T., Fitzgerald, W. M., Friel, S. N., & Phillips, E. D. (1996). *Connected mathematics project.* Palo Alto, CA: Seymour Publications.

Lovitt, C., & Lowe, I. (1993). *Chance and data investigations* (Vol. 1 & 2). Carlton, Vic., Australia: Curriculum Corporation.

Meira, L. R. (1991). *Explorations of mathematical sense-making: An activity-oriented view of children's use and design of material displays* (an unpublished Ph.D. dissertation). Berkeley: University of California Press.

Ministry of Education (1992). *Mathematics in the New Zealand curriculum.* Wellington, NZ: Author.

Monk, G. S. (1988). Students' understanding of functions in calculus courses. *Humanistic Mathematics Network Journal, 9*, 21–27.
Moore, D. S. (1990). Uncertainty. In Lynn Steen (Ed.), *On the shoulders of giants: A new approach to numeracy* (pp. 95–137). National Academy of Sciences.
Moore, D. S. (1992). Teaching statistics as a respectable subject. In F. & S. Gordon (Eds.), *Statistics for the 21st century* (pp. 14–25). Washington, DC: Mathematical Association of America.
Moore, D. S. (1997). New pedagogy and new content: The case of statistics. *International Statistical Review 65*(2), 123–165.
Moore, D. S. (2000). *The basic practice of statistics* (second edition). New York: Freeman.
Moore, D. S., & McCabe, G. P. (1993). *Introduction to the practice of statistics* (2nd ed.). New York: Freeman.
Moschkovich, J. D. (1989). *Constructing a problem space through appropriation: A case study of guided computer exploration of linear functions* (an unpublished manuscript, available from the author).
Mosteller, F., & Tukey, J. (1977). *Data analysis and regression*. Boston: Addison-Wesley.
National Council of Teachers of Mathematics (1989). *Curriculum and evaluation standards for school mathematics*. Reston, VA: Author.
National Council of Teachers of Mathematics (2000). *Principles and standards for school mathematics*. Reston, VA: Author.
Perkins, D., & Unger, C. (1999). Teaching and learning for understanding. In C. M. Reigeluth (Ed.), *Instructional-design theories and models* (pp. 91–114). Hillsdale, NJ: Erlbaum.
Resnick, L. (1988). Treating mathematics as an ill-structured discipline. In R. Charles & E. Silver (Eds.), *The teaching and assessing of mathematical problem solving* (pp. 32–60). Reston, VA: National Council of Teachers of Mathematics.
Resnick, T., & Tabach, M. (1999). *Touring the Land of Oz—Algebra with computers for grade seven* (in Hebrew). Rehovot, Israel: Weizmann Institute of Science.
Rubin, A., & Mokros, J. (1998). *Data: Kids, cats, and ads* (Investigations in number, data, and space Series). New York: Seymour Publications.
Schoenfeld, A. H. (1992). Learning to think mathematically: Problem solving, metacognition, and sense making in mathematics. In D. Grouws (Ed.), *Handbook of research on mathematics teaching and learning* (pp. 334–370). New York: Macmillan.
Schoenfeld, A. H. (1994). Some notes on the enterprise (research in collegiate mathematics education, that is). *Conference Board of the Mathematical Sciences Issues in Mathematics Education, 4*, 1–19.
Shaughnessy, J. M., Garfield, J., & Greer, B. (1996). Data handling. In A. J. Bishop, K. Clements, C. Keitel, J. Kilpatrick, & C. Laborde (eds.), *International handbook of mathematics education* (Vol. I, pp. 205–237). Dordrecht, The Netherlands: Kluwer.
TERC (2002). *Tabletop*. Geneva, IL: Sunburst Technology.
<http://www.terc.edu/TEMPLATE/products/item.cfm?ProductID=39>
Tukey, J. (1977). *Exploratory data analysis*. Reading, MA: Addison-Wesley.
Velleman, P. (2003). *Data Desk* (Version 6.2). Ithaca, NY: Data Description Inc.
<http://www.datadesk.com/products/data_analysis/datadesk/>
Velleman, P., & Hoaglin, D. (1981). *The ABC's of EDA: Applications, basics, and computing of exploratory data analysis*. Boston, MA: Duxbury.
Voigt, J. (1995). Thematic patterns of interaction and sociomathematical norms. In P. Cobb & H. Bauersfeld (Eds.), *Emergence of mathematical meaning: Interaction in classroom cultures* (pp. 163–201). Hillsdale, NJ: Erlbaum.
Vygotsky, L. S. (1978). *Mind in society: The development of higher psychological processes*. M. Cole, V. John-Steiner, S. Scribner, & E. Souberman (Eds.). Cambridge, MA: Harvard University Press.
Wild, C. J., & Pfannkuch, M. (1999). Statistical thinking in empirical enquiry. *International Statistical Review, 67*(3), 223–265.
Yackel, E., & Cobb, P. (1996). Socio-mathematical norms, argumentation and autonomy in mathematics. *Journal for Research in Mathematics Education, 27*(4), 458–477.
Yerushalmy, M., Chazan, D., & Gordon, M. (1990). Mathematical problem posing: Implications for facilitating student inquiry in classrooms. *Instructional Science, 19*, 219–245.

Chapter 7

LEARNING TO REASON ABOUT DISTRIBUTION

Arthur Bakker and Koeno P. E. Gravemeijer
Freudenthal Institute, Utrecht University, the Netherlands

OVERVIEW

The purpose of this chapter is to explore how informal reasoning about distribution can be developed in a technological learning environment. The development of reasoning about distribution in seventh-grade classes is described in three stages as students reason about different representations. It is shown how specially designed software tools, students' created graphs, and prediction tasks supported the learning of different aspects of distribution. In this process, several students came to reason about the shape of a distribution using the term bump along with statistical notions such as outliers and sample size.

This type of research, referred to as "design research," was inspired by that of Cobb, Gravemeijer, McClain, and colleagues (see Chapter 16). After exploratory interviews and a small field test, we conducted teaching experiments of 12 to 15 lessons in 4 seventh-grade classes in the Netherlands. The design research cycles consisted of three main phases: design of instructional materials, classroom-based teaching experiments, and retrospective analyses. For the retrospective analysis of the data, we used a constant comparative method similar to the methods of Glaser and Strauss (Strauss & Corbin, 1998) and Cobb and Whitenack (1996) to continually generate and test conjectures about students' learning processes.

DATA SET AS AN AGGREGATE

An essential characteristic of statistical data analysis is that it is mainly about describing and predicting aggregate features of data sets. Students, however, tend to conceive a data set as a collection of individual values instead of an aggregate that has certain properties (Hancock, Kaput, & Goldsmith, 1992; Konold & Higgins, 2002; Ben-Zvi & Arcavi, 2001; Ben-Zvi, Chapter 6). An underlying problem is that middle-grade students generally do not see "five feet" as a value of the variable

D. Ben-Zvi and J. Garfield (eds.),
The Challenge of Developing Statistical Literacy, Reasoning and Thinking, 147–168.

"height," but as a personal characteristic of, say, Katie. In addition to this view, students should learn to disconnect the measurement value from the object or person measured and consider data against a background of possible measurement values. They should furthermore develop a notion of distribution, since that is an organizing conceptual structure with which they can conceive the aggregate instead of just the individual values (Cobb, 1999; Petrosino, Lehrer, & Schauble, 2003).

These learning goals formed the motivation to explore the possibilities for students in early secondary education with little or no prior statistical knowledge to develop an informal understanding of distribution. Such understanding could then be the basis for more formal statistics in higher grades. The main question in this study is therefore: How can seventh-grade students learn to reason about distribution in an informal way?

DISTRIBUTION

To answer this question, we first analyze the relation between data and distribution. Distinguishing between data as individual values and distribution as a conceptual entity, we examine aspects of both data sets and distributions such as center, spread, density, and skewness (Table 1). Measures of center include mean, median, and midrange. Spread can be quantified with, for instance, range, standard deviation, and interquartile range. The aspects and measures in the table should not be seen as excluding each other; outliers and extreme values, for instance, influence skewness, density, spread, and even most measures of center.

Table 1. Between data and distribution

distribution (conceptual entity)			
center mean, median, midrange, …	**spread** range, standard deviation, interquartile range, …	**density** (relative) frequency, majority, quartiles	**skewness** position majority of data
data (individual values)			

This structure can be read upward and downward. The upward perspective is typical for novices in statistics: Students tend to see individual values, which they can use to calculate, for instance, the mean, median, range, or quartiles. This does not automatically imply that they see mean or median as a measure of center or as representative of a group (Mokros & Russell, 1995; Konold & Pollatsek, Chapter 8). In fact, students need a notion of distribution before they can sensibly choose between such measures of center (Zawojewski & Shaughnessy, 2000). Therefore, students need to develop the downward perspective as well: conceiving center, spread, and skewness as characteristics of a distribution, and looking at data with a notion of distribution as an organizing structure or a conceptual entity. Experts in statistics can easily combine the upward and downward perspectives. We might say that the upward perspective leads to a frequency distribution of a data set. In the downward perspective, we typically use probability distributions such as the normal distribution to model data.

The table shows that the concept of distribution has a complex structure, but this concept is also part of a larger structure consisting of big ideas such as variation and sampling (Reading & Shaughnessy, Chapter 9; Watson, Chapter 12). Without variation, there is no distribution, and without sampling there are mostly no data. We therefore chose to deal informally and coherently with all these big ideas at the same time with distribution in a central position. As Cobb (1999) notes, focusing on distribution as a multifaceted end goal of instruction might bring more coherence in the statistics curriculum. The question is how. Our answer is to focus on the informal aspects of shape.

The shape of a distribution is influenced by various statistical aspects. A high peak, for example, is caused by a high frequency of a certain class and long tails on the left or right with the hill out of center indicate skewed distributions. This implies that by reasoning with informal terms about the shape of a distribution, students may already reason with aspects of that distribution. And indeed, students in this study used informal words to describe density (crowded, empty, piled up, clumped, busy), spread (spread out, close together), and shape (hill, bump). If students compare the height distributions of two different grades, they might realize that the graphs have the same shape but are shifted in location (Biehler, 2001). And they might see that samples of different sizes still have similar shapes. We envisioned that reasoning with shapes forms the basis for reasoning about distributions.

METHODOLOGY AND SUBJECTS

To answer the main question of how students can develop a notion of distribution, we carried out developmental research, which is also called design research (Freudenthal, 1991; Gravemeijer, 1994; Edelson, 2002; Cobb & McClain, Chapter 16). Design research typically involves the design of instructional materials, teaching experiments, and retrospective analyses. In line with the principles of Realistic Mathematics Education (Freudenthal, 1991; Gravemeijer, 1994) and the National Council of Teachers of Mathematics (NCTM) Standards (2000), we looked for ways to guide students in being active learners dealing with increasingly sophisticated means of support.

To assist students in exploring data and developing the concept of distribution, we decided to use some specially designed Minitools (see Cobb, 1999). These web applets were developed by reasoning backward from the intended end goal of reasoning about distribution to possible starting points. One aspect of distribution, shape, can be inferred from stacked dot plots. To understand what dots in a dot plot stand for, students need to realize that a dot represents a value on some variable. One way to help students develop this insight is to let them start with case-value bars, which range from 0 to the corresponding value on the horizontal axis. We presume that bars representing values are closer to students' daily life reality than dots on an axis, because they are used to bar graphs and because horizontal bars are natural ways to symbolize certain variables such as the braking distance of cars, the life span of batteries, or the wingspan of birds. For that reason, each case in Minitool 1 (Figure 1) is signified by a bar whose relative length corresponds to the value of the case, and each case in Minitool 2 (Figure 2) is signified by a dot in a dot plot.

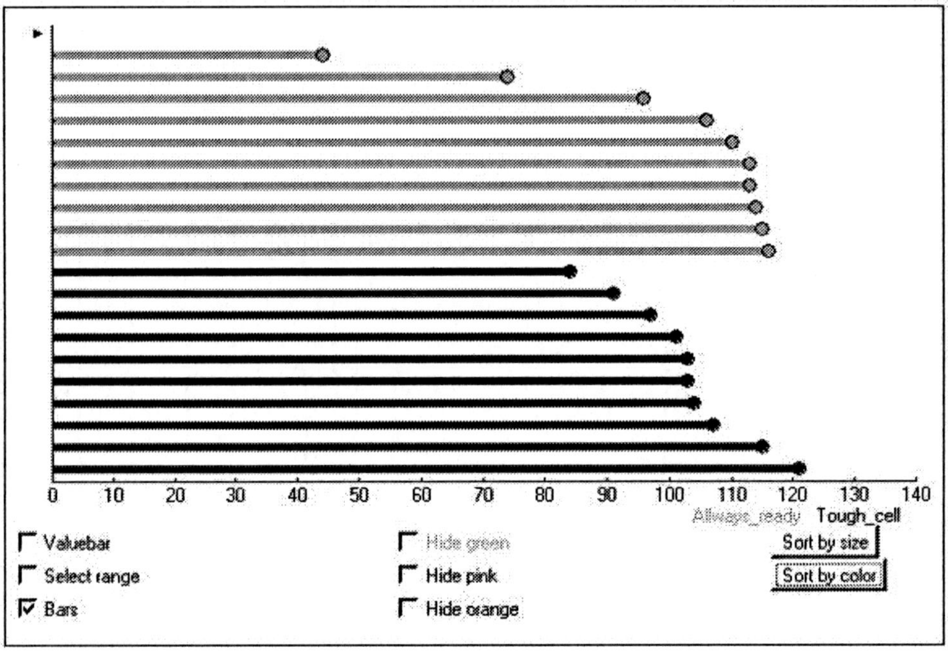

Figure 1. Minitool 1 (sorted by size and color).

To identify a baseline of what Dutch seventh-grade students already know about statistics and how easily they would solve statistical problems using the two Minitools, we interviewed 26 students about these issues. The students had encountered no statistics before except the arithmetic mean and bar graphs. They had almost no problems in reading off values from the Minitools, but they focused on individual data values (Section 2). We then did a small field test and conducted teaching experiments in 4 seventh-grade classes, which worked through a complete sequence of 12 to 15 lessons of 50 minutes each. The experiments were carried out during the school year 1999–2000, in a public school in a small town near Utrecht (the Netherlands) that prepared about 800 students for university (vwo) or higher vocational education (havo). At that time about 15% of the Dutch students went to the vwo level, 20% to the havo level, about 40% to the mavo level (for middle vocational education), and the remaining 25% to lower vocational education (in the meantime the last two levels have been merged). These percentages indicate that the learning abilities of the vwo and havo students of our teaching experiments were above average.

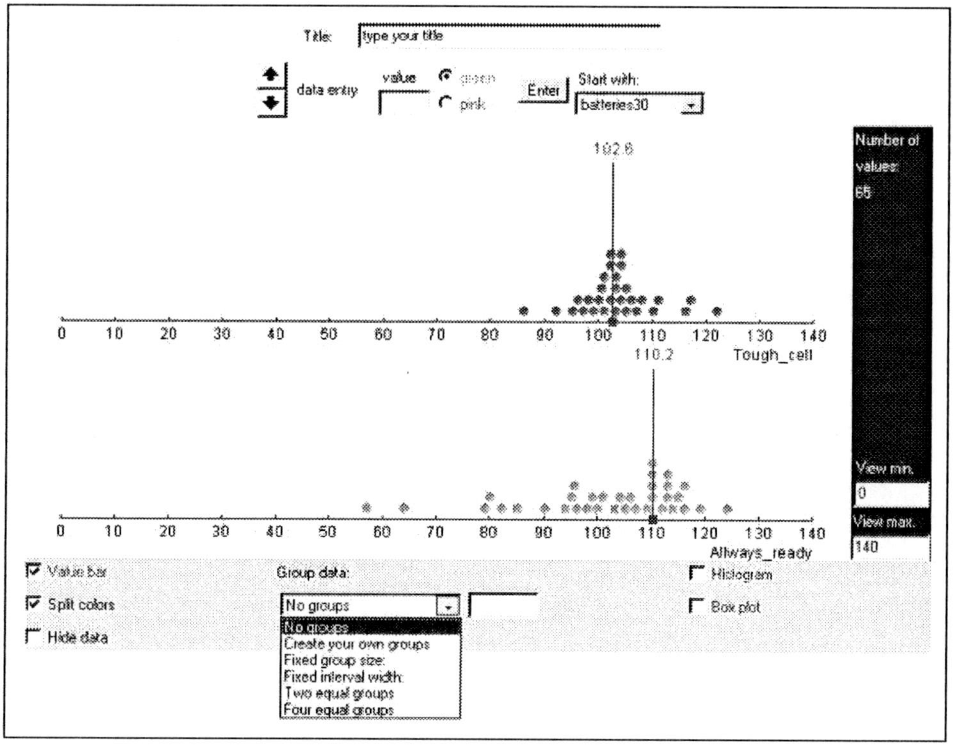

Figure 2. Minitool 2 (split colors and with vertical value bars).

The collected data include audio recordings, student work, field notes, and final tests in all classes, as well as videotapes and pretests in the last two experiments (see Table 2). The pretests were meant to find out if students already knew what we wanted them to learn (they did not).

An essential part of the data corpus was a set of mini-interviews that were held during lessons. Mini-interviews varied from about 20 seconds to 4 minutes and were meant to find out what concepts and graphs meant for the students. We realize that this influenced their learning, because the mini-interviews often stimulated reflection. In our view, however, the validity of the research was not in danger: Our aim was to find out how students could learn to reason with distribution, not whether teaching the sequence in other seventh-grade classes would lead to the same results.

For the retrospective analysis of the fourth teaching experiment, we have read the transcripts, watched the videotapes, and formulated conjectures on students' learning based on the transcript and video episodes. The generated conjectures were being tested at the other episodes and the rest of the collected data (student work, field observations, and tests) in the next round of analysis (triangulation). Then the whole generating and testing process was repeated. This method resembles Glaser and Strauss's constant comparative method (Strauss & Corbin, 1998; Cobb and Whitenack, 1996). Important transcript fragments, including those in this chapter, have been discussed with colleagues (peer examination).

Table 2. Overview of subjects, teaching experiments, data collection, number of lessons, and levels of education

Subjects (grade 7)	Type of Experiment	Data Collection	No. of Lessons	Level
26 students (1999)	Exploratory interviews (15 minutes for two students)	audio	—	mavo, havo, vwo
Class A (25)	Exploratory field test	student work, final test, field notes, audio	4	havo
Class F (27)	First teaching experiment		12	vwo
Class E (28)	Second teaching experiment		15	vwo
Class C (23) (2000)	Third teaching experiment	idem plus pretest and video	12	havo
Class B (23)	Fourth teaching experiment		12	havo
12 classes (2000–2002)	Implementation	e-mail reports of two teachers, field notes from incidental visits	144	havo and vwo

Furthermore, we have identified patterns of student answers that were similar in all teaching experiments, and categorized the evolving learning trajectory in three stages according to students' reasoning with the representations used. The sections describing stages 1 and 2 describe observations that were similar for all four observed classes. In the first stage, students worked with graphs in which data were represented by horizontal bars (Minitool 1, Figure 1). In the second stage, from lesson 5 to 12, students mainly worked with dot plots (Minitool 2, Figure 2). In the third stage students used both Minitools and came to reason with bumps; the examples stem from the second teaching experiment. The students in this class had good learning abilities (vwo) and had 15 lessons—three more than in the other classes. The specific stages began to overlap each other when we started to stimulate comparison of different graphs during the last two teaching experiments.

STAGE 1—DATA ARE REPRESENTED BY BARS

The aim of the first activities was to let students reason about different aspects of distributions in an informal way such as about majority, center, extreme values, spread-out-ness, and consistency. In the second lesson, for example, students had to prepare reports to Consumer Reports (a consumers' journal) on the quality of two battery brands. They were given a data set of 10 battery life spans of two brands in Minitool 1; using different computer options, they could sort the data and split the data, for instance of the two brands. In the beginning they used the vertical value bar (Figure 3) to read off values, but later sometimes to estimate the mean visually.

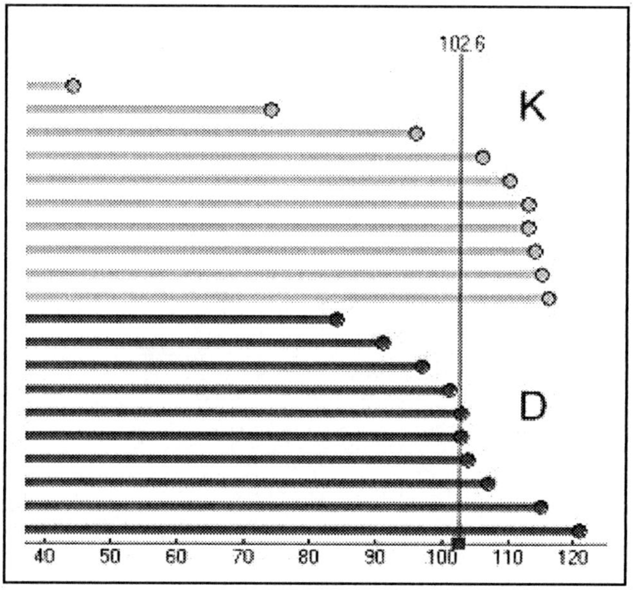

Figure 3. Estimating the mean of brand D with the movable vertical value bar (life span in hours).

During this battery activity, students in all teaching experiments could already reason about aspects of distributions. "Brand K has outliers, but you have more chance for a good one," was one answer. "Brand D is more reliable, since you know that it will last more than 80 hours," was another. This notion of reliability formed a good basis for talking about spread. Our observations resemble those of Cobb (1999) and Sfard (2000), who analyzed students' spontaneous use of the notion of "consistency."

The activities with Minitool 1 afforded more than informal reasoning about majority, outliers, chance, and reliability; they also supported the visual estimation of the mean (Figures 3 and 4). After this strategy had spontaneously emerged in the exploratory interviews, we incorporated instructional activities to evoke this strategy in other classes as well (Bakker, 2003). Minitool 1 supported the strategy with the movable vertical value bar. Students said that they cut off the longer bars, and gave the bits to the shorter bars. Several students in different classes could explain that this approach was legitimate: The total stays the same, and the mean is the total divided by the number. When students said that brand D is better because its mean is higher, they used the mean to say how good the brand is. In that case, the mean is not just a calculation on a collection of data, but refers to a whole subset of one brand. As we intended, they learned to use the mean as a representative value for a data set and to reason about the brand instead of the individual data values.

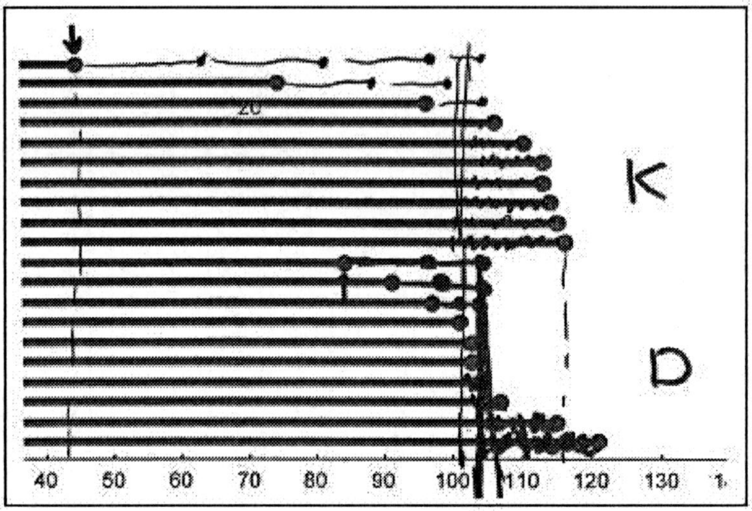

Figure 4. Scribblings on a transparency during class discussions after estimating means of both brands. The mean of brand D is slightly higher than that of K.

To assess students' understanding of distribution aspects and to establish a tighter relationship between informal statistical notions and graphs, we decided to "reverse" this battery task. In the last two teaching experiments, during the fourth lesson, we asked students to invent their own data according to certain characteristics such as "brand A is bad but reliable; brand B is good but unreliable; brand C has about the same spread as brand A, but it is the worst of all brands." Many students produced a graph similar to the one in Figure 5 (in this case, the variation of C is less than that of A). A sample response was:

> Why is brand A better. Because it lives long. And it has little spread. Brand B is good but unreliable. Because it has much spread. But it lives long. Brand C has little spread but the life span is not very long.

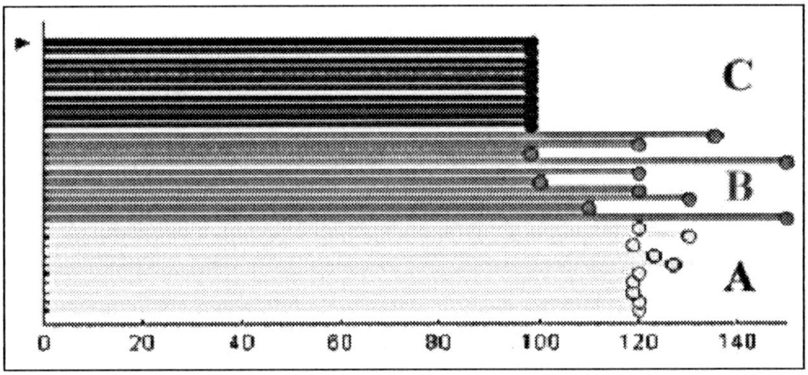

Figure 5. Invented data set according to certain features: Brand A is bad but reliable; brand B is good but unreliable; brand C has about the same spread as brand A, but it is the worst of all.

With hindsight, we have come to see this back-and-forth movement between interpreting graphs and constructing graphs according to statistical notions as an important heuristic for instructional design in data analysis, for a number of reasons:

- Students can express ideas with graphs that they cannot express in words (Lemke, 2003). If students invent their own data and graphs, teachers and researchers can better assess what students actually understand.
- If students think of characteristics such as "good but not reliable," the lack of data prevents them from focusing on individual data, because it is cognitively impossible to imagine many individual data points. With this reverse activity, we create the need for a conceptual unity that helps in imagining a collection of data with a certain property. The notion of distribution serves that purpose (Section 3).
- In many schoolbooks, students mainly interpret ready-made graphs (Friel et al., 2001; Moritz, Chapter 10). And if students have to make graphs, the goal is too often just to learn how to produce a particular graph. De Lange, Burrill, Romberg, & van Reeuwijk (1993) and Meira (1995) strongly recommend letting students invent their own graphs. We may assume that students' own graphs are meaningful and functional for them.
- The importance of the back-and-forth movement between data and graphs (or different graphs) is also indicated by the research on symbolizing. Steinbring (1997), for example, distinguishes *reference* systems and *symbol* systems. Students interpret a symbol system in the light of a better-known reference system. Reference systems are therefore relatively well known and symbol systems relatively unknown. In learning the relationship between a symbol system and a reference system, students must go back and forth between the two systems. A next step can then be that students use the symbol system they have just learned to reason with (Minitool 1, for example) as a reference system for a new symbol system (Minitool 2, for example), and so on.

From the examples of the first stage, it is clear that students informally reasoned about different aspects of distribution from the very start. They argued about the mean (how good the battery is), spread (reliability), chance for outliers or extreme values, and where the majority is (skewness). Without the bar representation the students would probably not have developed a compensating strategy for finding the mean. Their reasoning, however, was bound to one representation and two contexts.

STAGE 2—DOTS REPLACE BARS

Our next aim was to let students reason about shapes of distributions in suitable representations and in different contexts. Additionally, we strove for quantification of informal notions such as frequency and the majority and to prepare students for using conventional aggregate plots such as histograms and box plots.

As mentioned in the previous section, Minitool 1 can be seen as a reference system for the new symbol system of Minitool 2. When solving problems with Minitool 1, the students reasoned with the endpoints of the bars. In Minitool 1, students could hide the bars, which they sometimes preferred, because "it is better organized." The dot plot of Minitool 2 can be obtained by hiding the bars of Minitool 1 and imaginatively dropping the endpoints on the horizontal axis or on the other dots that prevent them from dropping further down (cf. Wilkinson, 1999). Note that the dots are stacked and do not move sideways to fill up white areas in the graph (Figure 6). The advantages of this dot plot representation are that it is easy to interpret, it comes closer to conventional representations of distributions than Minitool 1, and students can organize data in ways that come close to histogram and box plot, for instance.

Minitool 2 has more options to organize data than Minitool 1. Apart from sorting by size and by subgroup (color), students can also group data into their own groups, two equal groups (for the median), four equal groups (for a box plot, Figure 7a), equal interval width (for a histogram, Figure 7b), and fixed group size (Figure 6b). This last option turned out to be useful for stimulating reasoning about density.

A particular statistical problem that students solved with Minitool 2 was the one on jeans sizes. Students had to report to a factory the percentage of each size that should be made, based on a data set of the waist measurements (in inches) of 200 men. This activity, typically done during the ninth lesson, was meant to distract students' attention away from the mean and toward the whole distribution. Furthermore, it could be an opportunity to let students reason about absolute and relative frequencies.

We expected that students would reason about several aspects of distribution when comparing different grouping options. The option of fixed group size (Figures 6b and 6c) typically evoked remarks such as "with the thin ones [the narrow bins] you know that there are many dots together." We interpret such expressions as informal reasoning about density, which we see as a key aspect of distribution. Many students used the four equal groups option to support their conclusion that "you have to make a lot of jeans in sizes 34–36, and less of 44–46." Generally, a skeptical question was needed to provoke more exact answers: "If the factory hired you for $1,000, do you think the factory would be satisfied with your answer?" Most students ended up with the fixed interval option and a table with percentages, that is, relative frequencies.

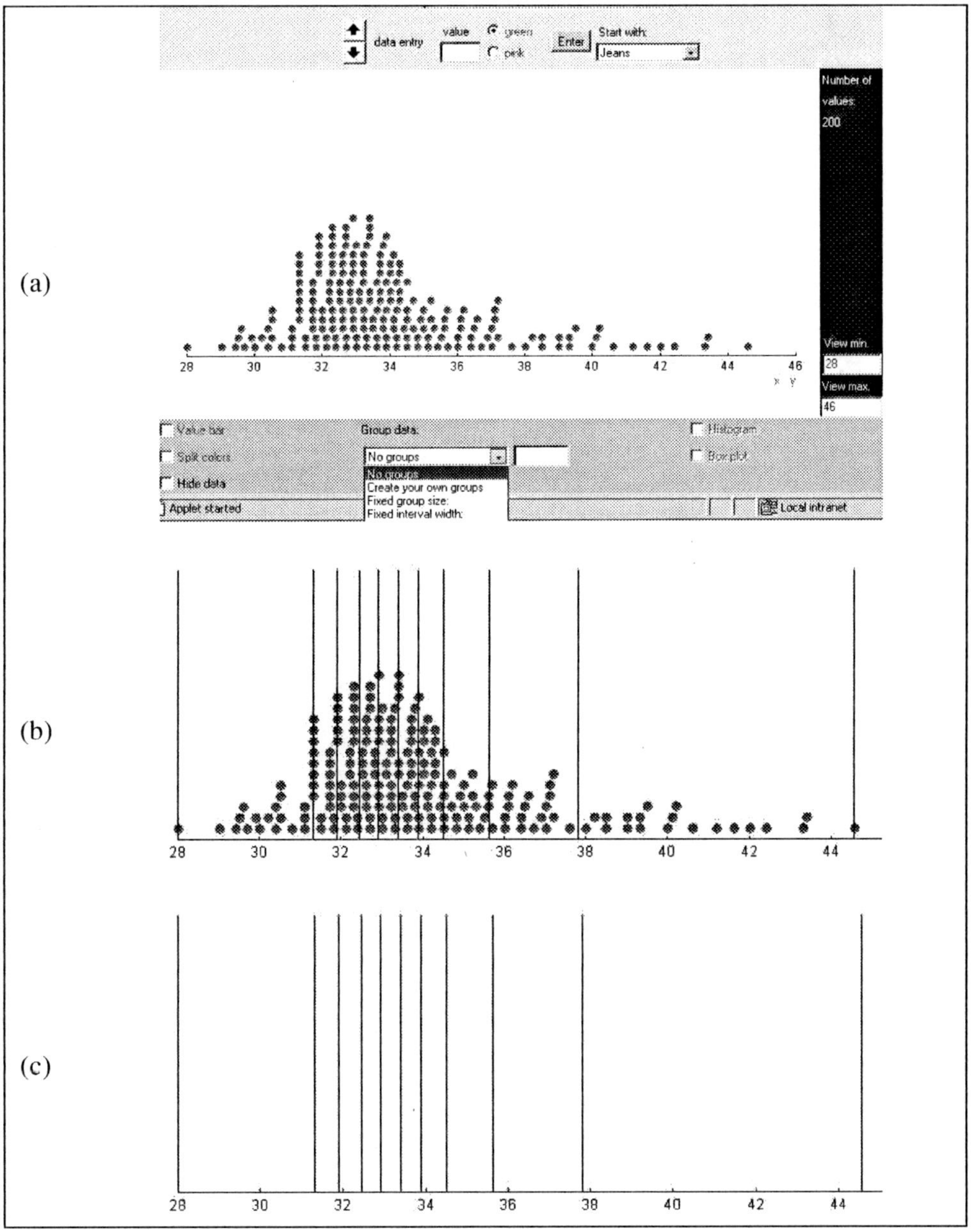

Figure 6. (a) Minitool 2 with jeans data set (waist size in inches, n = 200). (b) Fixed group size with 20 data points per group. (c) Minitool 2 with "hide data" function.

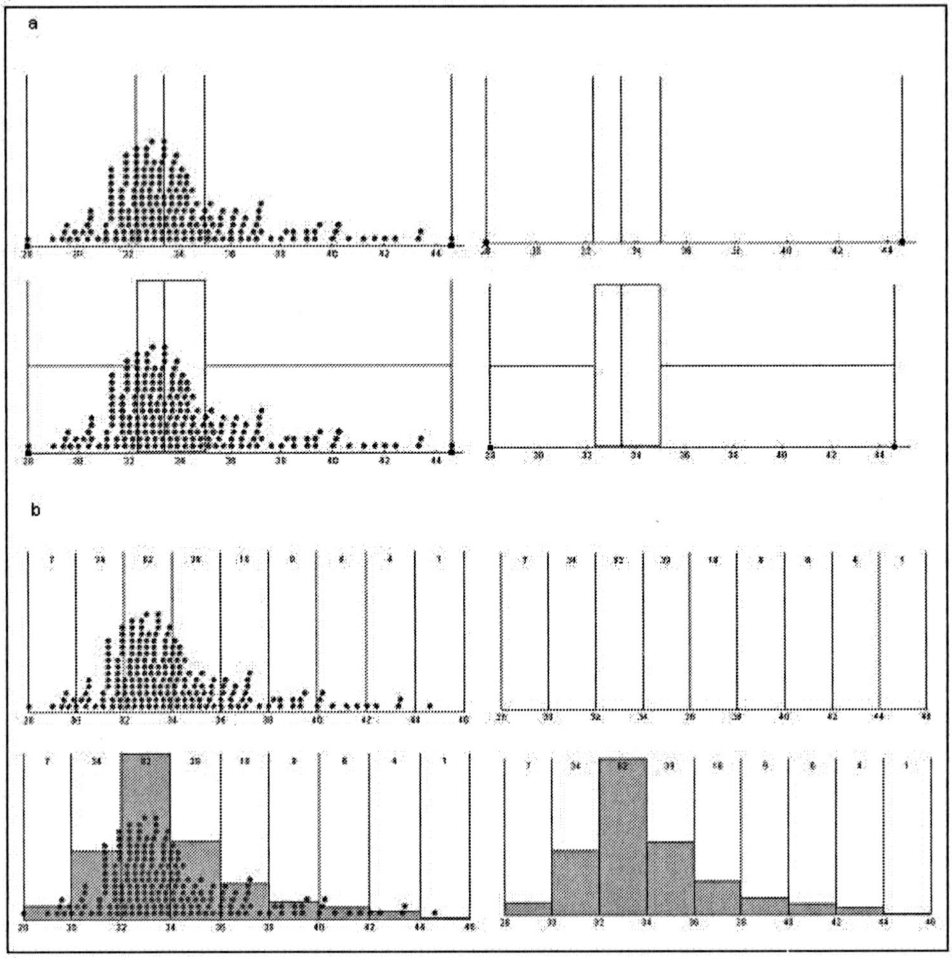

Figure 7. (a) Four equal group option with and without data. Box plot overlay was added after these seventh-grade teaching experiments. (b) Fixed interval width option with and without data. Histogram overlay was added after these seventh-grade teaching experiments.

An instructional idea that emerged during the last teaching experiment was that of "growing samples." Discussing and predicting what would happen if we added more data appeared to lead to reasoning about several aspects of distribution in a coherent way. For the background to this activity, we have to go back to a problem from the beginning of the instructional unit:

> In a certain hot air balloon basket, eight adults are allowed [in addition to the driver]. Assume you are going to take a ride with a group of seventh-graders. How many seventh-graders could safely go into that balloon basket if you only consider weight?

This question was meant to let students think about variation of weight, sampling, and representativeness of the average. A common solution in all classes was that students estimated an average or a typical weight for both adults and children. Some used the ratio of those numbers to estimate the number of children allowed, but most students calculated the total weight allowed and divided that by the average student weight. The student answers varied from 10 to 16.

This activity formed the basis for a class discussion on the reliability of the estimated weights, during which we asked for a method of finding more reliable numbers. A student suggested weighing two boys and two girls. The outcome of the discussion was that the students decided to collect weight data from the whole class. (In the second teaching experiment, they also collected height data.)

In the next lesson, we first showed the sample of four weight data in Minitool 2 (Figure 8a) and asked what students expected if we added the rest of the data. Students thought that the mean would be more precise. Because we did not want to focus on the mean, we asked about the shape and the range. Some students then conjectured that the range would be larger, and others thought the graph would grow higher. After showing the data for the whole class (Figure 8b), we asked what would happen if we added the data for two more classes (Figure 8c). In this way, extreme values, spread, and shape became topics of discussion. The graphs that students made to predict the shape if sample size were doubled tended to be smoother than the graphs students had seen in Minitool 2 (Figure 8d). In our interpretation, students started to see a pattern in the data—or in Konold and Pollatsek's words, a "signal in the noise" (Chapter 8). We concluded that stimulating reasoning about distribution by "growing samples" is another useful heuristic for instructional design in statistics education.

A conjecture about students' evolving notion of distribution that was confirmed in the retrospective analyses was that students tend to divide unimodal distributions into three groups of low, "average," and high values. We saw this conceptual grouping into three groups for the first time in the second teaching experiment when we asked what kind of graph students expected when they collected height data. Daniel did three trials (Figure 9). During his second trial, he said: "You have smaller ones, taller ones, and about average." After the third trial he commented: "There are more around the average." Especially in the third trial, we clearly see his conceptual organization into three groups, which is a step away from focusing on individual data points.

One step further is when students think of small, average, tall, *and* "in between." When in the final test students had to sketch their class when ordered according to height, Christa drew Figure 10 and wrote: "There are 3 smaller ones, about 10 average, 3 to 4 taller, and of course in between."

The "average" group, the majority in the middle, seems to be more meaningful to students than the single value of the mean. Konold and colleagues (2002) call these ranges in the middle of distributions *modal clumps*. Our research supports their view that these modal clumps may be suitable starting points for informal reasoning about center, spread, and skewness. When growing samples, students might even learn to see such aspects of distribution as stable features of variable processes.

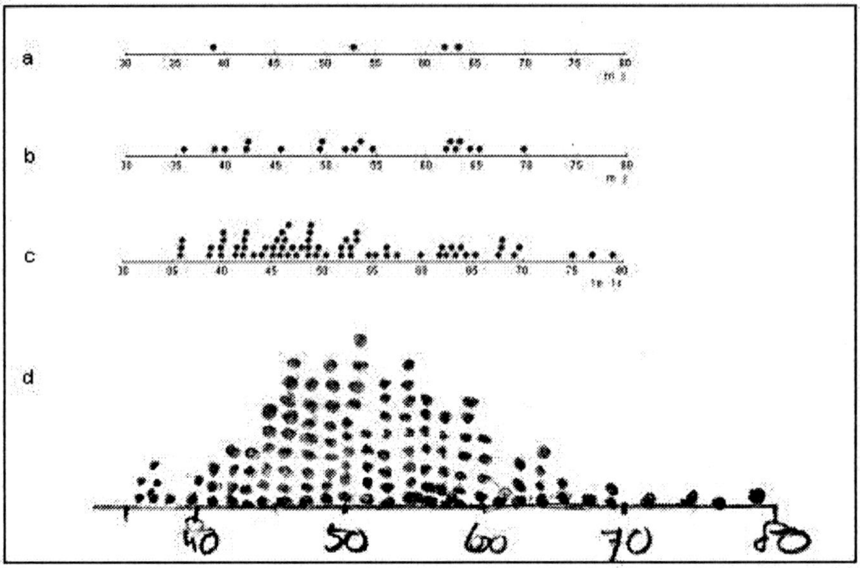

Figure 8. Growing samples (weight data in kg): (a) Four students; (b) one class; (c) three classes; (d) a student's smoother prediction graph of larger sample.

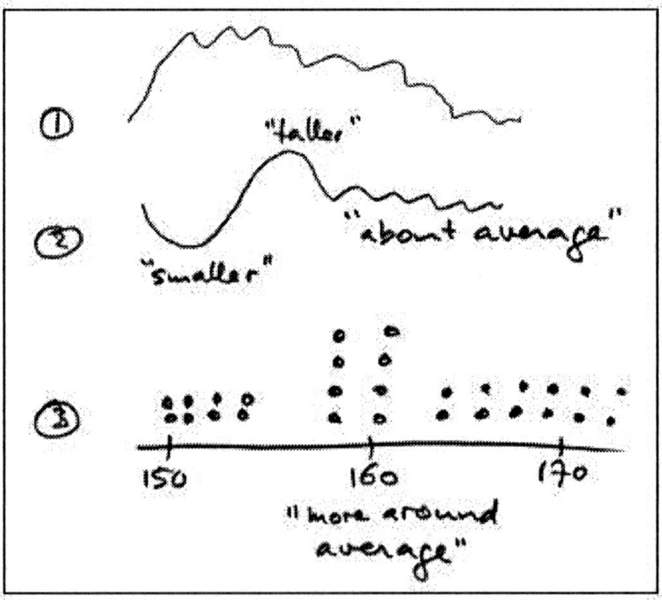

Figure 9. Three prediction trials of height data; the second and third show three groups.

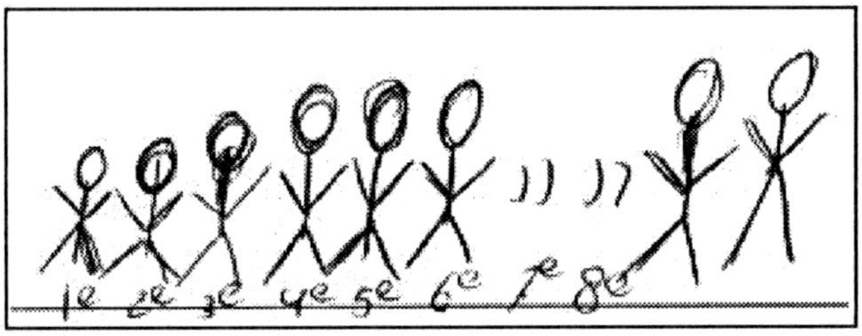

Figure 10. Class ordered by height. Christa's explanation: "There are 3 smaller ones, about 10 average, 3 to 4 taller, and of course in between."

STAGE 3—SYMBOLIZING DATA AS A "BUMP"

Though students in the first two teaching experiments started to reason with majorities and modal clumps in the second stage, they did not explicitly reason with shape. We had hoped that they would reason with "hills," as was the case in the teaching experiment of Cobb, Gravemeijer, and McClain (Cobb, 1999), but they did not. A possible reason is that their teaching experiment lasted 34 lessons, whereas ours lasted only 12 or 15 lessons. In the second teaching experiment, we decided to try something else. In line with the reasons to let students invent their own data (Section 5), we asked students to invent their own graphs of their own data. As a follow-up of the balloon activity mentioned earlier, the students had to make a graph for the balloon rider, which she could use in deciding how many students she could safely take on board.

The students of the second teaching experiment drew various graphs. The teacher focused the discussion on two graphs, namely, Michiel's and Elleke's (Figure 11).

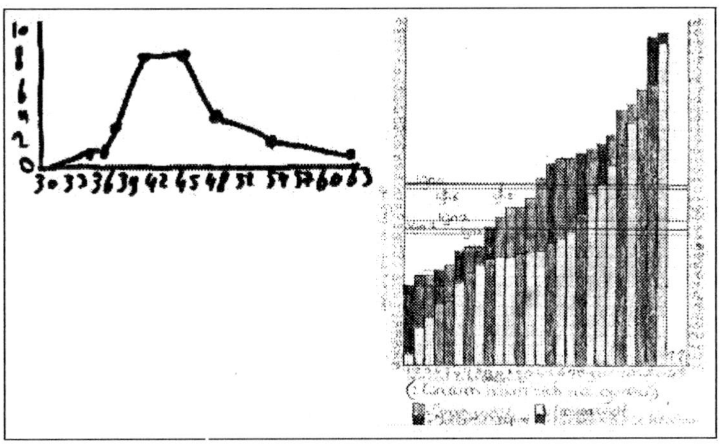

Figure 11. Michiel's graph (left) and Elleke's graph.

The shorter bars represent students' weights; the lightest bars signify girls' weights. Though all students used the same data set, Michiel's graph on a transparency does not exactly match the values in Elleke's graph on paper. Michiel's graph is more like a rough sketch.
Michiel's graph is especially interesting, since it offered the opportunity to talk about shape. Michiel explained how he got the dots as follows. (Please note that a translation of ungrammatical spoken Dutch into written English does not sound very authentic.)

Michiel: Look, you have roughly, averagely speaking, how many students had that weight and there I have put a dot. And then I have left [y-axis] the number of students. There is one student who weighs about 35 [kg], and there is one who weighs 36, and two who weigh 38 roughly.

And so on: the dot at 48, for example, signifies about four students with weights around 48. After some other graphs had been discussed, including that of Elleke, the teacher asked the following question.

Teacher: What can you easily see in this graph [by Michiel]?
Laila: Well, that the average, that most students in the class, uhm, well, are between 39 and, well, 48.
Teacher: Yes, here you can see at once which weight most students in this class roughly have, what is about the biggest group. Just because you see this bump here. We lost the bump in Elleke's graph.

It was the teacher who used the term bump for the first time. Although she had tried to talk about shapes earlier, this was the first time the students picked it up. As Laila's answer indicates, Michiel's graph helped her to see the majority of the data—between 39 and 48 kg. This "average" or group of "most students" is an instance of what Konold and colleagues (2002) call a modal clump. Teachers and curriculum designers can use students' informal reasoning with clumps as preparation for using the average as a representative value for the whole group, for example.

Here, the teacher used the term bump to draw students' attention to the shape of the data. By saying that "we lost the bump in Elleke's graph," she invited the students to think about an explanation for this observation. Nadia reacted as follows.

Nadia: The difference between … they stand from small to tall, so the bump, that is where the things, where the bars [from Elleke's graph] are closest to one another.
Teacher: What do you mean, where the bars are closest?
Nadia: The difference, the endpoints [of the bars], do not differ so much with the next one.

Eva added to Nadia's remarks:

Eva: If you look well, then you see that almost in the middle, there it is straight almost and uh, yeah that [teacher points at the horizontal part in Elleke's graph].
Teacher: And that is what you [Nadia] also said, uh, they are close together and here they are bunched up, as far as [...] weight is concerned.
Eva: And that is also that bump.

These episodes demonstrate that, for the students, the bump was not merely a visual characteristic of a certain graph. It signified a relatively large number of data points with about the same value—both in a hill-type graph and in a value-bar graph. For the students, the term bump signified a range where there was a relatively high density of data points. The bump even became a tool for reasoning, as the next episode shows, when students revisited the battery task as one of the final tasks.

Laila: But then you see the bump here, let's say [Figure 3].
Ilona: This is the bump [pointing at the straight vertical part of the lower 10 bars].
Researcher: Where is that bump? Is it where you put that red line [the vertical value bar]?
Laila: Yes, we used that value bar for it [...] to indicate it, indicate the bump. If you look at green [the upper ten], then you see that it lies further, the bump. So we think that green is better, because the bump is further.

The examples show that some students started to reason about density and shape in the way intended. However, they still focused on the majority, the modal clump, instead of the whole distribution. This seemed to change in the 13th lesson of the second teaching experiment

In that lesson, we discovered that asking students to predict and reason without available data was helpful in fostering a more global view of data. A first example of such a prediction question is what a graph of the weights of eighth-graders would look like, as opposed to one of seventh-graders. We hoped that students would shift the whole shape instead of just the individual dots or the majority.

Teacher: What would a graph of the weights of eighth-graders look like?
Luuk: I think about the same, but another size, other numbers.
Guyonne: The bump would be more to the right.
Teacher: What would it mean for the box plots?
Michiel: Also moves to the right. That bump in the middle is in fact just the box plot, which moves more to the right.

It could well be that Luuk reasoned with individual numbers, but he thought that the global shape would look the same. Instead of talking about individual data points, Guyonne talked about a bump, in singular, shifted to the right. Michiel related to the box plot as well, though he just referred to the box of the box plot.

Another prediction question also led to reasoning about the whole shape, this time in relation to other statistical notions such as outliers and sample size. Note that students used the term *outliers* for extreme values, not for values that are questionable.

Researcher: If you would measure all seventh-graders in the city instead of just your class, how would the graph change, or wouldn't it change?

Elleke: Then there would come a little more to the left and a little more to the right. Then the bump would become a little wider, I think. [She explained this using the term outliers.]

Researcher: Is there anybody who does not agree?

Michiel: Yes, if there are more children, than the average, so the most, that also becomes more. So the bump stays just the same.

Albertine: I think that the number of children becomes more and that the bump stays the same.

In this episode, Elleke relates shape to outliers; she thinks that the bump grows wider if the sample grows. Michiel argues that the group in the middle also grows higher, which for him implies that the bump keeps the same shape. Albertine's answer is interesting in that she seems to think of relative frequency: for her the shape of the distribution seems to be independent of the sample size. If she thought of absolute frequency she would have thought that the bump would be much higher. Apparently, the notion of a bump helped these students to reason about the shape of the distribution in hypothetical situations. In this way, they overcame the problem of seeing only individual data points and developed the notion of a bump, which served as a conceptual unity.

There are several reasons why predictions about shape in such hypothetical situations can help to foster understanding of shape or distribution. First, if students predict a graph without having data, they have to reason more *globally* with a property in their mind. Konold and Higgins (2002) write that with the individuals as the foci, it's difficult to see the forest for the trees. Our conclusion is that we should ask questions about the forest, or predict properties of other forests—which we consider another heuristic for statistics education. This heuristic relates to the cognitive limitations mentioned in Section 5: If there are no available data and students have to predict something on the basis of some conceptual characteristic, it is impossible to imagine many individual data points.

A second reason has to do with the *smoothness* of graphs. Cobb, McClain, and Gravemeijer (2003) assume that students can more easily reason about hills if the hills are smooth enough. We found evidence that the graphs students predict tend to be smoother than the graphs of real data, and we conjecture that reasoning with such smoother graphs helps students to see the shape of a distribution through the variation or, in other words, the signal through the noise (Konold & Pollatsek, Chapter 8). If they do so, they can model data with a notion of distribution, which is the downward perspective we aimed for (Section 3).

A last example illustrates how several students came to reason about distributions. These two girls were not disturbed by the fact that distributions did not look like hills in Minitool 1. The question they dealt with was whether the distributions of the battery brands looked normal or skewed, where *normal* was informally defined as "symmetrical, with the median in the middle and the majority close to the median." The interesting point is that they used the term *hill* to indicate the majority (see Figure 3), although it looked straight in the case-value bar graph. This indicates that the hill was not a visual tool; it had become a conceptual tool in reasoning about distributions.

Albertine:	Oh, that one [battery brand D in Figure 3] is normal [...].
Nadia:	That hill.
Albertine:	And skewed if like here [battery brand K] the hill [the straight part] is here.

DISCUSSION

The central question of this chapter was how seventh-grade students could learn to reason about distributions in informal ways. In three stages, we showed how certain instructional activities, supported by computer tool use and the invention of graphs, stimulated students to reason about aspects of distributions. After a summary of the results we discuss limitations of this study and implications for future research.

When solving statistical problems with Minitool 1, students used informal words such as *majority*, *outliers*, *reliability*, and *spread out*. The examples show that students reasoned about aspects of distribution from the very start of the experiment. The students invented data sets in Minitool 1 that matched certain characteristics of battery brands such as "good but not reliable." We argued that letting students invent their own data sets could stimulate them to think of a data set as a whole instead of individual data points (heuristic 1). The bar representation of Minitool 1 stimulated a visual compensation strategy of finding the mean, whereas many students found it easier to see the spread of the data in Minitool 2.

When working with Minitool 2, students developed qualitative notions of more advanced aspects of distribution such as frequency, classes, spread, quartiles, median, and density. The dot plot representation in combination with the options to structure data into two equal groups, four equal groups, fixed group size, and fixed interval width supported the development of an understanding of the median, box plot, density, and histogram respectively. Like Konold and colleagues (2002), we expect that modal clumps are useful to help students reason with center and other distribution aspects. Growing samples is a promising instructional activity to let students reason with stable features of variable processes (heuristic 2). The big ideas of sampling and distribution can thus be developed coherently, but how this could be done is a topic of future research.

In the third stage, students started to reason with bumps in relation to statistical notions such as majority, outliers, and sample size in hypothetical situations and in relation to different graphs. We argued that predictions about the shape and location of distributions in hypothetical situations are useful to foster a more global view and to let students see the signal in the noise (heuristic 3).

IMPLICATIONS

The results of this research study suggest that it is important to provide opportunities for students to contribute their own ideas to the learning process, which requires much discussion and interaction during class. We believe that formal measures such as median and quartiles should be postponed until intuitive notions about distribution have first been developed. We also encourage teachers to allow students to use less than precise statistical definitions as students develop their reasoning, and then make a transition to more specific definitions as students are able to comprehend these details. We are convinced that teachers should try to learn about how students are reasoning about distribution by listening and observing as well as by gathering assessment data. A type of assessment that we found useful asked students to create a graph representing statistical information. One such task that was very effective asked students to make graphs that were compatible with a short story with both informal and statistical notions related to running practice. There were no restrictions on the type of graph students could use. We had deliberately incorporated characteristics in the story that ranged from easy (the fastest runner needed 28 minutes) to difficult (the spread of the running times at the end was much smaller than in the beginning but the range was still pretty big). This is the item we used:

A seventh grade is going to train for running 5 km. To track their improvement they want to make three graphs. One before training starts, one halfway through, and one after ten training sessions. Draw the graphs that belong to the following story:

- Before training started some students were slow and some were already very fast. The fastest ran the 5 km in 28 minutes. The spread between the other students was large. Most of them were on the slow side.
- Halfway through, the majority of the students ran faster, but the fastest had improved his time only a little bit, as had the slowest.
- After the training sessions had finished, the spread of the running times was much smaller than in the beginning, but the range was still pretty big. The majority of the students had improved their times by about 5 minutes. There were still a few slow ones, but most of the students had a time that was closer to the fastest runner than in the beginning.

We found that students were able to represent many elements in their graphs and we learned more about their thinking and reasoning by examining their constructions.

Although we conclude that it is at least possible for seventh-graders to develop the kind of reasoning about distribution that is shown in this chapter, it should be stressed that the students in these experiments had above-average learning abilities and had been stimulated to reflect during mini-interviews. Other students probably need more time or need to be older before they can reason about distribution in a similar way.

Another limitation of this study is that the examples of the third stage were to a certain extent unique for the second teaching experiment. What would have happened if Michiel had not made his "bump" graph? This research does not completely answer that question (there was some reasoning with bumps in the third and fourth teaching experiment), but it shows what the important issues are and which heuristics might be useful for instructional activities.

In addition, we noticed that making predictions graphs without having data is not a statistical practice that automatically emerges from doing an instructional sequence such as the one described here. We concluded this from observations during the two subsequent school years, when two novice teachers used the materials in 12 other seventh-grade classes. When we asked prediction questions, the students seemed confused because they were not used to such questions. An implication for teaching is that establishing certain socio-mathematical norms and certain practices (Cobb & McClain, Chapter 16) are as important as suitable computer tools, carefully planned instructional activities, and skills of the teacher to orchestrate class discussions.

These teachers also reported that some of the statistical problems we had used or designed were too difficult and not close enough to the students' world of experience. The teachers also needed much more time than we used in the first year, and they found it difficult to orchestrate the class discussions. We acknowledge that the activities continually need to be adjusted to local contingencies, that the mini-interviews probably had a learning effect, and that the teachers needed more guidance for teaching such a new topic. Hence, another question for future research is what kind of guidance and skills teachers need to teach these topics successfully.

NOTE

We thank the teachers Mieke Abels, Maarten Jasper, and Mirjam Jansens for joining this project. The research was funded by the Netherlands Organization for Scientific Research under number 575-36-003B. The opinions expressed in this chapter do not necessarily reflect the views of the Organization.

REFERENCES

Bakker, A. (2003). The early history of average values and implications for education. *Journal of Statistics Education, 11*(1). Online: http://www.amstat.org/publications/jse/v11n1/bakker.html

Biehler, R (2001). Developing and assessing students' reasoning in comparing statistical distributions in computer-supported statistics courses. In C. Reading (Ed.), *Proceedings of the Second International Research Forum on Statistical Reasoning, Thinking, and Literacy (SRTL-2)*. Armidale, Australia: University of New England.

Ben-Zvi, D., & Arcavi, A. (2001). Junior high school students' construction of global views of data and data representations. *Educational Studies in Mathematics, 45*(1), 35–65.

Cobb, P. (1999). Individual and collective mathematical development: The case of statistical data analysis. *Mathematical Thinking and Learning, 1*(1), 5–43.

Cobb, P., McClain, K., & Gravemeijer, K. P. E. (2003). Learning about statistical covariation. *Cognition and Instruction 21*(1), 1–78.

Cobb, P., & Whitenack, J. W. (1996). A method for conducting longitudinal analyses of classroom videorecordings and transcripts. *Educational Studies in Mathematics, 30*(3), 213–228.

de Lange, J., Burrill, G., Romberg, T., & van Reeuwijk, M. (1993). *Learning and testing mathematics in context: The case—data visualization*. Madison, WI: University of Wisconsin, National Center for Research in Mathematical Sciences Education.

Edelson, D. C. (2002). Design research: What we learn when we engage in design. *Journal of the Learning Sciences, 11*(1), 105–121.

Freudenthal, H. (1991). *Revisiting mathematics education: China lectures*. Dordrecht, The Netherlands: Kluwer Academic Publishers.

Friel, S. N., Curcio, F. R., & Bright, G. W. (2001). Making sense of graphs: Critical factors influencing comprehension and instructional implications. *Journal for Research in Mathematics Education, 32*(2), 124–158.

Gravemeijer, K. P. E. (1994). *Developing realistic mathematics education*. Utrecht, The Netherlands: CD Bèta Press.

Hancock, C., Kaput, J. J., & Goldsmith, L. T. (1992). Authentic enquiry with data: Critical barriers to classroom implementation. *Educational Psychologist, 27*(3), 337–364.

Konold, C., & Higgins, T. (2002). Highlights of related research. In S. J. Russell & D. Schifter & V. Bastable (Eds.), *Developing mathematical ideas: Working with data* (pp. 165–201). Parsippany, NJ: Seymour.

Konold, C., Robinson, A., Khalil, K., Pollatsek, A., Well, A. D., & Wing, R., et al. (2002). Students' use of modal clumps to summarize data. In B. Phillips (Ed.), *Developing a Statistically Literate Society: Proceedings of the International Conference on Teaching Statistics* [CD-ROM], Cape Town, South Africa, July 7–12, 2002.

Lemke, J. L. (2003). Mathematics in the middle: Measure, picture, gesture, sign, and word. In M. Anderson, A. Sáenz-Ludlow, S. Zellweger, & V. V. Cifarelli (Eds.), *Educational perspectives on mathematics as semiosis: From thinking to interpreting to knowing* (pp. 215–234). Ottawa, Ontario: Legas Publishing.

Meira, L. (1995). Microevolution of mathematical representations in children's activity. *Cognition and Instruction, 13*(2), 269–313.

Mokros, J., & Russell, S. J. (1995). Children's concepts of average and representativeness. *Journal for Research in Mathematics Education, 26*(1), 20–39.

National Council of Teachers of Mathematics. (2000). *Principles and standards for school mathematics*. Reston, VA: NCTM.

Petrosino, A. J., Lehrer, R., & Schauble, L. (2003). Structuring error and experimental variation as distribution in the fourth grade. *Mathematical Thinking and Learning, 5*(2&3), 131-156.

Sfard, A. (2000). Steering (dis)course between metaphors and rigor: Using focal analysis to investigate an emergence of mathematical objects. *Journal for Research in Mathematics Education, 31*(3), 296–327.

Steinbring, H. (1997). Epistemological investigation of classroom interaction in elementary mathematics teaching. *Educational Studies in Mathematics, 32*(1), 49–92.

Strauss, A. & Corbin, J. (1998). *Basics of qualitative research: Techniques and procedures for developing grounded theory*. Thousand Oaks, CA: Sage.

Wilkinson, L. (1999). Dot plots. *American Statistician, 53*(3), 276–281.

Zawojewski, J. S., & Shaughnessy, J. M. (2000). Mean and median: Are they really so easy? *Mathematics Teaching in the Middle School, 5*(7), 436–440.

Chapter 8

CONCEPTUALIZING AN AVERAGE AS A STABLE FEATURE OF A NOISY PROCESS[1]

Clifford Konold and Alexander Pollatsek
University of Massachusetts, Amherst, USA

INTRODUCTION

Until recently, the study of statistics in the United States was confined to the university years. Following recommendations made by the National Council of Teachers of Mathematics (NCTM, 1989; 2000), and building on the ground-breaking Quantitative Literacy series (see Scheaffer, 1991), statistics and data analysis are now featured prominently in most mathematics curricula and are also appearing in the K–12 science standards and curricula (Feldman, Konold, & Coulter, 2000; National Research Council, 1996). Concurrently, university-level introductory statistics courses are changing (e.g., Cobb, 1993; Gordon & Gordon, 1992; Smith, 1998) in ways that pry them loose from the formulaic approach copied with little variation in most statistics textbooks published since the 1950s.[1] At all levels, there is a new commitment to involve students in the analysis of real data to answer practical questions. Formal inference, at the introductory levels, is taking a less prominent place as greater emphasis is given to exploratory approaches (à la Tukey, 1977) to reveal structure in data. This approach often capitalizes on the power of visual displays and new graphic-intensive computer software (Biehler, 1989; Cleveland, 1993; Konold, 2002).

Despite all the criticisms that we could offer of the traditional introductory statistics course, it at least has a clear objective: to teach ideas central to statistical

---

[1] This article originally appeared as "Data Analysis as the Search for Signals in Noisy Processes," in the *Journal for Research in Mathematics Education, 33* (4), 259–289, copyright 2002, and is reproduced here with the permission of the National Council of Teachers of Mathematics. All rights reserved. The writing of this article was supported by National Science Foundation (NSF) grants REC-9725228 and ESI-9818946. Opinions expressed are those of the authors and not necessarily those of NSF.

D. Ben-Zvi and J. Garfield (eds.),
The Challenge of Developing Statistical Literacy, Reasoning and Thinking, 169–199.

inference, including the Law of Large Numbers and the Central Limit Theorem. For the students now learning more exploratory forms of data analysis, the objective is less clear. There are various proposals about which core ideas we should target in early instruction in data analysis. Wild and Pfannkuch (1999), for example, view variation as the core idea of statistical reasoning and propose various subconstructs that are critical to learning to reason about data. Recently designed and tested materials for 12- to 14-year-olds aim at developing the idea of a distribution (Cobb, 1999; Cobb, McClain, & Gravemeijer, 2003). According to the supporting research, this idea entails viewing data as "entities that are distributed within a space of possible values," in which various statistical representations—be they types of graphical displays or numerical summaries—are viewed as different ways of structuring or describing distributions (see Cobb, 1999, pp. 10–11). Others have argued the centrality of the idea of data as an aggregate—an emergent entity (i.e., distribution) that has characteristics not visible in any of the individual elements in the aggregate (Konold & Higgins, 2003; Mokros & Russell, 1995).

In this article, we build on these ideas of variation, distribution, and aggregate to offer our own proposal for the core idea that we believe should guide statistics and data analysis instruction, beginning perhaps as early as age 8. In short, that idea involves coming to see statistics as the study of noisy processes—processes that have a signature, or signal, which we can detect if we look at sufficient output.

It might seem obvious that a major purpose of computing statistics such as the mean or median is to represent such a "signal" in the "noise" of individual data points. However, this idea is virtually absent from our curricula and standards documents. Neither NCTM's *Principles and Standards for School Mathematics* (2000) nor the American Association for the Advancement of Science (AAAS), *Science for All Americans* (1989), explicitly describes an average as anything like a signal. Our search through several middle school and high school mathematics curricula has not uncovered a single reference to this idea. Nor does it appear in earlier research investigating students' ideas about averages and their properties (Mokros & Russell, 1995; Pollatsek, Lima, & Well, 1981; Strauss & Bichler, 1988). The idea is evident, however, in a few recent studies. In their investigation of statistical reasoning among practicing nurses, Noss, Pozzi, and Hoyles (1999) refer briefly to this interpretation; one nurse the authors interviewed characterized a person's average blood pressure as "what the normal range was sort of settling down to be." The idea of signal and noise is also evident in the work of Biehler (1994), Wild and Pfannkuch (1999), and Wilensky (1997).

OVERVIEW

We begin by describing how statisticians tend to use and think about averages as central tendencies. We then contrast this interpretation with various other interpretations of averages that we frequently encounter in curriculum materials. Too frequently, curricula portray averages as little more than summaries of groups of values.[2] Although this approach offers students some rationale for summarizing group data (for example, to see what is "typical"), we will argue that it provides little conceptual basis for using such statistical indices to characterize a set of data, that is, to represent the whole set. To support this claim, we review research that has demonstrated that although most students know how to compute various averages such as medians and means, few use averages to represent groups when those averages would be particularly helpful—to make a comparison between two groups. We recommend beginning early in instruction to help students develop the idea of central tendency (or data as a combination of signal and noise). To explore the conceptual underpinnings of the notion of central tendency, we briefly review its historical development and then examine three types of statistical processes. For each process, we evaluate the conceptual difficulty of regarding data from that process as a combination of signal and noise. Finally, we outline some possible directions for research on student thinking and learning.

In this article, we focus our discussion on averages, with an emphasis on means (using the term *average* to refer to measures of center collectively, including the mean, median, and mode). By focusing on averages, we risk being misunderstood by those who have recently argued that instruction and public discourse have been overemphasizing measures of center at the expense of variability (e.g., Shaughnessy, Watson, Moritz, & Reading, 1999; also see Gould, 1996). A somewhat related but more general critique comes from proponents of Tukey's (1977) exploratory data analysis (EDA) who advocate that, rather than structure our curricula around a traditional view of inferential statistics, we should instruct young students in more fluid and less theory-laden views of analysis (e.g., Biehler, 1989; 1994).

Those concerned that measures of center have been overemphasized as well as proponents of EDA may misread us as suggesting that instruction should aim at teaching students to draw conclusions by inspecting a limited number of simple summaries such as means. In fact, we agree wholeheartedly with Shaughnessy et al. (1999) and with EDA proponents that we should be teaching students to attend to general distributional features such as shape and spread, and to look at distributions in numerous ways for insights about the data. We do not view the decision to focus our analysis here on measures of center as being at odds with their concerns. Our decision is partly pragmatism and partly principle.

On the pragmatic side, we wanted to simplify our exposition. Almost all statistical measures capture group properties, and they share an important property with good measures of centers: They stabilize as we collect more data. These measures include those of spread, such as the standard deviation, interquartile range, percentiles, and measures of skewness. But switching among these different measures would needlessly complicate our exposition.

The deeper reason for focusing our discussion on measures of center is that we believe such measures do have a special status, particularly for comparing two sets of data. Here, some proponents of teaching EDA may well disagree with us. Biehler (1994), for example, maintained that the distribution should remain the primary focus of analysis and that we should regard an average, such as the mean, as just one of many of its properties. We will argue that the central idea should be that of searching for a signal and that the idea of distribution comes into better focus when it is viewed as the "distribution around" a signal. Furthermore, we claim that the most basic questions in analyzing data involve looking at *group differences* to determine whether some factor has produced a difference in the two groups. Typically, the most straightforward and compelling way to answer these questions is to compare averages. We believe that much of statistical reasoning will elude students until they understand when a comparison of two averages makes sense and, as a corollary, when such a comparison is misleading. If they do not understand this, students' explorations of data (i.e., "data snooping") will almost certainly lack direction and meaning.

SIGNALS IN NOISY PROCESSES

A statistician sees group features such as the mean and median as indicators of stable properties of a variable system—properties that become evident only in the aggregate. This stability can be thought of as the certainty in situations involving uncertainty, the signal in noisy processes, or, the descriptor we prefer, central tendency. Claiming that modern-day statisticians seldom use the term *central tendency,* Moore (1990, p. 107) suggests that we abandon the phrase and speak instead of measures of "center" or "location." But we use the phrase here to emphasize conceptual aspects of averages that we fear are often lost, especially to students, when we talk about averages as if they were simply locations in distributions.

By central tendency we refer to a *stable* value that (a) represents the signal in a variable process and (b) is better approximated as the number of observations grows.[3] The obvious examples of statistics used as indicators of central tendency are averages such as the mean and median. Processes with central tendencies have two components: (a) a stable component, which is summarized by the mean, for example; and (b) a variable component, such as the deviations of individual scores around an average, which is often summarized by the standard deviation.

It is important to emphasize that measures of center are not the only way to characterize stable components of noisy processes. Both the shape of a frequency distribution and global measures of variability, for example, also stabilize as we collect more data; they, too, give us information about the process. We might refer to this more general class of characteristics as *signatures* of a process. We should point out, however, that all the characteristics that we might look at, including the shape and variability of a distribution, are close kin to averages. That is, when we look at the shape of a particular distribution, we do not ordinarily want to know precisely how the frequency of values changes over the range of the variable. Rather, we tame the distribution's "bumpiness." We might do this informally by visualizing a smoother underlying curve or formally by computing a best-fit curve. In either case, we attempt to see what remains when we smooth out the variability. In a similar manner, when we employ measures such as the standard deviation or interquartile range, we strive to characterize the *average* spread of the data in the sample.

Implicit in our description of central tendency is the idea that even as one speaks of some stable component, one acknowledges the fundamental variability inherent in that process and thus its probabilistic nature. Because of this, we claim that the notion of an average understood as a central tendency is inseparable from the notion of spread. That average and variability are inseparable concepts is clear from the fact that most people would consider talking about the average of a set of identical values to be odd. In addition, it is hard to think about why a particular measure of center makes sense without thinking about its relation to the values in the distribution (e.g., the mean as the *balance point* around which the sum of the deviation scores is zero, or the median as the point where the number of values above equals the number of values below).

Not all averages are central tendencies as we have defined them above. We could compute the mean weight of an adult lion, a Mazda car, and a peanut, but no clear process would be measured here that we could regard as having a central tendency. One might think that the mean weight of all the lions in a particular zoo would be a central tendency. But without knowing more about how the lions got there or their ages, it is questionable whether this mean would necessarily tell us anything about a process with a central tendency. Quetelet described this distinction in terms of *true* means of distributions that follow the law of errors versus *arithmetic* means that can be calculated for any assortment of values, such as our hodgepodge above (see Porter, 1986, p. 107).

Populations versus Processes

In the preceding description, we spoke of processes rather than populations. We contrast these two ways of thinking about samples or batches of data, as shown in Figure 1. When we think of a sample as a subset of a population (see the left graphic), we see the sample as a piece allowing us to guess at the whole: The average and shape of the sample allow us perhaps to estimate the average and shape of the population. If we wanted to estimate the percentage of the U.S. population favoring gun control, we would imagine there being a population percentage of some unknown value, and our goal would be to estimate that percentage from a well-chosen sample. Thinking in these terms, we tend to view the population as static and to push to the background questions about why the population might be the way it is or how it might be changing.

From the process perspective (as depicted in the right graphic of Figure 1), we think of a population or a sample as resulting from an ongoing, dynamic process, a process in which the value of each observation is determined by a large number of causes, some of which we may know and others of which we may not. This view moves to the foreground questions about why a process operates as it does and what factors may affect it. In our gun control example, we might imagine people's opinions on the issue as being in a state of flux, subject to numerous and complex influences. We sample from that process to gauge the net effect of those influences at a point in time, or perhaps to determine whether that process may have changed over some time period.

For many of the reasons discussed by Frick (1998), we have come to prefer thinking of samples (and populations, when they exist) as outputs of processes.[4] One reason for this preference is that a process view better covers the range of statistical situations in which we are interested, many of which have no real population (e.g., weighing an object repeatedly). Another reason for preferring the process view is that when we begin thinking, for example, about how to draw samples, or why two samples might differ, we typically focus on factors that play a role in producing the data. That is, we think about the *causal processes* underlying the phenomena we are studying. Biehler (1994) offered a similar analysis of the advantages of viewing data as being produced by a probabilistic mechanism—a mechanism that could be altered to produce predictable changes in the resultant distribution. Finally, viewing data as output from a process highlights the reason that we are willing to view a collection of individual values as in some sense "the same" and thus to reason about them as a unity: We consider them as having been generated by the same process.

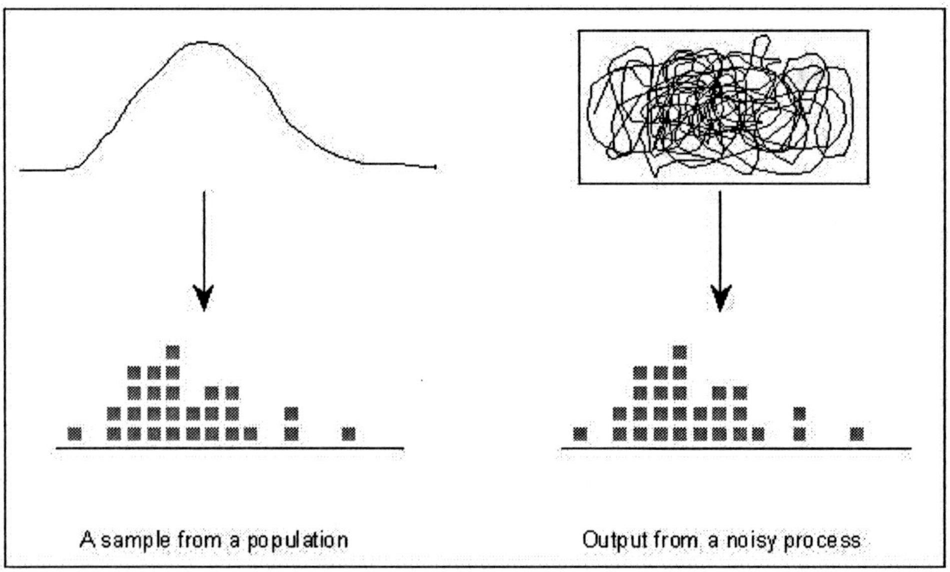

Figure 1. Data viewed as a sample of a population (left) versus data viewed as output of a noisy process (right).

This notion of process is, of course, inherent in the statistician's conception of a population, and we expect that most experts move between the process and population perspectives with little difficulty or awareness.[5] However, for students new to the study of statistics, the choice of perspective could be critical. To illustrate more fully what we mean by reasoning about processes and their central tendencies, we discuss recent results of the National Assessment of Educational Progress (NAEP).

NAEP Results as Signals of Noisy Processes

NAEP is an assessment of student capabilities in Grades 4, 8, and 12, conducted every 4 years in the United States. On the 1998 assessment, eighth graders averaged 264 on the reading component.[6] What most people want to know, of course, is how this compares to the results from previous assessments. In this case, the mean had increased 4 points since the 1994 assessment. The 12th graders had also gained 4 points on average since 1994, and the fourth graders, 3 points. Donahue, Voelkl, Campbell, and Mazzeo (1999) interpreted these differences as evidence that children's reading scores were improving.

Reports such as this are now so commonplace that we seldom question the logic of this reasoning. But what is the rationale in this case for comparing group means and for taking the apparently small difference between those means seriously? We will argue that to answer these questions from a statistical perspective requires a well-formed idea of a central tendency.

Interpreted as a central tendency, the mean of 264 is a measure of a complex process that determines how well U.S. children read at a given point in time. An obvious component of this process is the reading instruction that children receive in school. Another component of the process is the behavior of adults in the home: their personal reading habits, the time they spend reading to their children, and the kind and quantity of reading material they have in the home. A third component consists of factors operating outside the home and school, including determinants of public health and development, such as nutrition levels and the availability and use of prenatal care; genetic factors; and the value placed on literacy and education by local communities and the society at large.

Using a statistical perspective, we often find it useful to regard all these influences together (along with many others that we may be unaware of) as a global process that turns out readers of different capabilities. In the sense that we cannot know how these various factors work together in practice to produce results, the global process is a probabilistic one, unpredictable at the micro level. However, even though readers produced by this process vary unpredictably in their performance, we can regard the entire process at any given time as having a certain stable capability to produce competent readers. The average performance of a large sample of readers produced by this process is one way to gauge the power of that process (or its propensity) to produce a literate citizenry. As Mme. de Staël explained in 1820, "events which depend on a multitude of diverse combinations have a periodic recurrence, a fixed proportion, when the observations result from a large number of chances" (as quoted in Hacking, 1990, p. 41). And because of the convergence property of central tendencies, the larger the data set, the better the estimate we expect our sample average to be of the stable component of the process.

Given the huge sample size in the reading example (about 11,000 eighth graders) and assuming proper care in composing the sample, we expect that the sample mean of 264 is very close to this propensity. Assuming that the 1994 mean is of equal quality, we can be fairly certain that the difference between these two means reflects a real change in the underlying process that affects reading scores. Note that the important inference here does not concern a sampling issue in the narrow sense of randomly sampling from a fixed known population. That is, assuming no changes in the system, we would expect next year's mean to come out virtually the same even though the population of eighth graders would consist of different individuals. Focusing on the process rather than the population helps make the real intent of our question clear.

The mean is not necessarily the best single number to serve as an index of such a change. The median is also a good index, and changes in the 25th percentile, the percent above some minimal value, the standard deviation, or the interquartile range could also be valid indicators of changes in the underlying educational process. As long as a process remains stable, we expect the mean, or any of these other statistical indices obtained from that process, to remain relatively unchanged from sample to sample. Conversely, when a statistic from a large sample changes appreciably, we assume that the process has changed in some way. Furthermore, these expectations are crucial in our attempts to evaluate efforts to alter processes. In the case of reading, we might introduce new curricula, run an advertising campaign encouraging parents to read to their children, expand the school free lunch program in disadvantaged areas, and upgrade local libraries. If we do one or more of these things and the mean reading scores of an appropriate sample of children increases, we have grounds for concluding that we have improved the process for producing readers. Again, we emphasize that though we have specified the mean in this example, we might be as happy using the median or some other measure of center.

The above example, however, indicates a way in which a measure of center is often special. That is, the practical issue in which we are usually interested is whether, overall, things are getting better or worse, a question most naturally phrased in terms of a change of center. It is much harder to think of examples where we *merely* want to increase or decrease the variability or change the shape of the distribution. We could imagine an intervention that tried only to narrow the gap between good and poor readers, in which case we would compare measures of spread, such as the standard deviation. Although there are questions that are naturally phrased in terms of changes in variability or distribution shape, such questions are typically second-order concerns. That is, we usually look at whether variability or shape have changed to determine whether we need to qualify our conclusion about comparing measures of center. Even in situations where we might be interested in reducing variability, such as in income, we are certainly also interested in whether this comes at the expense of lowering the average.

DIFFERENT INTERPRETATIONS OF AVERAGES

We have argued that statisticians view averages as central tendencies, or signals in variable data. But this is not the only way to think about them. In Table 1, we list this interpretation along with several others, including viewing averages as data reducers, fair shares, and typical values. We consider an interpretation to be the goal that a person has in mind when he or she computes or uses an average. It is the answer that a person might give to the question, "Why did you compute the average of those values?" Some of these interpretations are described in Strauss and Bichler (1988) as "properties" of the mean. Mokros and Russell (1995) described other interpretations as "approaches" that they observed elementary and middle school students using.[7] In Table 1, we also provide an illustrative problem context for each interpretation. Of course, any problem could be interpreted from a variety of perspectives. But we chose these particular examples because their wording seemed to suggest a particular interpretation.

Table 1. Examples of contexts for various interpretations of average

Interpretation/ meaning	**Example context**
Data reduction	Ruth brought 5 pieces of candy, Yael brought 10 pieces, Nadav brought 20, and Ami brought 25. Can you tell me in one number how many pieces of candy each child brought? (From Strauss & Bichler, 1988)
Fair share	Ruth brought 5 pieces of candy, Yael brought 10 pieces, Nadav brought 20, and Ami brought 25. The children who brought many gave some to those who brought few until everyone had the same number of candies. How many candies did each girl end up with? (Adapted from Strauss & Bichler, 1988)
Typical value	The numbers of comments made by eight students during a class period were 0, 5, 2, 22, 3, 2, 1, and 2. What was the typical number of comments made that day? (Adapted from Konold & Garfield, 1992)
Signal in noise	A small object was weighed on the same scale separately by nine students in a science class. The weights (in grams) recorded by each student were 6.2, 6.0, 6.0, 15.3, 6.1, 6.3, 6.2, 6.15, 6.2. What would you give as the best estimate of the actual weight of this object? (Adapted from Konold & Garfield, 1992)

Data Reduction

According to this view, averaging is a way to boil down a set of numbers into one value. The data need to be reduced because of their complexity—in particular, due to the difficulty of holding the individual values in memory. Freund and Wilson (1997) draw on this interpretation to introduce averages in their text: "Although distributions provide useful descriptions of data, they still contain too much detail for some purposes" (p. 15). They characterize numerical summaries as ways to further simplify data, warning that "this condensation or data reduction may be accompanied by a loss of information, such as information on the shape of the distribution" (p. 16). One of the high school students interviewed by Konold,

Pollatsek, Well, and Gagnon (1997) used this as a rationale for why she would look at a mean or median to describe the number of hours worked by students at her school:

> We could look at the mean of the hours they worked, or the median. ... It would go through a lot to see what every, each person works. I mean, that's kind of a lot, but you could look at the mean. ... You could just go through every one ... [but] you're not going to remember all that.

Fair Share

The computation for the mean is often first encountered in elementary school in the context of fair-share problems, with no reference to the result being a mean or average. Quantities distributed unevenly among several individuals are collected and then redistributed evenly among the individuals. The word *average,* in fact, derives from the Arabic *awariyah,* which translates as "goods damaged in shipping." According to Schwartzman (1994), the Italians and French appropriated this term to refer to the financial loss resulting from damaged goods. Later, it came to specify the portion of the loss borne by each of the many people who invested in the ship. Strauss and Bichler (1988) provided 11 problems as examples of tasks that they used in their research, and we would regard all but three of them as involving the idea of fair share. We can view many commonly encountered rates, such as yearly educational expenditure per student, as based on the fair-share idea, since we tend to think most naturally about these rates as distributing some total quantity equally over some number of units. In such cases, we do not ordinarily think of the computed value in relation to each individual value; nor do we worry, when computing or interpreting this fair share, about how the component values are distributed or whether there are outliers.

Typical Value

Average as a typical score is one of the more frequently encountered interpretations in current precollege curricula. What appears to make values typical for students are their position (located centrally in a distribution of values) and/or their frequency (being the most frequent or even the majority value). Younger students favor the mode for summarizing a distribution, presumably because it can often satisfy both of these criteria (Konold & Higgins, 2003). Mokros and Russell (1995) speculated that those students they interviewed who used only modes to summarize data may have interpreted *typical* as literally meaning the most frequently occurring value. Researchers have also observed students using as an average a range of values in the center of a distribution (Cobb, 1999; Konold, Robinson, Khalil, Pollatsek, Well, Wing, & Mayr, 2002; Mokros & Russell, 1995; Noss, Pozzi, & Hoyles, 1999; Watson & Moritz, 1999). These "center clumps" are located in the heart of the distribution and often include a majority of the observations. In this respect, these clumps may serve as something akin to a mode for some students.

Signal in Noise

According to this perspective, each observation is an estimate of an unknown but specific value. A prototypical example is repeatedly weighing an object to determine its actual weight. Each observation is viewed as deviating from the actual weight by a measurement error, which is viewed as "random." The average of these scores is interpreted as a close approximation to the actual weight.

Formal Properties of Averages

Many school tasks involving averages seem unrelated to any of the particular interpretations we describe above. For example, finding the average of a set of numbers out of context seems intended only to develop or test students' computational abilities. Other school tasks explore formal properties of averages, which we also would not view as directly related to particular interpretations. Such tasks include those meant to demonstrate or assess the idea that (a) the mean of a set of numbers is simply related to the sum of those numbers, (b) the mean is a balance point and the median a partition that divides the cases into two equal-sized groups,[8] (c) the mean and median lie somewhere within the range of the set of scores, and (d) the mean or median need not correspond to the value of an actual observation. In their longitudinal study of the development of young students' understandings of average, Watson and Moritz (2000) focused in particular on these relations, asking students, for example, how the mean number of children per family could possibly be 2.3 rather than a whole number. We consider most of the properties enumerated by Strauss and Bichler (1988, p. 66) to be formal relations of this sort. We are not arguing that these are unimportant or trivial ideas, but rather that they are usually not tied to particular interpretations of averages.

Applying Interpretations to the Problem of Group Comparison

In the NAEP example, we explored the notion of central tendency and showed how it provides a basis for using averages—means, in that case—to compare groups. Because the mean is a very stable estimator in large samples, we can use it to track changes in a process even though the output from that process is variable and unpredictable in the short run.

What bases do the other interpretations of average provide for evaluating the two NAEP results by comparing means? Consider the data reduction interpretation: Data are distilled to a single value, presumably because of our inability to consider all the values together. We argue that nothing in this interpretation suggests that any new information emerges from this process; indeed, a considerable loss of information seems to be the price paid for reducing complexity. By this logic, it would seem that as a data set grows larger, any single-value summary becomes less representative of the group as increasingly more information is lost in the reduction process.

The typical-value interpretation is nearer to the central tendency interpretation since it may involve the idea that the value, in some sense, represents much of the data in the group. However, as with the data reduction interpretation, it is not clear why one ideally would like to have typical values from large samples rather than from small ones. Indeed, it would seem as reasonable to regard a typical score as becoming less (rather than more) representative of a group as that group became larger and acquired more deviant values.

The fair-share interpretation may provide some basis for using means to compare groups. One could think of the mean in the 1998 NAEP data as the reading score that all students sampled that year would have if reading ability were divided evenly among all the students sampled. Based on this reasoning, one might reasonably conclude that the 1998 group had a higher reading score than the 1994 group. Cortina, Saldanha, and Thompson (1999) explored the use of this notion by seventh- and eighth-grade students and concluded that these students could use the idea of fair share to derive and compare means of unequal groups. However, we would guess that many students would regard such reasoning skeptically unless it were physically possible to reallocate quantities in the real-world situation. If, for example, we were thinking about the number of boxes of cookies sold by different scout troops (as in the study by Cortina et al.), redistributing the cookie boxes evenly makes some sense. In contrast, if we were reasoning about mean weight, height, or IQ of a number of individuals, we would have to think of these pounds, inches, or IQ points being shared metaphorically.[9]

Furthermore, we are skeptical about whether the fair-share interpretation is a statistical notion at all. It seems to ignore, in a sense, the original distribution of values and to attend only to the total accumulation of some amount in a group. Consider, for example, the value we would compute to decide how the different numbers of candies brought by various children to a party could be equally redistributed among the children (see Table 1). In this context, the particulars about how the candies were originally distributed seem irrelevant. That is, the number that constitutes a fair share is not viewed as a representation or summary of the original distribution but rather as the answer to the question of how to divide the candies equitably.

In conclusion, whereas some of the interpretations may be useful to summarize a group of data, it is quite another thing to take a statistic seriously enough as to use it to represent the *entire* group, as one must do when using averages to compare groups. We claim that viewing an average as a central tendency provides a strong conceptual basis for, among other things, using averages to compare two groups, whereas various other interpretations of average, such as data reducers and typical values, do not.

We acknowledge that our analysis of these alternative interpretations has been cursory and that it should thus be regarded skeptically. However, our primary purpose is to highlight some of the questions that should be asked in exploring different approaches to introducing students to averages. Furthermore, there is good evidence that whatever interpretations students do have of averages, those interpretations usually do not support using averages to compare one group to another. Many studies have demonstrated that even those who know how to compute and use averages in some situations do not tend to use them to compare groups.

Students' Tendency Not to Use Averages to Compare Groups

Gal, Rothschild, and Wagner (1990) interviewed students of ages 8, 11, and 14 about their understanding of how means were computed and what they were useful for. They also gave the students nine pairs of distributions in graphic form and asked them to decide whether the groups were different or not. Only half of the 11- and 14-year-olds who knew how to compute the mean of a single group (and, also, to some extent, how to interpret it) went on to use means to compare two groups. Hancock, Kaput, and Goldsmith (1992) and, more recently, Watson and Moritz (1999) have reported similar findings.

This difficulty is not limited to the use of means. Bright and Friel (1998) questioned 13-year-old students about a stem-and-leaf plot that showed the heights of 28 students who did not play basketball. They then showed them a stem-and-leaf plot that included these data along with the heights of 23 basketball players. This latter plot is shown in Figure 2. Heights of basketball players were indicated in bold type, as they are here. Students had learned how to read this type of display and had no difficulty reading values from it. Asked about the "typical height" in the single distribution of the non–basketball players, the students responded by specifying middle clumps (e.g., 150–160 cm), a reasonable group summary. Yet, shown the plot with both distributions, they could not generalize this method or find another way to determine "How much taller are the basketball players than the students who did not play basketball?"

We found similar difficulties when we interviewed four high school seniors (ages 17–18) who had just completed a yearlong course in probability and statistics (Biehler, 1997; Konold et al., 1997). During the course, the students had frequently used medians (primarily in the context of box plot displays) as well as means to make group comparisons. However, during a postcourse interview in which they were free to use whatever methods of comparison seemed appropriate, they seldom used medians or means for this purpose. Instead, they tended to compare the number of cases in each group that had the same value on the dependent variable. For example, to decide if males were taller than females, they might inspect the sample for all individuals who were 6 feet tall and argue that males were taller because there were more males than females of that height. In making these comparisons, students typically did not attend to the overall number of individuals in the two groups (in this case, to the overall number of males vs. females). Other researchers, including Cobb (1999) and Watson and Moritz (1999), have reported students using this same "slicing" technique over a range of different problems to compare two groups.

10	
11	
12	
13	8 8 8 9
14	1 2 4 7 7 7
15	0 0 1 1 1 1 2 2 2 2 3 3 5 6 6 7 8
16	
17	1
18	**0 3 5**
19	**0 2 5 7 8 8**
20	**0 0 2 3 5 5 5 5 7**
21	**0 0 0 5**
22	**0**
23	

Note. The row headed by 13 (the stem) contains four cases (leaves)—three students of 138 centimeters and a fourth student of 139 centimeters.

Figure 2. Stem-and-leaf plot of heights of students and basketball players (boldface) from "Helping Students Interpret Data," by G. Bright and S. N. Friel, in Reflections on Statistics: Learning, Teaching, and Assessment in Grades K–12 (p. 81), edited by S. P. Lajoie, 1998, Mahwah, NJ: Lawrence Erlbaum Associates. Copyright 1998 by Lawrence Erlbaum Associates.

In short, even though instruction in statistics usually focuses on averages, many students do not use those measures of central tendency when they would be particularly helpful—to make comparisons between groups composed of variable elements. We suggest that this pattern is symptomatic of students' failure to interpret an average of a data set as saying something about the entire distribution of values. To address this problem instructionally, we believe that we should be encouraging students early in statistics instruction to think of averages as central tendencies or signals in noisy processes. We acknowledge that this is a complex idea and one that is particularly difficult to apply to the type of processes that we often have students investigating. We explore these conceptual difficulties below.

THREE TYPES OF PROCESSES AND THEIR CONCEPTUAL CHALLENGES

Hints about the cognitive complexity of central tendency are found in the historical account of its development. It was Tycho Brache in the late 1500s who introduced the use of means as central tendencies to astronomy (Plackett, 1970). He used them to address a problem that had long troubled astronomers: What to take as the position of a star, given that the observed coordinates at a particular time tended to vary from observation to observation. When early astronomers began computing means of observations, they were very cautious, if not suspicious, about whether and when it made sense to average observations. In fact, before the mid-eighteenth century, they would never combine their own observations with those obtained from another astronomer. They were fearful that if they combined data that had anything but very small errors, the process of averaging would multiply rather than reduce the effect of those errors (Stigler, 1986, p. 4). Taking the mean of multiple observations became the standard solution only after it had been determined that the mean tended to stabilize on a particular value as the number of observations increased.

It was another hundred years before Quetelet began applying measures of central tendency to social and human phenomena (Quetelet, 1842). The idea of applying means to such situations was inspired partly by the surprising observation that national rates of birth, marriage, and suicides—events that at one level were subject to human choice—remained relatively stable from year to year. Some, including Arbuthnot and De Moivre, had taken these stable rates as evidence of supernatural design. Quetelet explained them by seeing collections of individual behaviors or events as analogous to repeated observations. Thus, he regarded observing the weights of 1,000 different men—weights that varied from man to man—as analogous to weighing the same man 1,000 times, with the observed weight varying from trial to trial. The legitimacy of such an analogy, of course, has been a heated controversy in statistics. Even at the time, Quetelet's ideas brought stiff rebukes from thinkers such as Auguste Comte, who thought it ludicrous to believe that we could rise above our ignorance of values of individual cases simply by averaging many of them (Stigler, 1986, p. 194). To Comte, statistics applied to social phenomena was computational mysticism.

We think that the way these early thinkers reacted to different applications of the mean is not merely a historical accident but instead says something about the "deep structure" of these different applications. To explore the challenges of learning to think about data as signal and noise, we examine the metaphor in the context of three types of statistical processes: repeated measures, measuring individuals, and dichotomous events.

Repeated Measures

Consider weighing a gold nugget 100 times on a pan balance, a prototypical example of repeated measurement. It almost goes without saying that the purpose of weighing the nugget is to determine its weight. But how does one deal with the fact that the observed weight varies from trial to trial? We assume that statisticians and nonstatisticians alike would regard these fluctuations as resulting from errors in the measurement process. But given this variation, how should we use the 100 measurements to arrive at the object's weight? Should all the measurements be used? Perhaps not, if they are all not equally accurate. A novice might attempt to deal with this question by trying to separate the 100 measurements into two classes: those that are truly accurate versus those that are not. The problem then becomes how to tell which observations are truly accurate, because the actual weight is not known.

One aspect of this situation that makes using a mean of the observations particularly compelling is that, conceptually, we can separate the signal from the noise. Because we regard an object as having some unknown but precise weight, it is not a conceptual leap to associate the mean of several weighings with this actual weight, while attributing the trial-by-trial variations to a distinctly different thing: chance error produced by inaccuracies of the measurement instrument and by the process of reading values from it. Indeed, we can also regard each individual weighing as having two components—a fixed component determined by the actual weight of the nugget and a variable component attributable to the imperfect measurement process.

The relative clarity of this example hinges on our perception that the weight of the nugget is a real property of the nugget. A few philosophers might regard it (possibly along with the nugget itself) as a convenient fiction. But to most of us, the weight is something real that the mean weight is approximating closely and that individual weighings are approximating somewhat less closely. Another reason that the idea of central tendency is compelling in repeated measurement situations is that we can easily relate the mean to the individual observations as well. To help clarify why this is so, we will make some of our assumptions explicit.

We have been assuming that the person doing the weighing is careful and that the scale is unbiased and reasonably accurate. Given these assumptions, we expect that the variability of the weighings would be small and that the frequency histogram of observations would be single-peaked and approximately symmetric. If instead we knew that the person had placed the nugget on different parts of the balance pan, read the dial from different angles, or made errors in transcribing the observations, we would be reluctant to treat the mean of these numbers as a central tendency of the process. We would also be hesitant to accept the mean as a central tendency if the standard deviation was extremely large or if the histogram of weights was bimodal. In the ideal case, most observations would be *close* to the mean or median and the distribution would peak at the average, a fact that would be more apparent with a larger data set because the histogram would be smoother. In this case, we could easily interpret the sample average as a good approximation to a signal or a central tendency and view the variability around it as the result of random error.

These assumptions about the procedure and the resulting data may be critical to accepting the mean of the weighings as a central tendency, but they are not the only things making that interpretation compelling. As indicated earlier, we maintain that the key reason the mean observation in this example is relatively easy to accept as a central tendency is that we can view it as representing a property of the object while viewing the variability as a property of a distinctly independent measurement process. That interpretation is much harder to hold when—rather than repeatedly measuring an attribute of a single object—we measure an attribute of many different objects, taking one measurement for each object and averaging the measurements.

Measuring Individuals

Consider taking the height of 100 randomly chosen adult men in the United States. Is the mean or median of these observations a central tendency? If so, what does it represent? Many statisticians view the mean in this case as something like the *actual* or *true* height of males in the United States (or in some subgroup). But what could a statement like that mean?

For several reasons, an average in this situation is harder to view as a central tendency than the average in the repeated measurement example. First, the gold nugget and its mass are both perceivable. We can see and heft the nugget. In contrast, the population of men and their average height are not things we can perceive as directly. Second, it is clear why we might want to know the weight of the nugget. But why would we want to know the average height of a population of men? Third, the average height may not remain fixed over time, because of factors such as demographic changes or changes in diet. Finally, and perhaps most important, we cannot easily compartmentalize the height measurements into signal and noise. It seems like a conceptual leap to regard each individual height as partly *true height,* somehow determined from the average of the population, and partly *random error* determined from some independent source other than measurement error.

For all of these reasons, it is hard to think about the average height of the group of men as a central tendency. We speculate, however, that it is somewhat easier to regard *differences* between the averages of two groups of individual measurements as central tendencies. Suppose, for example, we wanted to compare the average height of U.S. men to the average height of (a) U.S. women or (b) men from Ecuador. We might interpret the difference between averages as saying something in the first case about the influence of genetics on height and in the second, about the effects of nutrition on height. When making these comparisons, we can regard the difference in averages as an indicator of the "actual effect" of gender or of nutrition, things that are easier to imagine wanting to know about even if they are difficult to observe directly.[10]

Some support for this speculation comes from Stigler (1999), who claims that Quetelet created his infamous notion of the "average man" not as a tool to describe single distributions, but as a method for comparing them: "With Quetelet, the essential idea was that of *comparison*—the entire point was that there were different average men for different groups, whether categorized by age or nationality, and it was for the study of the nature and magnitude of those differences that he had introduced the idea" (p. 61). Although we concede that the notion of a "true" or "actual" value is still a bit strained in these comparison cases, we believe that one needs some approximation to the idea of true value to make meaningful comparisons between two groups whose individual elements vary. To see why, let us look more closely at the comparison of men versus women.

Suppose we compute a mean or median height for a group of U.S. men and another for a group of U.S. women. Note that the act of constructing the hypothesis that gender partly determines height requires us to conceive of height as a process influenced by various factors. Furthermore, we cannot see how comparing the two groups is meaningful unless we have (a) an implicit model that gender may have a *real* genetic effect on height that is represented by the difference between the average for men and the average for women, and (b) a notion that other factors have influences on height that we will regard as *random error* when focusing on the influences of gender on height.[11] Thus, we claim that the concept of an average as approximating a signal, or true value, comes more clearly into focus when we are considering the influence of a particular variable on something (in this case, gender on height). Such a comparison scheme provides a conceptual lever for thinking about signal (gender influences) and noise (other influences). We return to this point later.

Discrete Events

Another measure that is often used as an index of central tendency is the rate of occurrence of some event. As a prototypical example, consider the rate of contracting polio for children inoculated with the Salk vaccine. Even though individual children either get the disease or do not, the rate tells us something about the ability of inoculated children, as a group, to fight the disease.

How can we view a rate (or probability) as a measure of central tendency? First, a probability can be formally viewed as a mean through what some would regard as a bit of trickery. If we code the event "polio" as a 1, and the event "no polio" as a 0, then the probability of getting polio is merely the mean of these Boolean values. Producing a formal average, however, does not automatically give us a measure of central tendency. We need to be able to interpret this average as a signal related to the causes of polio. Compare the distribution of values in the dichotomous case to the ideal case of the weighing example. In the dichotomous case, the mean is not a value that can actually occur in a single trial. Rather than being located at either of the peaks in the distribution, the mean is located in the valley between, typically quite *far* from the observed values. Thus, it is nearly impossible to think about the rate or probability as the *true-value* component of any single observation and the occurrence or nonoccurrence of an individual case of polio as the sum of a true value and a *random error* component. We suspect this is largely why the idea of a central tendency in dichotomous situations is the least tangible of all.

It might help in reasoning about this situation to conceive of some process about which the rate or probability informs us. In the disease example, the conception is fairly similar to the earlier height example: A multitude of factors influence the propensity of individuals to get polio—level of public health, prior development of antibodies, incident rate of polio, age—all leading to a rate of getting the disease in some population. So even though individuals either get polio or do not, the propensity of a certain group of people to get polio is a probability between 0 and 1. That value is a general indicator of the confluence of polio-related factors present in that group.

As with our height example, although an absolute rate may have some meaning, we think it is much easier to conceptualize the meaning of a signal when we are comparing two rates. In the polio example, this might involve comparing the rate in an inoculated group to the rate in a placebo control group. Here, as with the height example, most people would consider the difference in rates (or the ratio of the rates) to be a valid measure of the efficacy of the vaccine or as a reasonable way to compare the efficacy of two different vaccines.

The Role of Noise in Perceiving a Collection as a Group

We have argued that the idea of central tendency, or data as signal and noise, is more easily applied to some types of processes than to others. But other factors, to which we have alluded, may affect the difficulty of applying this idea. Consider the case of comparing the heights of men and women. We would expect that the shape and the relative spread of the distributions would affect how easy it is to conceive of each distribution as a coherent group and, consequently, to be able to interpret each group's average as an indicator of a relatively stable group characteristic.

Indeed, perhaps the most critical factor in perceiving a collection of individual measurements as a group is the nature of the variability within a group and how it relates to the differences between groups. In general, we expect that these individual measurements are easier to view as belonging to a group (and thus as having a central tendency) when the variability among them is relatively small. To explain what we mean by *relatively small,* we find the idea of *natural kinds* helpful. According to Rosch and Mervis (1975), people often mentally represent real-world concepts as prototypes and judge particular instances as "good" or "bad" depending on how closely those instances match the category prototype. For example, a prototypical bird for most North Americans is a medium-sized songbird, something like a robin. The closer an instance is to the category prototype, the less time it takes to identify that instance as a member of the category. North Americans can categorize a picture of a starling as a bird faster than they can a picture of an ostrich.

In this theory of natural kinds, prototypes function much as averages do: Instances of the category are single observations that can be some distance from the average (or prototype). In fact, some competing theories of natural kinds (e.g., Medin & Schaffer, 1978) claim there is no actual instance that functions as a prototype, but that the effective prototype is simply a mean (in some multidimensional feature space) of all the instances in memory. What makes some categories, such as birds, natural kinds is that there is little variability across features within the category relative to the variability of those features between various animal categories. So, even though there are some non-prototypical instances of birds, such as penguins and ostriches, the distributions of features of birds overlap little with those of other natural kinds such as mammals, so that the groups cohere. This research suggests that it might be easier to accept, for example, the mean heights of the men and women as representing group properties if there were no overlap in heights of the men and women, or if at least the overlap were small relative to the spread of the distributions.[12]

Applying Central Tendency to Nonstandard Cases

In the foregoing examples, we focused on relatively ideal cases. We tacitly assumed that our histograms of people's heights, for example, were single-peaked, approximately symmetric, and, configured as two histograms, had approximately equal spread. In such cases, most experts would accept some average as a meaningful measure of central tendency. Is the idea of central tendency applicable only to these ideal cases, or is it more generalizable than that? In this section, we consider several nonstandard examples to make the case that we can and do apply the idea of central tendency to less ideal situations, in which there is some doubt about whether a single measure of center is adequate to describe the data. We argue that statistical reasoning in these situations still rests to a large extent either on the conception of an average as a central tendency or on its cousin, a single measure that describes the variability of a group of observations.

Distributions with Outliers

Consider cases where there are outliers that we decide should be removed from the data set. In the case of weighing, suppose a typical observation differs from the mean weight by something like 1 mg. If one of our observations was 5 mg away from the mean, most people might think it sensible to omit that value in calculating the mean. Two ideas seem implicit in this thinking: (a) that "true" measurement error is associated with weighing on that scale and (b) that some *different* process can sometimes generate observations with unusually high measurement error. Only with such an implicit model can we consider, let alone decide, that an extremely deviant observation must have been due to nonrandom error (e.g., misrecording the observation or having a finger on the pan). Similarly, if we had one or two height observations that were 60 cm from the mean, we might disregard them in certain analyses as resulting from a process different from the process producing the rest of the data (e.g., from a mutation or birth defect). Here again, this makes sense only if we have some implicit model of a *typical* (male or female) height from which individual observations differ by something like "random genetic and/or environmental variation." We can then regard extremely tall or short people as not fitting this model—as resulting from a somewhat different process and therefore calling for a different explanation. For these same reasons, Biehler (1994, p. 32) suggested that "symmetrical unimodal distributions are something distinctive," and deviations from them require additional modeling.

Distributions with Unusual Shape

Continuing with the example of men's heights, consider the case perhaps furthest from the ideal, where the histogram of men's heights is bimodal. We would be reluctant in this case to interpret any average as a central tendency of men's heights. Why? With a bimodal histogram, we would be doubtful that the men we were looking at comprised a simple process, or "natural kind." Rather, we would suspect that our batch of men consisted of two distinct groups and that we could not make any useful statements unless we uncovered some underlying variable that distinguished the two. A similar but somewhat less severe problem would result if the histogram was unimodal but the variability in the group seemed enormous (e.g., if men's heights from an unknown country varied from 60 cm to 900 cm with a mean of 450 cm). Given the huge variability in this case, we would question whether the data came from a coherent process and whether it made sense, therefore, to use an average to represent it. Of course, people's intuitions about whether variability is enormous may differ and are likely to depend on the model they have of *typical* variability (or indeed whether they have any conceptual model for thinking about sources of variability).

Comparing Groups with Skewed or Differently Shaped Distributions

When comparing two histograms, say of men's and women's heights, we run into difficulties when the histograms are of different shape. Imagine, for example, that the men's heights were positively skewed and the women's heights negatively skewed. Because there is clearly something different about the variability in each group, we would be reluctant to compare the two groups using their averages. That is, unless we could generate a model of why the groups' histograms differed in shape and, as a result, conclude that the different shapes were just two versions of random error, we would probably be wary of viewing the difference between the two averages as representing something like the "gender effect on height."

Consider the comparison of differences in income from one decade to another, where both histograms are highly skewed with long tails out to the right. If the histograms have the same variance and the same shape, we claim it is reasonable to accept the shift in central tendency as an estimate of the *actual* change in income for the group, even though we might have misgivings about using the average for either group as the best measure of *actual* income. That is, even though the variability in each group may not match our ideal view of "noise," we can at least convince ourselves that it is the same noise process in both groups. Of course, even though one histogram is a horizontal translation of the other, it does not necessarily mean that income has improved the same amount for each individual (or each type of individual), give or take random error. Indeed, a finer analysis could indicate that certain groups have become better off while other groups have not changed or have even become worse off. It is worth noting, however, that many such arguments about why looking at the differences between group averages is inappropriate or misleading rely on the perception that the groups are, in some sense, not "natural kinds" (e.g., that the processes determining incomes of poor people are different from those determining incomes of rich people). Nonetheless, these arguments are usually most compelling when we can identify natural subgroups in the larger group and can show that the changes in the averages in these subgroups differ from each other (e.g., the rich got richer and the poor got poorer, or different things happened to Blacks and Whites).

Another classic difficulty involves comparing two averages when the distributions differ in spread. For example, what if Country A not only has a higher mean income than Country B but also has a higher standard deviation? This would call for more serious modeling of the variability. A special case that would make it conceptually easier to compare the averages of the two groups would be the situation in which the difference in standard deviations was commensurate with the difference in means (ideally, the ratio of standard deviations would be equal to the ratio of the means). In that case, we could view the effect as multiplicative rather than additive, since Country A's typical income would be equal to Country B's multiplied by a factor that represents the effect(s) that distinguish A from B. And it would be reasonable to assume that the same multiplicative factor also applied to the noise process.

Summary of Analyses of Nonstandard Cases

As we have implied in our argument above, we do not necessarily see these nonstandard cases as problems for the type of framework that we are advocating. Indeed, we think that the idea of central tendency of a process allows us to (a) decide to eliminate an outlier or break data into suitable subsets, (b) come up with a conceptual model that explains why the groups are asymmetric or differ in spread or shape, or (c) decide that there is little we can sensibly conclude about the differences between the two sets of data.

Let us summarize by asking what we could conclude about the difference in men's and women's heights from the distributions we described earlier that were skewed in opposite directions. We assert that we could conclude nothing without some conceptual model. If we were trying to make a statement about genetic gender differences, for example, we would have to be convinced that everything else was *random* and that, for instance, we could not explain the mean height difference as resulting from gender differences in diet. In other words, there is virtually nothing about analyzing data that is model-free. Some may regard this as a radical proposal, but we claim that a mean or median has little heuristic value (and is likely to have little meaning or heuristic value for the student) unless we can conceive of the data coming from some coherent process that an average helps to elucidate.

IMPLICATIONS FOR STATISTICS EDUCATION

The idea of noisy processes, and the signals that we can detect in them, is at the core of statistical reasoning. Yet, current curricula do not introduce students to this idea, instruments meant to assess student reasoning about data do not include items targeting it, and statistics education researchers have not given it much attention. If our argument is valid, then critical changes are called for in education research, the formulation of education objectives, curriculum materials, teacher education, and assessment. These are tightly interrelated components of educational reform. If we fail to advance our efforts on all these fronts, we run the risk of continuing to lose the small ground gained on any one of them.

Accordingly, we describe here what we see as essential components of a signal-versus-noise perspective and offer suggestions about how we might help students (and future teachers) develop these ideas. We do not aim our speculations at curriculum designers or teachers in the hope that that they will implement them. Instead, we intend them for researchers and others who are considering what big ideas should guide our standards and curriculum objectives, for those designing and running teacher institutes, and for those developing assessment frameworks and instruments.

Using Repeated Measures

According to our analysis, processes involving repeated measures are easier than other types of statistical processes to view as part signal and part noise. This suggests that to establish the signal-versus-noise interpretation of various statistical measures, we initially involve students in investigations of repeated measures.

Current curricula make little use of repeated measures. Perhaps this is because many of the prototypical situations, such as our weighing example, can be somewhat boring and seemingly pointless unless they are introduced in meaningful ways. There are many suitable and potentially interesting contexts.[13] In the later grades, these include a number of high-stakes scientific and political issues. For informed public policy, we need good estimates of the thickness of the ozone layer, of dissolved oxygen in rivers, of concentrations of atmospheric CO_2. Statistical control of manufacturing processes provides another context in which it is relatively clear why we need to track a process by looking at its outputs. Of course, time-series analyses are complex, and we need more research to help determine the kinds of questions regarding them that introductory students can fruitfully explore.

Lehrer, Schauble, and their colleagues have employed some interesting repeated measure contexts with younger students. For example, students in a second-grade class designed cars to race down a track (Lehrer, Schauble, Carpenter, & Penner, 2000). During trial runs, students became unhappy about a decision to base a claim about a car's speed on a single trial. Frequently, something would happen to impede a car—for example, it would run up against the track's railing. The agreed-on remedy was to race each car five times. Not surprisingly, the students could not agree later on how to get a single measure of speed from the five trials. However, their proposal of multiple trials was, by itself, suggestive of some notion of signal (a car's actual top speed on that track) and noise (its observed times resulting from unpredictable events).

This classroom episode suggests an important distinction. That is, a student might perceive data as comprising signal and noise and yet not necessarily view a statistical measure such as an average as an acceptable indicator of signal. We would expect that with processes involving repeated measures, students would tend to think of each measurement as a combination of signal and noise, particularly if sources of measurement error were easy to identify, as in measuring length with a ruler. But these same students might not be likely to think of an average of repeated measures as indicative of signal (any more than the early astronomers were). Thus, the instructional challenge is how to help students interpret measures such as averages as indicators of central tendency. Taking a clue from the historical development of the concept, it would seem fruitful to have students explore the relative stability of various indicators in different samples.

Explorations of Stability

The idea of stability is closely related to the idea of signal. If the weight of an object is not changing from trial to trial, it seems reasonable to expect that a good indicator of its weight should also not vary much from sample to sample. Recall that it was observing the stability from year to year of such things as birth and death rates that led Quetelet to begin regarding these rates as indicators of prevailing and relatively stable societal conditions, and to make the analogy to means of repeated measures. Similar investigations by students could set the stage for interpreting averages as indicators of signal.

A method frequently used to demonstrate stability is to draw multiple samples from a known population and evaluate particular features, such as the mean, across these replications. However, we expect that these demonstrations are often conducted prematurely—before students have understood why one is interested in the mean. Furthermore, in real sampling situations we never do these repeated samplings, which leaves many students confused about what we can possibly learn from this hypothetical exercise. The following three alternative methods of exploring stability appear promising on the basis of their use in classrooms with students as young as 8 years old.

Comparing Different Measures

In this approach, students compare the relative accuracy of different measurement methods. Lehrer, Schauble, Strom, and Pligge (2001) used this approach with third and fifth graders, who measured weights and volumes as part of a study of densities of different materials. The students explored several different ways to measure each attribute. They did this by using each method repeatedly to measure the same object. The students came to favor those methods that produced less variability in these repeated measures. Having established what measurement technique they would use, students then considered various proposals of what to use as, for example, the volume of a particular object. The problem, of course, was that even with the same measurement method, repeated measuring gave the students a range of values. They ultimately decided to discard outliers and compute the means of the remaining observations as their "best guess" of the weights and volumes of these objects.

Observing Growing Samples

Another way of exploring stability is to have students observe a distribution as the sample gets larger. We tested this approach recently in a seventh-grade mathematics class. Students had conducted an in-class survey to explore whether boys and girls were paid similar allowances. While comparing the two distributions, one student expressed reservations about drawing conclusions, arguing that she had no idea what the distributions might look like if they collected more data. Her classmates agreed.

To help the class explore this issue, we constructed an artificial pond filled with two kinds (colors) of paper fish. According to our cover story, a farmer wanted to determine whether a new type of genetically engineered fish grew longer, as claimed, than the normal fish he had been using. Students "captured" fish from the pond, reading off fish type and length (which was written on the fish.) On an overhead display, we constructed separate stacked dot plots for each type of fish as students read off their data. After about 15 fish had been sampled, we asked students what the data showed so far. Students observed that the data for the normal fish were clustering at 21–24 cm, whereas the data for the genetically engineered fish were clustering at 25–27 cm. Then we asked them what they thought would happen as we continued to sample more fish, reminding them of their earlier reservations with the allowance data. Some said that the stacks would become higher and the range would get bigger, without mentioning what would happen to such features as the general shape or the location of the center clump. However, other students did anticipate that the center clusters would "grow up" but would nevertheless maintain their approximate locations along the horizontal axis. The latter, of course, is what they observed as they continued to add more fish to the sample distributions. After the sampling, we showed them both population distributions along with their sample data, calling their attention to the fact that the centers of their sample distributions were quite good predictors of the centers of the population distributions—that these stable features of the samples were signals.

Simulating Processes

A third way to explore stability is to investigate why many noisy processes tend to produce mound-shaped distributions. Wilensky (1997) described a series of interviews that he conducted with graduate students who were exploring this question through computer simulations. We conducted a similar investigation with fifth-grade students in an after-school program on data analysis. In analyzing a data set on cats (from Rubin, Mokros, & Friel, 1996), students noticed that many frequency distributions, like tail length and body weight, were mound shaped. As part of exploring why this might be, students developed a list of factors that might cause a cat's tail to be longer or shorter. Their list included diet, being in an accident, and length of father's and mother's tails. Using this list, we constructed a spinner to determine the value of each factor for a particular cat's tail. One student might spin +2 inches for diet, +3 inches for mother's contribution, –2 inches for an accident, and so on (Of course, each student wanted his or her cat to have the longest tail.) Before they began spinning, students predicted that if they built 30 cat tails in this way, they would get about equal numbers of cats with short, medium, and long tails. After several trials they noticed they were tending to get medium tails, which they explained by pointing out that you would have to be "real lucky" to get a big number every spin, or "real unlucky" to get a small number every spin. As this was our last session with these students, we could not explore what they might have generalized from this experience; but we believe that understanding why such processes produce normal-shaped distributions is a critical part of coming to trust how process signals rise up through the noise.

Group Comparison

We have speculated that it is often easier to regard the difference between two averages as a central tendency than it is to think of a single average that way. This suggests, perhaps somewhat counterintuitively, that rather than beginning instruction by having students explore single distributions of individual values, we instead might fruitfully start with questions involving group comparison. Some support for the benefit of having even young students grapple with comparison problems comes from accounts from teachers of data analysis in the elementary grades (Konold & Higgins, 2003). Similarly, all the problems in the middle-school materials developed by Cobb, McClain, and Gravemeijer involve group comparison (Cobb, 1999; Cobb, McClain, & Gravemeijer, 2003). As Watson and Moritz (1999) pointed out, some of the benefits of comparison contexts are undoubtedly related to their being more interesting and allowing students to see more clearly why the question matters and why averages might be useful. But in addition, we expect that in a comparison situation, students can more easily view averages of the individual groups as summary measures of processes and can readily perceive the difference between those measures as some signal rising through the din of variability.

Conducting Experiments

Many educators have touted the benefits of students' collecting their own data (e.g., Cobb, 1993). Among the expected advantages are increased student interest and the rich source of information that students can draw on as they later analyze and reason about the data. There may be additional benefits to having students design and run simple, controlled experiments. One benefit derives from the fact that experimental setups involve group comparison. In addition, we speculate that data from experiments are easier than observational data to view as coming from a process. As experimenters, students take an active role in the process—for example, by fertilizing one group of plants and comparing their growth to that of an unfertilized group of plants. Even quite young students can understand the importance in such cases of treating both groups of plants the same in all other respects (Lehrer, Carpenter, Schauble, & Putz, 2000; Warren, Ballenger,

Ogonowski, Rosebery, & Hudicourt-Barnes, 2001). They then observe firsthand that not every plant in the fertilized group responds the same and that the effect of the fertilizer becomes evident, if at all, only when comparing the two groups. With observational data, students must reason backwards from observed differences to possible explanations for those differences, and their tendency in explaining the data is to offer different causal accounts for each individual value. With the experimental setup, students first see the process and then the data resulting from it, a difference in perspective that may help them focus on the class of causes that apply uniformly at the group, as opposed to the individual, level.

CONCLUSIONS

We fear that some readers will hear in our analysis and recommendations a call to abandon the teaching of noninferential exploratory methods of data analysis and to eschew data from other than well-defined samples. In fact, we believe that we should begin teaching informal methods of data analysis in the spirit of EDA to students at a young age. Moreover, we are not recommending that the teaching of data analysis should be grounded in, or necessarily headed toward, the technical question of drawing formal inferences from carefully constructed samples.

We agree with Tukey (1977) that we should not, as a rule, approach data with the knee-jerk desire to model them mathematically. Rather, our objective should be more general—to learn from them. For this purpose, being able to display data flexibly and in various ways can lead to interesting insights and hypotheses, some of which we may then choose to model more formally (Cleveland, 1993). It is this sensible approach to the general enterprise—not only to *how* but also to *why* we collect and explore data—that we believe is most important to convey to students in early introductions to statistics.

It is important that we keep in mind, however, that most of us who regularly use exploratory methods of data analysis have strong backgrounds in inferential methods. When we approach data exploration with fewer assumptions, we often set aside, for the moment, much of the power of the mathematical models of statistics. But to play data detective, we have a host of tools and experiences to draw on, many of which stem from our knowledge of the mathematical models of statistics. As Cleveland (1993) observes, "Tools matter (p. 1)." The tools that he was referring to were methods of displaying data. We would add that underlying the skillful use of such graphical tools is the skillful use of conceptual ones, which matter even more.

Our references to the pioneering work of Quetelet were meant to point out that the early users of means did not regard them simply as ways to describe centers of distributions, which is how some today (misleadingly) characterize them. Recent histories of the development of statistics (Hacking, 1990; Porter, 1986; Stigler, 1986) portray the early innovators of statistics as struggling from the beginning with issues of interpretation. In this regard, Quetelet's idea of the "average man" was a way to take the interpretation of a mean as a "true value" of repeated measures and bootstrap it to a new domain—measurements of individuals—for which the mean did not initially make much intuitive sense. We believe that learning to reason about data requires students to grapple with the same sorts of interpretation issues; in the process, they need to develop conceptual (not necessarily mathematical) models of data that can guide their explorations. The idea of data as signal and noise, physically embodied in the workings of the Galton Board (see Biehler, 1994), is perhaps the most fundamental conceptual model for reasoning statistically. Future research should help us learn how the idea develops and how we can foster that development in our students.

NOTES

1. As George Cobb (1993) remarked, "If one could superimpose maps of the routes taken by all elementary books, the resulting picture would look much like a time-lapse night photograph of car taillights all moving along the same busy highway" (p. 53).
2. David Krantz (personal communication, December 13, 2001) shared with us his response to the question, "Do we really need the mean in descriptive stats?" which had appeared on a data analysis listserv. "I'm not very clear on what is meant by 'descriptive statistics.' To be honest, I don't think there is any such thing, except as a textbook heading to refer to the things that are introduced prior to consideration of sampling distributions. Any description must have a purpose if it is to be useful—it is supposed to convey something real. The line between 'mere description' and suggesting some sort of inference is very fuzzy."
3. Many use the term *central tendency* as a synonym for *average* or *center.* When referring to central tendency in this article, we have in mind the particular definition specified here.
4. Adopting this perspective, we will generally refer to *processes* rather than to *populations,* to *signals* or *central tendencies of processes* rather than to *population parameters,* and to *estimates of signals* rather than to *sample statistics.* We use the term *process* to refer both to processes that remain relatively stable over time as well as to stochastic processes, which can change quickly over time.
5. However, Frick (1998) argues that the difference between processes and populations is more than terminology, claiming that the tension between theoretical descriptions of random sampling and what we actually do in practice could be resolved if we thought explicitly of sampling from processes rather than from populations.
6. The maximum score on the reading component was 500, and the standard deviation was 50.
7. See Bakker (2001) for a review of the historical origins of various types of averages and a discussion of parallels between these ideas and the development of student thinking.
8. There are good grounds for considering the idea of mean as balance point as an interpretation. This interpretation figures centrally in mechanics, where the mean is a measure of center of mass. But in the statistics texts that we examined, the idea of mean as balance point seemed to be used solely as a way to visualize the location of the mean in a distribution of values and not as an interpretation as we have defined it.
9. We have to be careful using this logic. For example, mean income would be a different, and probably better, indicator of the power of the economic system to take care of its citizens if the wealth were in fact distributed equally.

10. Of course, both differences may reflect both *nature* and *nurture*.
11. It is possible that genetic differences may also (or instead) be reflected by differences in variability in the groups. Thinking about such differences, however, also requires thinking about some sort of measure (e.g., the standard deviation or the interquartile range) as a signal reflecting the typical variability in a group.
12. However, we should note that in the Bright and Friel (1998) study cited earlier, the two distributions were non-overlapping, yet students did not use averages to compare them.
13. For several good examples of activities written around such processes, see Erickson (2000).

REFERENCES

American Association for the Advancement of Science (AAAS). (1989). *Science for all Americans*. Washington, D.C.: American Association for the Advancement of Science (AAAS).

Bakker, A. (2001). Historical and didactical phenomenology of the average values. In P. Radelet-de Grave (Ed.), *Proceedings of the Conference on History and Epistemology in Mathematical Education* (Vol. 1, pp. 91–106). Louvain-la-Neuve and Leuven, Belgium: Catholic Universities of Louvain-la-Neuve and Leuven.

Biehler, R. (1989). Educational perspectives on exploratory data analysis. In R. Morris (Ed.), *Studies in mathematics education* (Vol. 7, pp. 185–201). Paris: UNESCO.

Biehler, R. (1994). Probabilistic thinking, statistical reasoning, and the search for causes—Do we need a probabilistic revolution after we have taught data analysis? In J. Garfield (Ed.), *Research papers from ICOTS 4* (pp. 20–37). Minneapolis: University of Minnesota.

Biehler, R. (1997). Students' difficulties in practicing computer-supported data analysis: Some hypothetical generalizations from results of two exploratory studies. In J. B. Garfield & G. Burrill (Eds.), *Research on the role of technology in teaching and learning statistics: Proceedings of the 1996 IASE Round Table Conference* (pp. 169–190). Voorburg, The Netherlands: International Statistical Institute.

Bright, G. W., & Friel, S. N. (1998). Helping students interpret data. In Lajoie, S. P. (Ed.), *Reflections on statistics: Learning, teaching, and assessment in grades K–12* (pp. 63–88). Mahwah, NJ: Erlbaum.

Cleveland, W. S. (1993). *Visualizing data*. Summit, NJ: Hobart Press.

Cobb, G. (1993). Reconsidering statistics education: A National Science Foundation conference [Electronic version]. *Journal of Statistics Education, 1*(1), Article 02.

Cobb, P. (1999). Individual and collective mathematical development: The case of statistical data analysis. *Mathematical Thinking and Learning, 1*(1), 5–43.

Cobb, P., McClain, K., & Gravemeijer, K. (2003). Learning about statistical covariation. *Cognition and Instruction, 21*, 1-78.

Cortina, J., Saldanha, L., & Thompson, P. (1999). Multiplicative conceptions of the arithmetic mean. In F. Hitt & M. Santos (Eds.), *Proceedings of the 21st Meeting of the North American Chapter of the International Group of the Psychology of Mathematics Education* (Vol. 2, pp. 466–472), Cuernavaca, Mexico: Centro de Investigaci6n y de Estudios Avanzados.

Donahue, P. L., Voelkl, K. E., Campbell, J. R., & Mazzeo, J. (1999) *NAEP 1998 Reading Report Card for the Nation and the States*. Document No. NCES 1999-500. Washington, DC: National Center for Educational Statistics, U.S. Department of Education. Available at http://nces.ed.gov/pubsearch/pubsinfo.asp?pubid=1999500

Erickson, T. (2000). *Data in depth: Exploring mathematics with Fathom*. Emeryville, CA: Key Curriculum Press.

Feldman, A., Konold, C., & Coulter, R. (2000). *Network science, a decade later: The Internet and classroom learning*. Mahwah, NJ: Erlbaum.

Freund, R. J., & Wilson, W. J. (1997). *Statistical methods*. Boston: Academic Press.

Frick, R. W. (1998). Interpreting statistical testing: Process and propensity, not population and random sampling. *Behavior Research Methods, Instruments, & Computers, 30*(3), 527–535.

Gal, I., Rothschild, K., & Wagner, D. A. (1990). *Statistical concepts and statistical reasoning in school children: Convergence or divergence.* Paper presented at the annual meeting of the American Educational Research Association, Boston, MA.

Gordon F. S., & Gordon S. P. (1992). *Statistics for the twenty-first century* (MAA Notes, no. 26). Washington, D.C.: Mathematical Association of America.

Gould, S. J. (1996). *Full house.* New York: Harmony Books.

Hacking, I. (1990). *The taming of chance.* Cambridge, UK: Cambridge University Press.

Hancock, C., Kaput, J. J., & Goldsmith, L. T. (1992). Authentic inquiry with data: Critical barriers to classroom implementation. *Educational Psychologist, 27*(3), 337–364.

Konold, C. (2002). Teaching concepts rather than conventions. *New England Journal of Mathematics, 34*(2), 69–81.

Konold, C., & Garfield, J. (1992). *Statistical reasoning assessment: Intuitive thinking.* Unpublished Manuscript. Amherst: University of Massachusetts.

Konold, C., & Higgins, T. (2003). Reasoning about data. In J. Kilpatrick, W. G. Martin, & D. E. Schifter (Eds.), *A research companion to principles and standards for school mathematics* (pp.193-215). Reston, VA: National Council of Teachers of Mathematics (NCTM).

Konold, C., Pollatsek, A., Well, A., & Gagnon, A. (1997). Students analyzing data: Research of critical barriers. In J. B. Garfield & G. Burrill (Eds.), *Research on the role of technology in teaching and learning statistics: Proceedings of the 1996 IASE Round Table Conference* (pp. 151–167). Voorburg, The Netherlands: International Statistical Institute.

Konold, C., Robinson, A., Khalil, K., Pollatsek, A., Well, A., Wing, R., & Mayr, S. (2002). Students' use of modal clumps to summarize data. Paper presented at *the Sixth International Conference on Teaching Statistics*, Cape Town, South Africa.

Lehrer, R., Carpenter, S., Schauble, L., & Putz, A. (2000). Designing classrooms that support inquiry. In J. Minstrell & E. V. Zee (Eds.), *Inquiring into inquiry learning and teaching in science* (pp. 80–99). Washington, DC: AAAS.

Lehrer, R., Schauble, L., Carpenter, S., & Penner, D. (2000). The inter-related development of inscriptions and conceptual understanding. In P. Cobb, E. Yackel, & K. McClain (Eds.), *Symbolizing and communicating in mathematics classrooms: Perspectives on discourse, tools, and instructional design* (pp. 325–360). Mahwah, NJ: Erlbaum.

Lehrer, R., Schauble, L., Strom, D., & Pligge, M. (2001). Similarity of form and substance: Modeling material kind. In D. Klahr & S. Carver (Eds.), *Cognition and instruction: 25 years of progress.* (pp. 39–74). Mahwah, NJ: Erlbaum.

Medin, D. L., & Schaffer, M. M. (1978). Context theory of classification learning. *Psychological Review, 85,* 207–238.

Mokros, J., & Russell, S. J. (1995). Children's concepts of average and representativeness. *Journal for Research in Mathematics Education, 26,* 20–39.

Moore, D. S. (1990). Uncertainty. In L. A. Steen, (Ed.), *On the shoulders of giants: New approaches to numeracy* (pp. 95–137). Washington, DC: National Academy Press.

National Council of Teachers of Mathematics (NCTM). (1989). *Curriculum and evaluation standards for school mathematics.* Reston, VA: NCTM.

National Council of Teachers of Mathematics (NCTM). (2000). *Principles and standards for school mathematics.* Reston, VA: NCTM.

National Research Council. (1996). *National science education standards.* Washington, DC: National Academy Press.

Noss, R., Pozzi, S., & Hoyles, C. (1999). Touching epistemologies: Meanings of average and variation in nursing practice. *Educational Studies in Mathematics, 40,* 25–51.

Plackett, R. L. (1970). The principle of the arithmetic mean. In E. S. Pearson and M. G. Kendall (Eds.), *Studies in the history of statistics and probability* (pp. 121–126). London: Charles Griffen.

Pollatsek, A., Lima, S., & Well, A. (1981). Concept or computation: Students' misconceptions of the mean. *Educational Studies in Mathematics, 12,* 191–204.

Porter, T. M. (1986). *The rise of statistical thinking, 1820-1900.* Princeton, NJ: Princeton University Press.

Quetelet, M. A. (1842). *A treatise on man and the development of his faculties.* Edinburgh, Scotland: William and Robert Chambers.

Rosch, E., & Mervis, C. B. (1975). Family resemblances: Studies in the internal structure of categories. *Cognitive Psychology, 8,* 382–439.

Rubin, A., Mokros, J., & Friel S. (1996). *Data: Kids, cats, and ads. Investigations in number, data, and space.* Palo Alto, CA: Seymour.

Scheaffer, R. (1991). The ASA-NCTM Quantitative Literacy Program: An overview. In D. Vere-Jones (Ed.), *Proceedings of the Third International Conference on Teaching Statistics* (pp. 45–49). Voorburg, The Netherlands: International Statistical Institute Publications.

Schwartzman, S. (1994). *The words of mathematics: An etymological dictionary of math terms used in English.* Washington, DC: Mathematical Association of America.

Shaughnessy, J. M., Watson, J., Moritz, J., & Reading, C. (1999). School mathematics students' acknowledgment of statistical variation. Paper presented at *the 77th annual meeting of the National Council of Teachers of Mathematics*, San Francisco.

Smith, G. (1998). Learning statistics by doing statistics [Electronic version]. *Journal of Statistics Education, 6*(3), Article 04.

Stigler, S. M. (1986). *The history of statistics: The measurement of uncertainty before 1900.* Cambridge, MA: Harvard University Press.

Stigler, S. M. (1999). *Statistics on the table: The history of statistical concepts and methods.* Cambridge, MA: Harvard University Press.

Strauss, S., & Bichler, E. (1988). The development of children's concepts of the arithmetic average. *Journal for Research in Mathematics Education, 19,* 64–80.

Tukey, J. W. (1977). *Exploratory data analysis.* Reading, MA: Addison-Wesley.

Warren, B., Ballenger, C., Ogonowski, M., Rosebery, A., & Hudicourt-Barnes, J. (2001). Rethinking diversity in learning science: The logic of everyday languages. *Journal of Research in Science Teaching, 38,* 1–24.

Watson, J. M., & Moritz, J. B. (1999). The beginning of statistical inference: Comparing two data sets. *Educational Studies in Mathematics, 37,* 145–168.

Watson, J. M., & Moritz, J. B. (2000). The longitudinal development of understanding of average. *Mathematical Thinking and Learning, 2*(1), 9–48.

Wild, C. J., & Pfannkuch, M. (1999). Statistical thinking in empirical enquiry. *International Statistical Review, 67*(3), 223–265.

Wilensky, U. (1997). What is normal anyway? Therapy for epistemological anxiety. *Educational Studies in Mathematics, 33,* 171–202.

Chapter 9

REASONING ABOUT VARIATION

Chris Reading[1] and J. Michael Shaughnessy[2]
University of New England, Australia [1], and Portland State University, USA[2]

OVERVIEW

"Variation is the reason why people have had to develop sophisticated statistical methods to filter out any messages in data from the surrounding noise" (Wild & Pfannkuch, 1999, p. 236). Both variation, as a concept, and reasoning, as a process, are central to the study of statistics and as such warrant attention from both researchers and educators. This discussion of some recent research attempts to highlight the importance of reasoning about variation. Evolving models of cognitive development in statistical reasoning have been discussed earlier in this book (Chapter 5). The focus in this chapter is on some specific aspects of reasoning about variation.

After discussing the nature of variation and its role in the study of statistics, we will introduce some relevant aspects of statistics education. The purpose of the chapter is twofold: first, a review of recent literature concerned, directly or indirectly, with variation; and second, the details of one recent study that investigates reasoning about variation in a sampling situation for students aged 9 to 18. In conclusion, implications from this research for both curriculum development and teaching practice are outlined.

NATURE OF VARIATION

Perusal of recent research literature suggests that the terms *variation* and *variability* are at times used interchangeably. Although some researchers do hold this view, a closer investigation of terms unfolds a distinction. A survey of various dictionaries demonstrated that *variation* is a noun used to describe the act of varying or changing condition, and *variability* is one noun form of the adjective *variable*, meaning that something is apt or liable to vary or change (see for example Pearsall

D. Ben-Zvi and J. Garfield (eds.),
The Challenge of Developing Statistical Literacy, Reasoning and Thinking, 201–226.

& Trumble, 2001, p. 1598). In the world of educators and researchers these two terms have come to have more specific usages.

In this chapter these two terms will not be treated as interchangeable, although some of the referenced research uses them interchangeably. The term *variability* will be taken to mean the characteristic of the entity that is observable, and the term *variation* to mean the describing or measuring of that characteristic. Consequently, the following discourse, relating to "reasoning about variation," will deal with the cognitive processes involved in describing the observed phenomena in situations that exhibit variability, or the propensity for change. Moore (1997) points out that both variability and the measuring and modeling of that variability are important. It is in this measuring and modeling that variation will become the focus of this chapter.

Patterns and relationships between variables in data indicate variability. The search for the source of such variability may result in explanations being found for the variability, or it may result in the need to estimate the extent of unexplained, or random, variation. Wild and Pfannkuch (1999, pp. 240–242) discuss the modeling of variation and the importance of considering both explained and unexplained variation when exploring data. They point out that while many will agree with those who view all variation as "caused" those who believe in "uncaused" variation should consider the possibility that unexplained variation may be due to sources as yet undiscovered. This leads one to question the notion of unexplained, or random, variation. If the concept of random variation is puzzling even to statisticians and researchers, how much more puzzling must it be to those just embarking on their data handling careers?

Possible confusion over the nature of variation may well influence the approach taken to data handling and to a description of variability. How do people react to variation in data? There appear to be three broad categories of reaction: those who ignore variation as if it does not exist; those who investigate existing patterns of variation and work to fit in with them; and those who try to change the pattern of variation to something more desirable (Wild & Pfannkuch, 1999, p. 236). The latter is possible only if manipulable causes of variation can be isolated. Isolation and modeling of variation allows prediction, explanation and control, as well as questioning of why variation occurs, resulting in looking for causes. In fact, students who are presented with numbers that vary will often seek a "real" explanation for why they are not the same without being too concerned about actually describing the variation. This is especially so when students have some contextual knowledge about a situation. Even after some instruction, the randomness ideas are still much weaker in students than the impulse to postulate causes (Wild & Pfannkuch, 1999, p. 238).

THE ROLE OF VARIATION IN STATISTICS

Why is reasoning about variation so important? Variation, or variability, is featured in the American Statistics Association definitions of statistical thinking, and so any serious discussion on statistical thinking must examine the role of variation. Meletiou (2002) cites many references that discuss the importance of variation and the role of variation in statistical reasoning. Two of these, Moore (1997) and Wild and Pfannkuch (1999), are critical to appreciating the role of variation in statistics. Moore emphasized the omnipresence of variability, and the importance of measuring and modeling variability, while Wild and Pfannkuch put variation at the heart of their model of statistical thinking when consideration of variation emerged from interviews with statisticians and students, as one of the five types of fundamental statistical thinking. There are four aspects of variation to consider: noticing and acknowledging, measuring and modeling (for the purposes of prediction, explanation or control), explaining and dealing with, and developing investigative strategies in relation to variation (Wild & Pfannkuch, 1999, pp. 226–227). We, the authors, also suggest two important aspects of variation—describing and representing—that need to be considered. Much of the uncertainty that needs to be dealt with when thinking statistically stems from omnipresent variation and from these six aspects of variation that form an important foundation for statistical thinking.

RELEVANT ASPECTS OF STATISTICS EDUCATION

What is missing in the study of statistics? Both in curriculum design, and statistics education research, variation has not been given the attention that is warranted given the general acknowledgment of the importance of variation to statistics. Two of the principal statistical concepts in the teaching and learning of statistics, or data handling as it appears in curricula, are measures of central tendency and measures of dispersion. The latter is often referred to as variability or spread. Whenever statistics are discussed there is an overemphasis on the measurement of central tendency and lack of attention to the measurement of variability. Research shows that there is a conceptual gap among students in the concept of variability (Shaughnessy, 1997, p. 3) that needs to be addressed both in the area of curriculum design and in statistics education research.

Since statistics is a recent addition to the mainstream school mathematics curriculum (at least in the United States, for example, National Council of Teachers of Mathematics [NCTM], 1989, 2000), one might suspect some gaps in student learning of statistical concepts. There is ample evidence from the 1996 National Assessment of Educational Progress (NAEP) in the United States data that students' have weak conceptions of measures of central tendency, and even weaker conceptions of the role and importance of variation and spread in statistical thinking (Zawojewski & Shaughnessy, 2000). Students' current lack of understanding of the nature of variability in data and chance may be partly due to the lack of emphasis of variation in our traditional school mathematics curriculum and textbooks. It may also be partly due to teachers' inexperience in teaching statistical concepts.

In the United States, for example, most school mathematics textbooks do not encourage students to identify potential sources of variation in data sets. Neither do they provide opportunities for students to visualize variability, nor to investigate ways of measuring variability, nor what such measures actually mean. Teachers and students may know the procedure for computing standard deviation; but they may be unable to explain what it means, or why or when it is a good measure for expected variation (Green, 1993). Exceptions to this trend of a lack of exploration of sources of variation can be found in some of the *Quantitative Literacy* materials and in the *Data Driven Mathematics* series, both written by teams of classroom teachers and statistics educators (Landwehr & Watkins, 1985, 1995; Scheaffer et al., 1999).

The variety of models for centers that have been researched and used in teaching students (Russell & Mokros, 1996) is not matched by a correspondingly rich array of models for students' conceptions of spread or variability. Shaughnessy (1997) speculates on the reasons for this absence of research about variation. One reason may be that research often mirrors the emphases in curricular materials which, to date, has lacked a variation focus.

Another reason may be that statisticians have traditionally been very enamored with standard deviation as *the* measure of spread or variability; teachers and curriculum developers may tend to avoid dealing with spread because they feel they would have to introduce standard deviation, which is computationally complex and perhaps difficult to motivate in school mathematics. Still another reason may be that centers, or averages, are often used for prediction; and comparison and the incorporation of spreads, or variation, into the process only confounds the issue. People are comfortable predicting from centers—it feels like firm ground compared to variability issues. Finally, the whole concept of variability may just be outside of many people's comfort zone, perhaps even outside their zone of belief.

If this imbalance in research focus is to be addressed, then more research on reasoning about variation needs to be undertaken to assist educators to better equip future students in measuring and modeling variability as they reason about variation. The focus of this chapter is research involving students aged 9 to 18 years. These students are living in a world where from an early age they are surrounded by variability in their everyday life. But when it comes to collecting, representing, reducing, and interpreting data, all too often their learning experiences are focused on the central tendency of data and lack opportunities to describe the variation that occurs. Educators need to modify learning experiences so that students can move comfortably from identifying variability; to describing, representing, and sifting out causes for; and finally, to measuring variation. The research described in the following sections discusses aspects of students' reasoning that may be used to inform the evolution of both curriculum and teaching practice. First, some of the research on reasoning about variation is discussed. Next we review some research on students' understanding of samples and sampling in general. Finally, we investigate recent research on the more specific area of students' reasoning about variation in a sampling environment.

RECENT RESEARCH INTO REASONING ABOUT VARIATION

Recently, research involving reasoning about variation in a diverse range of statistical situations has emerged. This research reflects some changing expectations of students. The research includes investigations into the role of variation in correlation and regression (Nicholson, 1999), graphical representation (Meletiou & Lee, 2002), probability sample space (Shaughnessy & Ciancetta, 2002), comparison of data sets (Watson & Moritz, 1999; Watson, 2001) and chance, data and graphs, and sampling situations (Watson & Kelly, 2002a, 2002b, 2002c).

Some researchers are now developing hierarchies to describe various aspects of variation and its understanding. Watson, Kelly, Callingham, and Shaughnessy (2003) investigated three contexts for variation—chance, data, and sampling—and described four levels of reasoning: *prerequisites of variation, partial recognition of variation, applications of variation,* and *critical aspects of variation.* The description of each level is based on various aspects of the three types of variation. Of most interest to the present discussion is the shift from Level 2 (partial recognition of variation) to Level 3 (applications of variation). Responses at Level 2 do not reflect an understanding of chance and variation, with students likely to make flawed interpretations. It is only responses at Level 3, or above, that demonstrate a focus on appropriate aspects of the concepts while ignoring irrelevant aspects.

A variety of research situations suggest that reasoning about variation could be more of a natural instinct than is catered to in the present structure of learning environments. When students were responding to an open-ended data reduction question, Reading and Pegg (1996, p. 190) found that although many students took the expected option of reducing data based on measures of central tendency, nearly a quarter of them preferred reductions based on measures of dispersion. When designing computer minitools, which had "bar(s)" for partitioning the data, McClain, Cobb, and Gravemeijer (2000, p. 181) anticipated that students would place one bar at the mean of the data set. Instead, some students adapted the partitioning feature to determine which of two data sets had more consistent values, suggesting a higher regard for the spread of the data than the central tendency.

More recently, some researchers have focused on investigating reasoning about variation in sampling situations. But before discussing this research, we consider some recent findings on conceptions of sampling.

RECENT RESEARCH INTO CONCEPTIONS OF SAMPLING

Statistical analysis often relies on studying a *part* (sample) to gain information about the *whole* (population). Sampling is the process of selecting this part, or sample, of the population (Moore & McCabe, 2003, p. 225) to provide the reliable and relevant information, and as such is a core concept in statistics. Some research studies have identified hierarchies of student thinking on sampling tasks in probability settings. Watson, Collis, and Moritz (1997) used tasks with grades 3 to 9 involving dice, drawing names from a hat, and Piagetian marble tasks; then they identified hierarchies of student reasoning about those tasks, based on the theoretical underpinnings of the structure of observed learning outcomes (SOLO) taxonomy (Biggs & Collis, 1991). These results were later extended to grade 11 (Watson & Moritz, 1998). In a similar way, levels of justification have been found in students' reasoning in a sampling task (Jones et al., 1999; Reading, 1999; Shaughnessy, Watson, Moritz, & Reading 1999; Torok & Watson, 2000; Zawojewski & Shaughnessy, 2000), while other researchers have focused on students' perceptions of samples and sampling in data handling settings (Wagner & Gal, 1991; Rubin, Bruce, & Tenney, 1991; Jacobs, 1997, 1999; Watson & Moritz, 2000).

Jacobs (1999) found that while some children were aware of potential bias issues in surveys, favoring random sampling, other children preferred a quasi-stratified random sampling, preoccupied with issues of fairness. Reading (2001) found similar results among secondary students, discussing data collection, who they tended to create selection criteria based on variables that they perceived would improve the range of responses in their sample. Thus, in constructing a "sample," students demonstrated a desire to make absolutely sure that anything *can* happen. Reading's and Jacob's results may provide more evidence of what has been called the "equi-probability" bias (Lecoutre, 1994), that all things *can* happen, and so they all should have an equal chance of happening. These results are also reminiscent of Konold's "outcome approach" (Konold, 1989; Konold, Pollatsek, Well, Lohmeier, & Lipson, 1993).

The previous research on students' understanding of sampling suggests that there may be conceptual ties between students' understanding of variation in samples, and students' understanding of sample space in a probability experiment. The probability question is: What is the range of all possible outcomes, and which ones are more likely to occur than others (i.e., what is the sample space)? The statistical question is: If we repeat a probability experiment many times, what sort of variation in outcomes do we observe, and what is the range of the more likely outcomes (i.e., what interval captures most of our trials)?

RECENT RESEARCH INTO REASONING ABOUT VARIATION IN SAMPLING SITUATIONS

Sampling at random attempts to avoid the biases that may occur when a sample is drawn. Such sampling is based on the principle that each unit in the population has an equal chance of being selected in the sample (Moore & McCabe, 2003, pp. 225–227). Variation occurs in all sampling situations, but the equal likelihood principle on which random sampling is based allows calculation of the likely size of errors that occur (Wild & Seber, 2000, pp. 6–9). This sampling variability dictates that the value of the parameter of interest, for example the number of a specific color of lollies out of a sample chosen, will vary with repeated random samplings (Moore & McCabe, 2003, pp. 260–261). Given that such variation occurs, two important issues arise—the size of the sample and how many samples should be taken.

Several problems associated with reasoning about variation in sampling situations have been identified. First, the long-held notion that small samples should provide reliable representations of the parent population from which they were drawn, which leads to estimates based on what is commonly called the representativeness heuristic (Tversky & Kahneman, 1974) continues to be supported by more recent research (Shaughnessy 1992, 1997; Watson & Moritz 2000; Watson 2000). This line of research suggests that people may focus more on issues of "centers or averages or population proportions" than on issues of spread or variation when making estimates for the likelihood of chance events, and that they have minimal conceptions and weak intuitions of how outcomes are distributed "around a center" in a binomial distribution.

Second, although a sample is generally considered heterogeneous, in certain contexts it may have homogeneous connotations, thus influencing notions of variation in the sample. When asked what the word sample meant to them, some students (Grades 3, 6, and 9) said "a little bit," like a shampoo sample, or a taste of food, or a blood sample (Watson & Moritz, 2000). Jacobs (1997) reported similar findings. Intuitive notions of statistical variation would be unlikely to arise in connection with such samples, where the issue of variation in a "sample" could actually be troublesome.

Third, a tension has been found to exist in secondary students between accounting for variability and a desire for representativeness in samples (Rubin et al., 1991). Acknowledging the possibility of too much variation conflicts with whether the sample really is representative of a population. Of course, this question permeates many real applied statistics situations: When do we have enough evidence to predict from a sample (is it truly representative)? Could two different samples really be from different populations (is there too much variance across groups)? This tension between representativeness and variability always exists in a sampling situation and needs to be carefully considered.

Finally, when subjects are given a question that involves estimating the likelihood of a single event, they may actually superimpose a sampling setting on the question where none was there to begin with, in order to establish a "center" from which to predict (Shaughnessy, 1997).

In various protocols based around sampling situations, some researchers have identified students' reasoning about variation, both from analysis of students' descriptions of possible outcomes in the sampling situation and their explanations of why the values were chosen. In particular, Torok and Watson (2000) described a hierarchy of four levels of developing concepts of variations: *weak appreciation of variation* (Level A), *isolated appreciation of aspects of variation and clustering* (Level B), *inconsistent appreciation of variation and clustering* (Level C) and *good, consistent appreciation of variation and clustering* (Level D). These were based on responses to a variety of tasks including some situations with isolated random variation and others with real-world variation. During analysis of the tasks, two features emerged as important when differentiating between students. One was the acknowledgment of variation and description of clustering, and the other was the use of proportion. Similar levels were also identified by Watson et al. (2003) when they analyzed responses to a wider range of chance and data tasks and developed a hierarchy that included aspects of centrality, as well as aspects of spread.

In summary, prior research that provides information about students' conceptions of variation has predominantly come indirectly from investigations of students' understandings of sampling in either a probability experiment or in a data collection setting. Following are some of the principal findings:

- Students may be strongly attracted to averages or population proportions, and focus on them to the neglect of issues involving spread or variation.
- The issue of "fairness" in creating samples in a survey setting is a prominent one, especially for younger children. They wish to control variation, or to allocate variation evenly across groups.
- The word *sample* can have both heterogeneous (e.g., random stratified sample) and homogeneous (e.g., food, blood) connotations for students.
- Students may superimpose a sampling environment on a problem when none was there to begin with, in order to justify their thinking by representativeness. Also, there is a tension between variability and representativeness.
- Students' reasoning about variation may depend on an understanding of centering and clustering.

CONTEXT FOR A RECENT STUDY: VARIABILITY IN REPEATED SAMPLES

In a secondary analysis of the statistics items from the 1996 National Assessment of Educational Progress (NAEP), the predominance of single-value responses given by students to a sampling task was intriguing (Zawojewski & Shaughnessy, 2000). Students were told the number of red-, yellow-, and blue-colored gumballs in a bowl, and then asked how many red gumballs they would expect to get in a handful of 10. Grade 4 students consistently gave a single number for their response—5, 10, 3, and so forth, with only *one* student in a convenience sample of 232 students (from a possible 1,302 responses) actually giving a *range* of possible numbers for the number of red gumballs that would be pulled. That student wrote 4–6.

This suggests that Grade 4 students tend to give "point value" answers for sampling problems, and that they do not normally, in such a "test" situation, give consideration to a "range of likely values" in their responses. This is troubling because it suggests that students do not recognize the role that variability plays in a sampling task. However, point-value responses do mirror the prototypical responses to the most frequent types of questions about data and chance posed in classrooms and textbooks, namely: "What is the probability that …?" Probability questions just beg students to provide a point-value response and thus tend to mask the issue of the variation that can occur if experiments are repeated.

What would happen if the sampling question were asked in a different way? What would students say about repeated samples of 10 gumballs? How many reds would they expect? Would they expect the same number every time? Or, would students acknowledge that variation exists in the number of reds in repeated handfuls? What sorts of "likely ranges" for the numbers of reds would students select? These questions gave birth to what we have come to call the lollie task (in Australia, a "lollie" is a hard sweet; in the United States, we called this the candy task): a sampling task involved pulling lollies from a bowl with a known mixture.

The lollie tasks were given in several written forms to over 400 students in Grades 4–6, in Australia, New Zealand, and the United States and to over 700 secondary students in Grades 9–12 in the United States. In one version, students were presented with a mixture of 100 lollies—50 red, 30 yellow, and 20 blue—and were asked how many reds they would expect if a handful of 10 lollies were pulled out. Then they were asked, "If this experiment were repeated six times, what would the likely numbers of reds be?" Students were told that after each sample pull, the lollies were returned to the bowl and thoroughly mixed up again. Six repetitions of the sampling were chosen, for two reasons. First, to be small enough that students would see the task as not too daunting and second, large enough that students had an opportunity to demonstrate variability that they considered might occur.

Some clear categories of student reasoning emerged in the lollie task. Students often make predictions that we might characterize as "wide" or "narrow," or "high" or "low" from what would be expected according to probability and statistics theory. For example, when given a 50% red mixture, some students expect a very wide range of numbers of reds to occur in 6 repeated samples of 10, such as 0,1,4,7,9,10 reds. While these students acknowledged that variability exists, they felt that *all* possible outcomes for the number of reds should show up. These students may believe in an equi-probability model for this sampling problem, that all numbers of reds have the same chance of occurring. Many of the younger students questioned indicated that everything should have a "fair" chance of occurring, as Jacobs has found (1997, 1999). Still other students' reasoning suggested that they think "anything can happen" in a chance experiment. Students who reasoned in any of these ways gave "wide" ranges for the numbers of reds in the repeated samples.

A surprising number of students predicted 5,5,5,5,5,5, suggesting no variability at all in the sampling results. This tendency is stronger among older mathematics students, who like to predict "what should happen" every time. This indicates that some students tend to think in terms of point values rather than a range of likely values, even when they are directed to provide a range of likely values.

Other students predicted high for the numbers of reds in each attempt, predicting numbers like 6,7,8,8,7,9, and reasoned "because there are a lot of reds in that mixture." There were students who in fact did recognize that outcomes in the lollie experiment are more likely to be distributed symmetrically around 5, such as "from 3 to 7 reds." However, in these initial investigations, less than 30% of all the students surveyed or interviewed were able to successfully integrate the roles of both centers and spreads in sampling scenarios like the lollie sampling problem (Shaughnessy et al., 1999).

In summary:

- There was a tendency for students to be preoccupied with the large number of reds in the population, rather than the proportion of reds, or a likely range of reds that would be pulled in a sample. This led students to overestimate the numbers of reds in samples.
- There was a tendency for some students to go wide in their predictions of the range of the numbers that would be pulled in repeated samples. This may be due to thinking that reflects aspects of the outcome approach, or to beliefs in equi-probability.
- A proportion of students changed their minds and produced a more normative response to the lollie problem after actually doing the experiment and seeing the variation. Thus, there is potential for student learning in such a sampling task, although some of the students did not change their minds even when confronted with evidence that conflicted with their original prediction.
- There was evidence of potential interference at higher grades (11–12) from recent experiences with calculating probabilities. These older students were more often the ones who would predict 5,5,5,5,5,5. It is our conjecture that since these students have normally been asked questions in the form of "what is the probability that" or "what is the most likely outcome," they do not recognize a situation that calls for dealing with a spread of outcomes.

Students' reasoning about variation can be investigated in a variety of situations; Torok and Watson (2002, p. 152) consider it important to include situations with isolated random variation as well as situations with real-world variation. In this chapter we will focus on isolated random variation in order to build on, and deepen, our understanding of students' reasoning about variation in a sampling environment. In the following sections of this chapter, we will present findings from one of the studies in the lollie research that conducted a qualitative analysis of explanations given by students. The study consisted of interviews based on the lollie task, conducted in Australian schools, designed to address the following research questions: What aspects of reasoning in a sampling task indicate consideration of variation? Is there a hierarchy of reasoning about variation in responses to a sampling situation?

METHODOLOGY

Twelve students, six from primary school and six from secondary school, were interviewed to expand on explanations given in written surveys, previously administered to other students. The primary students were from Grades 4 (Millie), 5 (Kate, Donna), and 6 (Jess, Alice, Tim); secondary students were from Grades 9 (Jane, Prue, Brad) and 12 (Max, Rick, Sue). Although these names are fictitious, they are used to refer to specific students during the discussions. The schools were asked to suggest students who were reasonably articulate and had average mathematical ability. The interviews were audiotaped, and students were given response forms on which to record as they were being interviewed. All interviews were transcribed. Students were encouraged to articulate all responses, but could choose whether to record aspects of the response.

Students were asked to respond to two different sampling situations: a mixture with 50 red, 30 blue, and 20 yellow and another with 70 red, 10 blue, and 20 yellow. A bowl containing the correct, relevant proportions of wrapped lollies was placed in full view. Students were told that the lollies were well mixed and the sampling was blind. In each case the students were asked how many red lollies could be expected in a handful of 10 lollies. They were then asked to report on the number of reds that would be drawn by six people in a handful of 10 lollies, with the lollies being returned to the bowl after each draw and thoroughly remixed.

The interviews were conducted by the researchers in the school setting familiar to the students. The student response form (condensed) for the 50 red (50R) situation is shown in Figure 1 as question 1. A suitably adapted question 2 was used for the 70 red (70R) situation. The interview protocol followed the wording of the student response sheet, with prompting-style encouragement given to students who hesitated when responding to the "why" questions.

Initially students were asked how many reds could be expected and whether that would happen every time (parts 1A, 2A). Then responses to the sampling task were sought in three different forms: LIST (parts 1B, 2B), CHOICE (parts 1C, 2C) and RANGE (parts 1D, 2D). Students were also asked why they gave the responses they did and then given the chance to alter their responses after having actually drawn 6 samples of 10 from the bowl.

Two conceptually difficult notions related to sampling were addressed in the extended questions for larger sample size (parts 1E, 2E—selecting 50 lollies instead of 10) and for increased repetitions (parts 1F, 2F—40 draws instead of 6). Taking a larger sample results in less variability in terms of providing accurate information about the population, and more repetitions of sampling can help to provide more detail about the sampling variability (Moore & McCabe, 2003, pp. 265–266). Both these notions were considered too difficult for primary students and hence were presented only to the secondary students.

Responses to question 1 and question 2 in Figure 1 were analyzed both qualitatively and quantitatively. Quantitatively, performance of the particular 12 students in this study is discussed in Reading and Shaughnessy (2000) based on a coding scheme developed in Shaughnessy et al. (1999). Qualitatively, the explanations given were arranged hierarchically depending on increasing appreciation of aspects of reasoning about variation, similar to those identified by Torok and Watson (2002). It is the results of this qualitative investigation that are reported here. Detailed case studies of four of the interviews, one in each of Grade 4 (Millie), Grade 6 (Jess), Grade 9 (Jane), and Grade 12 (Max), can be found in Reading and Shaughnessy (2000).

Student Response Form

1A) Suppose we have a bowl with 100 lollies in it. 20 are yellow, 50 are red, and 30 are blue.
Suppose you pick out 10 lollies.
How many reds do you expect to get? __
Would this happen every time? Why?

1B) Altogether six of you do this experiment.
What do you think is likely to occur for the numbers of red lollies that are written down? Please write them here.

_____, _____, _____, _____, _____, _____
Why are these likely numbers for the reds?

1C) Look at these possibilities that some students have written down for the numbers they thought likely. Which one of these lists do you think best describes what might happen? Circle it.
a) 5,9,7,6,8,7
b) 3,7,5,8,5,4
c) 5,5,5,5,5,5
d) 2,3,4,3,4,4
e) 7,7,7,7,7,7
f) 3,0,9,2,8,5
g) 10,10,10,10,10,10
Why do you think the list you chose best describes what might happen?

1D) Suppose that 6 students did the experiment—pulled out ten lollies from this bowl, wrote down the number of reds, put them back, mixed them up.
What do you think the numbers will most likely go from? From ____ (low) to ____ (high) number of reds.
Why do you think this?
**(After doing the experiment)
Would you make any changes to your answers in 1B–1D?
If so, write the changes here.

1E) Suppose that 6 students each pulled out 50 lollies from this bowl, wrote down the number of reds, put them back, mixed them up.
What do you think the numbers will most likely go from this time?
From ______ (low) to ______ (high) number of reds.
Why do you think this?

1F) Suppose that 40 students pulled out 10 lollies from the bowl, wrote down the number of reds, put them back, mixed them up. Can you describe what the numbers would be, what they'd look like?
Why do you think this?

Figure 1. Student Response Form (condensed).

RESULTS

As responses were analyzed they indicated important aspects of reasoning about variation, as acknowledged by Torok and Watson (2002) in their hierarchy. Two of these characteristics, one based on the description of the variation and the other looking for the cause of the variation, were considered important enough to warrant the development of two separate hierarchies in the present study. The *description hierarchy,* based around students' descriptions of the variation occurring, developed from aspects of students' responses such as *it is more spread out, there's not the same number each time* (Jess G6). The *causation hierarchy,* based around students' attempts to explain the source of the variation, developed from aspects of student responses such as *because there's heaps of red in there* (Jane G9). A categorization of student responses as giving causation (C) or description (D) or both (C&D) as part of the explanations is presented (Table 1) for both the 50R and 70R sampling situations. As trends were similar for primary and secondary students, the data were combined. Any question that does not have a total of 12 students coded indicates that the explanations given in the responses were absent or did not contain description of variation or a mention of causation.

Table 1. Categorization of Student Responses

		50 Red				70 Red		
		C	*D*	*C & D*		*C*	*D*	*C & D*
Every time?	1A	10	1	1	2A	6	0	0
LIST	1B	5	1	3	2B	6	3	1
CHOICE	1C	3	8	1	2C	4	4	1
RANGE	1D	5	3	1	2D	2	6	0

Although frequencies are too small for rigorous statistical analysis, trends can be observed. Discussions of causation were usually given for the explanation of whether the given answer will occur "every time" (1A, 2A) and when students were asked to LIST all outcomes (1B, 2B). For the CHOICE question (1C, 2C), a descriptive answer was more likely for the 50R situation; and a mixture of responses occurred in the 70R situation, which students were generally finding more difficult to deal with. For the RANGE question (see 1D, 2D), more descriptive-type explanations were given for the 70R situation and more causation-type explanations for the 50R situation. This was the only question where there were noticeable differences between primary and secondary student responses, with causation explanations for the 50R situational being from primary students and all but one of the descriptive explanations for the 70R situation being from secondary students. Given that the 70R situation was generally more difficult for students to explain, it is understandable that mainly secondary students chose to describe the variation while primary students took the option to look for cause.

These results suggest that the form (LIST, CHOICE, RANGE) in which the question is asked may influence whether a student chooses to describe the variation or look for a cause. Also, student responses indicated that the 50R situation, dealing with a more familiar proportion of 50%, was conceptually easier to understand than the 70R situation.

The details of the two hierarchies—description (coded D1 to D4) and causation (coded C1 to C4)—follow, together with typical responses from the students. When students' responses are quoted directly, either in part or in full, they appear in *italics*.

Description Hierarchy

As identified in previous research with the sampling task (Shaughnessy et al. 1999), the responses given usually indicated a notion of reasonable spread; but the actual way that the spread was described varied considerably. The description hierarchy was developed based on increasing sophistication in the way that students referred to notions of spread. The interviews under consideration here indicated that at a less sophisticated level, two different approaches appear to be taken. Some students chose to concentrate on the middle values, while others were more preoccupied with extreme values. The more sophisticated explanations by the students gave consideration to both extreme values as well as what is occurring between them.

It should be noted that the responses to the more complex situations of larger sample size (questions 1E, 2E), and increased repetitions (1F, 2F) appeared to play a more significant role in the development of this hierarchy than the causation hierarchy. Students found it challenging trying to describe the variation in these more complex situations, and their explanations of the questions brought deeper insight into how they reasoned about variation.

D1—Concern with Either Middle Values or Extreme Values

In this sense "extremes" are used to indicate data items that are at the uppermost end or the lowest end of the data, while "middle values" are used to indicate those data items that are between the extremes. Typical of responses concerned mainly with the extremes are those that justify values selected by explaining why it was not possible to get more extreme values. Jess (G6) chose to explain why she had excluded certain values rather than why she had included the ones that she did, because *there's 50 red I don't think you could get one or two*. Such responses were most likely for the RANGE questions but not exclusively. For example, Rick (G12) expressed his desire to exclude certain extreme values when eliminating some CHOICE options for 50R, deciding that *a 0, 8, 9 or 10 are not likely*.

Typical of those responses indicating more concern with the middle values were those that explained why specific values were important and those demonstrating more concern about the relationship of the numbers to each other. Jess (G6) explained her LIST by saying that the numbers *need to be all mixed up but it is hard to explain,* showing specific concern for the middle values and the variety that was needed within those numbers. Similarly, Prue (G9) *wanted a lot of different numbers,* and Kate (G6) stated that *size doesn't matter just different.* On the other hand, Sue (G12) showed concern for the actual values of the middle numbers when explaining in the 70R LIST that she was wavering around 6 and 7 *because it won't be around the 5 anymore because there's a larger number of reds.* Such discussions of middle values were more likely to occur when students were asked to LIST the values, but they also occurred when students gave reasons for CHOICE responses. For example, Jess (G6), choosing (b), explained: *because its more spread out there's not the same number each time* while Jane (G9), choosing (a), explained: *because it has got the most difference in it.*

An interesting explanation came from Jess (G6), who said you *have to pick them all in the higher half of 5.* She meant that all the numbers she was selecting needed to be between 5 and 10. This is an unusual way to express the range of numbers, but it probably contains no more information than a response that deals with extremes by stating an upper and lower value.

There is no attempt here to claim that a student is more or less likely to discuss extreme values or middle values, just that some responses contain information about extreme values and that others will contain information about middle values. In fact, some students dealt with extreme values in one response and then middle values in another. Perhaps the choice of students' focus, middles or extremes, is influenced by the types of questions asked or the order in which the questions were asked. Explaining her choice of RANGE in the 50R situation, Donna (G5) showed concern for the extremes: *2 and 1 might not come out because they are lower numbers than 3 and 3 is a bit more higher and usually don't get 2 reds,* but then when explaining her CHOICE for 70R showed more concern for the middle: *there's 5 and 7 and two 7s and then in between 6 and 8 there's 7.*

D2—Concern with Both Middle Values and Extreme Values

These responses described both the extreme values and what is happening with the values between. Sue (G12), after describing the individual *numbers between 3 and 6 there might be one 5, a couple of 4s, maybe a 6 and maybe a 3 and a 5 she added you would be less likely to get the maximum of 6 or the minimum of 3 than you would the ones more like 5.* In describing the individual numbers, she showed concern for what is happening in the middle of the data; and in discussing the maximum and minimum, she showed concern for extremes. Sue more succinctly identified aspects of both the extreme values and the middle values when she explained her CHOICE for the 70R situation by stating: *not likely to get all of the same, a 0 or a 2 or even a 3.*

A response such as *want something more spread* from Max (G12) needs to be interpreted carefully. It can be interpreted to mean that a larger range is what is wanted, but when Max gave this as his reason for changing his CHOICE option from the 5,5,5,5,5,5 previously chosen what he meant was that he did not want all the numbers to be exactly the same. Although this would inadvertently also cause the range to increase, what Max is really saying was that he wanted some sort of variation in the numbers as opposed to them being all the same.

Definite concern was shown for both extreme values and what was happening between them when Rick (G12) explained, on choosing 2,3,4,3,4,4: *because you are unlikely to get two over 5 and you are not likely to get all 10s or all 7s or all 5s and these are all pretty high numbers and as I said [list] b has a 7 and an 8 and 0 doesn't seem to be likely or the 9,8.* However, given that Rick was responding to a CHOICE question, where options are designed to have higher and lower ranges and to show different amounts of change within the numbers, systematically justifying why he would not select the various options within the question, it is not surprising that he addressed both aspects. This reinforces the notion that the style of question influences the type of approach students take when responding.

D3—Discuss Deviations from an Anchor (not necessarily central)

These responses indicate that deviations from some value have been taken into consideration; but either the anchor for such deviations was not central, or it was not specifically identified as central. For example, Kate (G5) stated: *at least two away from the highest and lowest* when explaining her LIST in the 70R situation, suggesting consideration of deviations from extreme values rather than central values.

Responses not specifically mentioning a central value, even though it is obviously being used as the anchor, are transitional from D3 to D4. Rick (G12) explained: *there are three numbers on each side* to justify giving a range of 4 to 10 in the 70R situation. In this case Rick wanted a deviation of 3, but did not state the central value that he used. However, given that in an earlier response he had identified 7 as the most likely value in the 70R situation, there is the implication that he is comparing the deviations to the central value of 7.

D4—Discuss Deviations from a Central Anchor

These responses indicated that consideration had been given to both a center and what is happening about that center. No responses of this type were given by primary students. Generally, the better responses at this level were to the questions concerning more repetitions of the sampling (1F, 2F). The fact that the younger students were not given questions, considered too conceptually difficult, may well have deprived them of a chance to express their ideas about describing variation in this way.

A poorer-quality example of a response at this level that did not describe the deviations well came from Max (G12), who explained his LIST in the 70R situation by stating: *averaged around half each time so around there somewhere.* A better response came from Sue (G12) who, when responding to the larger sample question (1E) for the 50R situation, explained: *with a small number of people more spread out but with 50 people it would probably be around 5—probably a wave but not irregular but in closer and closer to 5.* This suggests that she was trying to indicate the variation that would exist but with convergence toward an expected number. Then for the 70R situation, she said basically the same thing would happen: *but higher, the other was around 5, it would be around 7.* This indicates she was considering both central tendency and spread.

Another form of response, indicating deviations from a central value have been considered, is one that suggests a distribution of values have been considered. Student responses that struggle to describe a distribution could be considered as just discussions of both extreme values and middle values (D2), such as Sue's (G12) response in the 70R situation: *get a couple around 6 and 7, probably get a 4 and a 5 and maybe an 8, waver around 6 and 7 because I like those numbers, oh, because it won't be around the 5 anymore because there's a larger number of reds.*

There were, however, other responses more clearly trying to indicate a distribution of some type. Max (G12) explained that for the 70R situation *it would follow the same pattern,* indicating that he expected similar things to happen for the 50 repetitions as had done for the 6 repetitions. He went on to explain: *it would be more spread,* indicating that he thought there would be more variation; he concluded with an indication that he had considered a distribution by discussing the possible occurrence of the various numbers: *but it would be around 4 to 9, same number of 5,6,7,8, and 9 would appear and average would be around 7 or something, 6 not sure, may even get two or three.* Interestingly, while Sue identified that for the smaller number of repetitions (6) there would be more variation, Max decided that for 50 repetitions there would be more variation. However, Max was possibly trying to convey that with more repetitions it was possible to more easily demonstrate the nature of the variation.

Causation Hierarchy

The causation hierarchy was developed to indicate aspects of responses that indicated increasing recognition of the relevant source of variation and the sophistication of the articulation of that source. It is interesting to note that these "causes" were discussed in responses, even though students were asked only to elaborate upon why they gave various responses. At no time were students actually asked to identify causes. The four levels of responses identified are now described.

C1—Identify Extraneous Causes of Variation

The extraneous sources of variation identified usually focused on physical aspects of the sampling. These sources included where the lollies were placed in the bowl (Tim, G6: *might have put them all over the place except the center*), where students decided to select from in the bowl (Prue, G9: *reds are in the middle and might pick at the edge*), and how the mixing was done (Tim, G6: *mix them up, they scatter, might have put them all over the place*). Some students even appeared to lose sight of the fact that 10 lollies were to be drawn each time, attributing variation to how big the handful is (Alice, G6: *would depend on how big the handful is*) and to the size of the hand (Kate, G5: *with me its different because of my hand size*), while others forgot that this was a blind sampling (Jess, G6: *got more red as you can see more of the red*). Some authors have referred to these types of responses as idiosyncratic (for example, Jones et al., 1999).

C2—Discuss Frequencies of Color(s) as Cause of Variation

Responses, with explanations based on the frequencies of specific colors, indicate that the composition of the population had been considered but either the effect was not fully appreciated or it could not be articulated. For example, Prue (G9) could only state that *there's a lot of reds* to explain her numerical response in the 70R situation, suggesting she was not able describe the effect of the 70% proportion. Although many students focused on the predominance of reds, some chose to acknowledge the need to allow for the occurrence of other colors. In the 50R situation, when explaining why she would expect numbers no higher than those she had selected, Millie (G4) said: *there are lots of other colors as well.* The better responses at this level are transitional to the C3 level. Although proportion is not specified, it is suggested by referring to the frequencies of more than one of the colors. Sue (G12), although not identifying the exact proportions, was able to observe that *there are more red and less of the other colors.*

Interviewing students helped to identify that there was a need to be careful with responses that made statements like there were "more red." Generally, such a comment would be taken to mean that there are more red than there are other colors, as meant by Sue's comment; but for the 70R situation, some students used this to mean more red than in the 50R situation. Such a reference is more likely to be indicating an appreciation of the importance of proportionality. This became obvious when Jess (G6), for her 70R CHOICE response, explained: *because some of them are higher than the ones I chose last time, want it higher because there are more reds.* By *last time* Jess meant the 50R situation, and now she wanted to give higher numbers in her RANGE response because there were more reds for the 70R situation.

C3—Discuss Proportion(s) of Colors as the Cause of Variation

Responses, with explanations based on the proportionality of color(s), demonstrated a realization that the proportion of red influences the number of reds in the sample drawn. Explanations given used the proportions of the colors (usually, but not always, red) to justify answers. Poorer responses at this level cannot succinctly articulate the use of the proportion. These responses showed attempts to describe the proportion but confusion when explaining the ratio in detail. Max (G12) explained, *as the ratio to yellow is about half* when justifying 5 out of 10 in the 50R situation. Other responses, such as Max (G12) stating that *you have to get over half as more than half of them are red*, are basing the explanation on proportions, even though the exact proportion is not stated or perhaps even known. Still, other responses acknowledged the importance of proportions by referring to the population from which the samples are drawn. For example, Max (G12) pointed out that *it comes back to the amount there are in the first place*, meaning the ratio of colors as described in the contents of the bowl.

Better responses actually articulate the proportion in some way. In some cases, the percentage was actually stated, as Rick (G12) explained: *50% red in there so if you take out 10 you would expect to get 50% of 10.* Other students just acknowledged the importance of the proportion but did not elaborate; for example, Brad (G9) explained: *that's the percentage that there are.*

C4—Discuss Likelihoods Based on Proportions

Responses at this level alluded to both proportion and likelihood when explaining their choices, with the references to proportion often being better articulated than the attempts to explain the likelihood concept. Students giving such responses are not only reasoning proportionally, but inferring sample likelihoods from population proportions. A poorer example came from the only primary student to respond at this level. Tim (G6), after a rather detailed but clumsy attempt to explain the proportion, attributed other aspects of the variation, the small possibility that something else could happen, to something that he could not quite put his finger on by stating: *no matter how slim the chance is still there, you just can't throw it away like you do a used tissue.* Even some senior students, who had a feel for this "unexplained" extra that needed to be accounted for when choosing how much variation to demonstrate, could not articulate what they were thinking. Max (G12) discussed ratios to explain the possible values in the sample he chose but then resorted to luck to account for the variation, suggesting: *about half but you never know you could have a bit of luck one time.*

A more statistically sophisticated response at this level came from Sue (G12). Having already demonstrated an appreciation of the importance of proportion in question 1A by explaining that *if you joined the yellow and the blue together it could be 50 too, so it's 50/50 and out of 10, 5 is half,* Sue indicated an appreciation of likelihoods by adding: *but you could get different answers.* This was confirmed in question 1B when Sue explained her LIST choice by saying: *you would be less likely to get the maximum of 6 or the minimum of 3 than you would the ones more like 5.* Another Grade 12 student, Rick—having already indicated an appreciation of the importance of proportions—demonstrated a good understanding of likelihood. When justifying the CHOICE he made for the 70R situation, he explained: *got two 7s which I think are most likely and the 8s and 6s which are probably second most likely I think and the 5 and the 9 which can do other times as well.* This explanation discusses not only "most likely" but also lesser degrees of likelihood, even though the 5 and 9 occurrences were not well expressed.

It fact, it was mostly Grade 12 students who used the words *likely* and *unlikely;* and in most instances, it was in situations with references such as *less, un, not,* and *not as.* Use of such vocabulary may reflect that Grade 12 students have undertaken more intensive study of probability than the younger students. Responses also indicated likelihood by using expressions discussing the "chances" of something happening. Max (G12) stated, as part of the explanation of the LIST he chose in the 70R situation 10, *chances of it getting to 10 would be fairly low.*

When interpreting responses for evidence of the discussion of likelihood, care needs to be taken with the use of words like *likely* and *unlikely.* The word *likely* actually appeared as part of the questions; so when using *likely* in their responses, students may just be reacting to its being used in the questions. Millie explained away the CHOICE 5,5,5,5,5,5 by saying, *unlikely to get all 5s.* There was nothing about this response, or others that she gave, to indicate that Millie really understood the concept of likelihood. There is a suggestion that even those students who do not appreciate the importance of the proportions in the parent population attempt to attribute the possible variation to chance with comments such as *you don't pick up the same handful* made by Donna (G5).

However, some responses at the C4 level suggest a possible conflict. Grade 12 students, who have been exposed to probability instruction, are able to calculate expected values. This causes them to gravitate toward saying 5 red (in the 50R situation) for all 6 draws, demonstrating no reference to the possible variation, but then intuitively realizing that this would not happen. Max (G12) chose all 5s for the 50R situation; but when asked to give a reason, talked himself into a response with some variation, explaining: *the chances of getting 5 every time would not be very high, but I think the average would be around 5.* Rick (G12) also experienced this conflict but was not able to articulate the situation as well as Max. Although having stated that 5 was the most common, Rick added: *unlikely to get 5 every time or 5 really often, I suppose, may ... though it could be anything.*

Responses to the more complex situations of larger sample size (questions 1E, 2E) and increased repetitions (1F, 2F) did not add significantly to the discussion of students' attempts to find causes for the variation. As mentioned previously, these more challenging situations were offered only to the senior students, who found them far more difficult to deal with. Although able to argue reasonable causes of the variation in the simpler situations, they were not able to articulate these as well in the more difficult ones. When considering a larger sample, drawing out 50 lollies instead of 10, students found it very difficult to give a RANGE and even more difficult to give an explanation. Only one of the Grade 12 students who had been able to give some reasonable responses for the sample size 10 questions mentioned color of the lollies as an issue in explaining her estimates.

Summary

Basically two main aspects of variation have come to light in these descriptions. One aspect is how spread out the numbers are. Students give responses that suggest that some indication of variation is being considered when they are dealing with extreme values using the range. The other aspect is what is happening with the numbers contained within that range. Responses considering the behavior of the middle values may give specific information about the numbers; or they may just give attributes that are necessary for the numbers, such as wanting them to be different. When these two aspects of variation description are brought together, deviations begin to become an issue; and when these deviations are anchored to a specific value, usually a center of some description, it will eventually become the focus of the student's description of a distribution.

These hierarchies were developed to code how responses may demonstrate reasoning about variation. Coding of student responses, according to a spread scale, was reported in Shaughnessy et al. (1999). These hierarchies are describing reasoning in relation to variation, from two perspectives: how students describe the variation and how they attribute cause to variation.

DISCUSSION

How then have these results addressed the research questions and enriched the understanding of reasoning about variation? Previously, aspects of the lollie-sampling task indicated some consideration of variation in the numerical responses where samples are described. This study has shown that a richer source of information is contained in the actual explanations that students give for those numerical responses. While investigating levels of reasoning about the variation, two important aspects were identified, suggesting the development of two separate hierarchies—one for actual description of the variation and another for consideration of causation. Four levels were identified in each hierarchy. The description hierarchy presents a developing sense of exploring extremes and middle values, leading to consideration of deviations anchoring around a central value and finally to the notion of a distribution. The causation hierarchy describes a developing sense of identifying the variables that are the source of the variation being described. These two hierarchies cover two important aspects of reasoning about variation, and depending on the task in which one is engaged, either or both might be of relevance.

Importantly, the tasks proposed, and the form of the question asked, can affect the reasoning about variation. Students demonstrated more of their developing notions of variation in those tasks that they found more difficult. The challenges of dealing with the 70% proportion of red lollies rather than 50%, drawing larger samplings and increasing the number of repetitions, all provided more insight into students' notions as they struggled with the complexities of the situations. Reading and Shaughnessy (2000) noticed the influence of the form of question when discussing the four case studies. The LIST form of the question restricted the demonstration of understanding of variation because the descriptions were based on only 6 repetitions of the sampling, but it did allow more flexibility than both the CHOICE and RANGE forms of the question. This deeper analysis of the interviews undertaken has identified further issues resulting from the various forms of a question. First, students' attempts to describe the variation or to look for a cause depended on the form of question asked. Although explaining a LIST was more likely to lead to a cause being sought, better descriptions of variation were more common for CHOICE questions. The better descriptions may have arisen because students were exposed to seven different lists—parts (a) to (g)—in the CHOICE question and needed to compare them in order to make a choice. Second, differences in the type of information gained from responses were noted within one hierarchy. When describing variation in the RANGE explanations, students focused more on discussion of extremes, but when describing variation in the LIST explanations, they concentrated more on the middle values.

Analysis of responses to the 50R lollie-sampling task presented to a different group of students by Kelly and Watson (2002) led to the development of a hierarchy of four levels that ranged from *intuitive, ikonic reasoning* (Level 1), through *"more red" but inconsistent reasoning* (Level 2) and *"more" and "half" red with centered reasoning* (Level 3) to *distributional reasoning* (Level 4). Kelly and Watson's hierarchy overlaps aspects of the two hierarchies we have identified in the study in this chapter. For example, Level 3 and Level 4 responses are distinguished by the acknowledgment of the proportion of reds. But most important, the notion of distributional thinking is evident in the Level 4 responses but not in Level 3. The Kelly and Watson hierarchy was concerned with the "correctness" of the numbers offered as possible results of the sampling task as well as the explanations of the variation. The description and causation hierarchies proposed in this chapter, while acknowledging some features of the Kelly and Watson levels, show more concern for the approaches to and notions of variation than do the Kelly and Watson levels.

The more sophisticated responses, identified in both the Torok and Watson (2002) and the present studies, are able to link together aspects of both center and spread, leading to notions of distribution. In particular, Level D4 responses, where students consider deviations from some central value, clearly showed that some students were giving careful consideration to the distribution of possible value around the center.

Although the 12 interviews have provided a rich basis for delving into reasoning about variation, one should not lose sight of the various limitations of this study. First and foremost, the sampling situation as used in this study is a restricted context, with isolated random variation and a known population. There are many situations in which students are expected to reason about variation, and sampling is but one of those situations. Second, only a small number of students were interviewed. Interviewing and qualitative analysis of responses is a time-consuming methodology and necessarily restricts the number of students to be included; but the researcher is usually rewarded with a depth of richness in the data that is not possible from just written responses. Finally, the style of question asked could have influenced the approach taken by students in responding. Although some researchers may see this as a limitation, providing recognition is given to this effect, in the future it may be useful as a tool not only for designing questions to elicit certain types of responses from students but also for helping to guide the development of intuitive notions of variation.

IMPLICATIONS FOR RESEARCH

The findings of this study unfold many possibilities for future research into reasoning about variation. However, three questions are particularly relevant. The approaches taken by students indicated the desire not only to describe the variation but also to discover causes for the variation. This suggests the first question for future research: Does a similar approach to reasoning about variation, comprising both description and causation components, arise in other situations (apart from sampling) in which students engage? For example, other possible contexts include reasoning about data, either in tables or graphs; reasoning about probability experiments; reasoning about information from the media. Whilst investigating these two hierarchies, causation and description, various immature notions of reasoning about variation have been identified. Hence, the second question: How can these intuitive notions be harnessed to develop a more sophisticated notion of reasoning about variation? An important part of these intuitive notions appears to be dealing with aspects of center and spread, and with their linking. This gives rise to a final question for further investigation: How can students be encouraged to link the concepts of central tendency and dispersion?

IMPLICATIONS FOR INSTRUCTION AND ASSESSMENT

The findings of this study also unfold a number of issues relevant to instruction in, and assessment of, reasoning about variation. Five key points encompass many of these issues. First, do not be afraid to give students more challenging tasks. With the integration of calculators and computers into learning environments, students are no longer burdened with the cumbersome calculations once synonymous with the study of statistics. Students should be allowed to deal with more detailed data sets since they allow more opportunity for discovering and attempting to explain the variation that occurs. Second, do not separate the study of central tendency and spread. Too often, learning situations totally neglect the study of spread or artificially separate it for the study of central tendency. Educators need to encourage discussion of more than just centering tendencies and to link, as much as possible, reasoning about variation with that of central tendency. Third, when learning situations involve reasoning about variation, allow students to have their untrained explorations into what is happening with extreme and middle values. These early explorations are laying a basis for future, more structured, reasoning about variation.

These first three points are mainly applicable in relation to instruction, while the following two points are equally applicable to both instruction and assessment. First, educators need to encourage students to explain their responses. Short responses in both learning and assessment tasks can produce some information on students' reasoning about variation, but far more is gained if students are asked to explain the responses given. Finally, whether instructing or assessing, use a variety of tasks or forms of questions. Different tasks or questions can encourage different aspects of students' reasoning; for a chance to develop all aspects of their reasoning, students need to be offered the opportunity to react to a variety of tasks and respond to a variety of questions.

Following is a final message to statistics educators about teaching and learning statistics. Students need to be encouraged to discuss variation in a variety of settings and to be questioned in a variety of ways. We, as educators, need to tap students' thinking, reasoning and explanations, in order to get a better hook for where to go next in instruction and assessment. Unless we know how our students are thinking about variability, we are apt to miss opportunities to build on what they already know—or do not know. Thus having students share their thinking, and encouraging them to discuss and argue about statistical situations, is critical for *our* pedagogical knowledge.

REFERENCES

Biggs, J., & Collis, K. (1991). Multimodal learning and the quality of intelligent behavior. In H. Rowe (Ed.), *Intelligence, Reconceptualization and Measurement* (pp. 57–76). New Jersey: Erlbaum.

Green, D. (1993). Data analysis: What research do we need? In L. Pereira-Mendoza (Ed.), *Introducing data analysis in the schools: Who should teach it?* (pp. 219–239). Voorburg, The Netherlands: International Statistical Institute.

Jacobs, V. R. (1997). Children's understanding of sampling in surveys. Paper presented at the annual meeting of the American Educational Research Association, Chicago.

Jacobs, V. R. (1999). How do students think about statistical sampling before instruction? *Mathematics in the Middle School, 5*, 240–263.

Jones, G., Langrall, C., Thornton, C., & Mogill, T. (1999). Students' probabilistic thinking in instruction. *Journal for Research in Mathematics Education, 30*, 487–519.

Kelly, B. A., & Watson, J. M. (2002). Variation in a chance sampling setting: The lollies task. In B. Barton, K. C. Irwin, M. Pfannkuch, & M. O. J. Thomas (Eds.), *Mathematics Education in the South Pacific Proceedings of the 25th annual conference of the Mathematics Education Research Group of Australasia, Auckland* (pp. 366–373). Sydney: Mathematics Education Research Group of Australasia (MERGA).

Konold, C. (1989). Informal conceptions of probability. *Cognition and Instruction, 6*, 59–98.

Konold, C., Pollatsek, A., Well, A., Lohmeier, J., & Lipson, A. (1993). Inconsistencies in students' reasoning about probability. *Journal for Research in Mathematics Education, 24*, 392–414.

Landwehr, J. M., & Watkins, A. E. (1985, 1995). *Exploring data*. Palo Alto: Seymour.

McClain, K., Cobb, P., & Gravemeijer, K. (2000). Supporting students' ways of reasoning about data. In M. Burke & F. Curcio (Eds.), *Learning mathematics for a new century, 2000 Yearbook* (pp. 175–187). Reston, VA: National Council of Teachers of Mathematics (NCTM).

Meletiou, M. (2002). Conceptions of variation: A literature review. *Statistics Education Research Journal, 1*(1), 46–52.

Meletiou, M., & Lee, C. (2002). Student understanding of histograms: A stumbling stone to the development of intuitions about variation. In B. Phillips (Ed.), *Proceedings of the Sixth International Conference on Teaching Statistics Developing a Statistically Literate Society* (CD-ROM). The Netherlands: International Association for Statistical Education (IASE).

Moore, D. (1997). New pedagogy and new content: The case for statistics. *International Statistical Review, 65*, 123–165.

Moore, D. S., & McCabe, G. (2003). *Introduction to the practice of statistics* (4th ed.). New York: Freeman.

National Council of Teachers of Mathematics (NCTM). (1989). *Curriculum and Evaluation Standards*. Reston, VA: Author.

National Council of Teachers of Mathematics (NCTM). (2000). *Principles and Standards for School Mathematics*. Reston, VA: Author.

Nicholson, J. (1999). Understanding the role of variation in correlation and regression. Presentation at *the First International Research Forum on Statistical Reasoning, Thinking and Literacy*, Be'eri, Israel.

Pearsall, J., & Trumble, B. (Eds.). (2001). *The Oxford English Reference Dictionary* (2nd ed.). Oxford, UK: Oxford University Press.

Reading, C. (1998). Reactions to data: Students' understanding of data interpretation. In L. Pereira-Mendoza, L. Kea, T. Kee, & W.-K. Wong (Eds.), *Proceedings of the Fifth International Conference on Teaching of Statistics*, Singapore (pp. 1427–1434). Netherlands: ISI Permanent Office.

Reading, C. (1999). Variation in sampling. Presentation at *the First International Research Forum on Statistical Reasoning, Thinking and Literacy*, Be'eri, Israel.

Reading, C., & Pegg, J. (1996). Exploring understanding of data reduction. In L. Puig & A. Gutierrez (Eds.), *Proceedings of the 20th Conference of the International Group for the Psychology of Mathematics Education*, Valencia, Spain, 4, 187–194.

Reading, C., & Shaughnessy, J. M. (2000). Student perceptions of variation in a sampling situation. In T. Nakahar, & M. Koyama (Eds.), *Proceedings of the 24th Conference of the International Group for the Psychology of Mathematics Education*, Hiroshima, Japan, 4, 89–96.

Rubin, A., Bruce, B., & Tenney, Y. (1991). Learning about sampling: Trouble at the core of statistics. In D. Vere-Jones (Ed.), *Proceedings of the Third International Conference on Teaching Statistics* (pp. 1 314–319). Voorburg, The Netherlands: International Statistical Institute.

Russell, S. J., & Mokros, J. (1996). What do children understand about average? *Teaching Children Mathematics, 2*, 360–364.

Scheaffer, R., Burrill, G., Burrill, J., Hopfensperger, P. Kranendonk, H., Landwehr, J., & Witmer, J. (1999). *Data-driven mathematics*. White Plains, NY: Seymour.

Shaughnessy, J. M. (1992). Research in probability and statistics: Reflections and directions. In D. A. Grouws (Ed.), *Handbook of research on mathematics teaching and learning* (pp. 465–494). New York: Macmillan.

Shaughnessy, M. (1997). Missed opportunities in research on the teaching and learning of data and chance. In F. Biddulph & K. Carr (Eds.), *Proceedings of the Twentieth Annual Conference of the Mathematics Education Research Group of Australasia* (pp. 6–22). Rotorua, NZ: University of Waikato.

Shaughnessy, J. M., & Ciancetta, M. (2002). Students' understanding of variability in a probability environment. In B. Phillips (Ed.), *Proceedings of the Sixth International Conference on Teaching Statistics Developing a Statistically Literate Society* (CD-ROM), South Africa. The Netherlands: IASE.

Shaughnessy, J. M., Watson, J., Moritz, J., & Reading, C. (1999). School mathematics students' acknowledgment of statistical variation. NCTM Research Pre-session Symposium: *There's More to Life than Centers*. Paper presented at the 77th Annual NCTM Conference, San Francisco, California.

Torok, R., & Watson, J. (2000). Development of the concept of statistical variation: An exploratory study. *Mathematics Education Research Journal, 12*(2), 147–169.

Wagner, D. A., & Gal, I. (1991). *Project STARC: Acquisition of Statistical Reasoning in Children* (Annual Report: Year 1, NSF Grant No. MDR90-50006). Philadelphia, PA: Literacy Research Center, University of Pennsylvania.

Watson, J. (2000). Intuition versus mathematics: The case of the hospital problem. In J. Bana & A. Chapman (Eds.), *Mathematics Education Beyond 2000: Proceedings of the 23rd Annual Conference of the Mathematics Education Research Group of Australasia*, Fremantle, (pp. 640–647). Sydney: MERGA.

Watson, J. M. (2001). Longitudinal development of inferential reasoning by school students. *Educational Studies in Mathematics, 47*, 337–372.

Watson, J. M., Collis, K. F., & Moritz, J. B. (1997). The development of chance measurement. *Mathematics Education Research Journal, 9*, 60–82.

Watson, J. M., & Kelly, B. A. (2002a). Can grade 3 students learn about variation? In B. Phillips (Ed.), *Proceedings of the Sixth International Conference on Teaching Statistics: Developing a Statistically Literate Society* (CD-ROM), South Africa. The Netherlands: IASE.

Watson, J. M., & Kelly, B. A. (2002b). Grade 5 students' appreciation of variation. In A. Cockburn & E. Nardi (Eds.), *Proceedings of the 26th Annual Conference of the International Group for the Psychology of Mathematics Education*, University of East Anglia, United Kingdom, 4, 386–393.

Watson, J. M., & Kelly, B. A. (2002c). Variation as part of chance and data in grades 7 and 9. In B. Barton, K. C. Irwin, M. Pfannkuch, & M. O. J. Thomas (Eds.), *Mathematics Education in the South Pacific: Proceedings of the 25th Annual Conference of the Mathematics Education Research Group of Australasia*, Auckland (pp. 682–689). Sydney: MERGA.

Watson, J. M., Kelly, B. A., Callingham, R. A., & Shaughnessy, J. M. (2003). The measurement of school students' understanding of statistical variation, *International Journal of Mathematical Education in Science and Technology, 34*, 1-29..

Watson, J. M., & Moritz, J. B. (1998). Longitudinal development of chance measurement. *Mathematics Education Research Journal, 10*(2), 103–127.

Watson, J. M., & Moritz, J. (1999). The beginning of statistical inference: comparing two data sets. *Educational Studies in Mathematics Education, 37*, 145–168.

Watson, J. M., & Moritz J. (2000). Developing concepts of sampling. *Journal for Research in Mathematics Education, 31*, 44–70.

Wild, C., & Pfannkuch, M. (1999). Statistical thinking in empirical enquiry. *International Statistical Review, 67*(3), 223–265.

Wild, C. J., & Seber, G. A. F. (2000). *Chance encounters: A first course in data analysis and inference.* New York: Wiley.

Zawojewski, J. S., & Shaughnessy, J. M. (2000). Data and chance. In E. A. Silver & P. A. Kenney (Eds.), *Results from the Seventh Mathematics Assessment of the National Assessment of Educational Progress* (pp. 235–268). Reston, VA: NCTM.

Chapter 10

REASONING ABOUT COVARIATION

Jonathan Moritz
University of Tasmania, Australia

OVERVIEW

Covariation concerns association of variables; that is, correspondence of variation. Reasoning about covariation commonly involves translation processes among raw numerical data, graphical representations, and verbal statements about statistical covariation and causal association. Three skills of reasoning about covariation are investigated: (a) speculative data generation, demonstrated by drawing a graph to represent a verbal statement of covariation, (b) verbal graph interpretation, demonstrated by describing a scatterplot in a verbal statement and by judging a given statement, and (c) numerical graph interpretation, demonstrated by reading a value and interpolating a value. Survey responses from 167 students in grades 3, 5, 7, and 9 are described in four levels of reasoning about covariation. Discussion includes implications for teaching to assist development of reasoning about covariation (a) to consider not just the correspondence of values for a single bivariate data point but the variation of points as a global trend, (b) to consider not just a single variable but the correspondence of two variables, and (c) to balance prior beliefs with data-based observations.

THE PROBLEM

Covariation, in broad terms, concerns correspondence of variation. The nature of the covariation may be categorized according to the variation possible in the measure of each variable involved. For logical variables, which can be either True or False, the logical statement A = NOT(B) expresses *logical covariation* between A and B, since varying the value of A from True to False entails a corresponding variation in the value of B from False to True to maintain the equation as true. The equation $y = 2x$ expresses *numerical covariation* between real-number variables x and y, since a variation in the value of either x or y entails a corresponding variation in the value of the other variable. Other polynomial and piecewise functions also express numerical covariation. In all of these cases, the values of the variables may

D. Ben-Zvi and J. Garfield (eds.),
The Challenge of Developing Statistical Literacy, Reasoning and Thinking, 227–255.

be said to involve some form of relationship, association, function, dependency, or correspondence.

Statistical covariation refers to the correspondence of variation of two statistical variables that vary along numerical scales. Such covariation is commonly represented in scatterplots using a Cartesian coordinate system that shows the correspondence of the ordination of each variable. The more general term *statistical association* may refer also to associations between two categorical variables, commonly represented in two-way frequency tables, and between one categorical and one interval variable, often formulated as the comparison of group. Statistical association involves more than just a relation of values, but a relation of measured quantities of distinct characteristics because data are "not merely numbers, but *numbers with a context*" (Moore, 1990, p. 96). Much work in the social and physical sciences concerns attempts to use statistical association as evidence of causal association between two characteristics, which may be used to enhance our prediction and control of one variable by knowledge or manipulation of the other variable. In most cases the statistical association does not perfectly fit the deterministic models of logical or numerical covariation just described; that is, there is variation from the model. Tests of statistical significance are required to measure the degree to which data fit or vary from one of these models. Formal measures of statistical covariation depend on the type of variation of the measures of each variable involved: χ^2 tests may be used to judge the significance of the association between categorical variables, and *t*-tests or analyses of variance are used to judge the significance of mean values of an interval variable across groupings of a categorical variable. For statistical covariation, which involves two numerical variables, Pearson correlation coefficients are commonly used to test the significance of the linear fit of covariation between the variables. Much of the discussion in this chapter focuses on covariation that might otherwise be termed statistical association or correlation, but in the restricted sense of being considered in relation to degree of fit to a linear function, as opposed to polynomial or piecewise models.

Reasoning about covariation commonly involves translation processes among raw numerical data, graphical representations, and verbal statements about statistical covariation and causal association. Other processes may include calculating and interpreting statistical tests of association, mathematical modeling to fit the data to a specific functional equation, and translating to and from symbolic expressions of algebraic functions. A comprehensive taxonomy of translations among words, graphs, tables of data, and algebraic formulae was described by Janvier (1978; Bell & Janvier, 1981; Coulombe & Berenson, 2001). Common translation processes associated with reasoning about covariation are shown in Figure 1. It is important that students know what is involved in these translation processes in order to be sensitive to the possibility of bias or error. Graph production and graph interpretation are frequently recommended for students in schools. In daily life such as reading the newspaper, however, adults rarely engage in the data analysis sequence of graph production, verbal graph interpretation, followed by causal inference. Many newspaper reports and advertisements make verbal statements that involve causal claims, but only some use graphs to illustrate the statistical data that lie behind the claims. More commonly, adults read a causal statement based on a statistical association, and in order to understand and evaluate it critically, they must imagine what statistical data lie behind it, that is, *speculative data generation*. Speculative data generation requires an understanding of numerical covariation, and a *contextual understanding* of data elements concerning how the data might have been collected and measured. Tasks of speculative data generation have some degree of freedom in the speculation of what was lost in the forward process of data interpretation to arrive at the verbal statement. For assessment purposes, this reverse type of task may be more informative of student understanding than interpretation, as students are required to supply more detail in their responses. Previous research of students' ability to deal with covariation in graphs has more often concerned graph production and numerical graph interpretation. Drawing a graph to illustrate a verbal statement of covariation requires both graph production and speculative data generation; such tasks are rarely found in curricula or research. This chapter focuses on reasoning about covariation for the processes of speculative data generation, verbal graph interpretation, and numerical graph interpretation.

LITERATURE AND BACKGROUND

Curriculum

As part of data handling, covariation appears in statistics curricula in Australia (Australian Education Council [AEC], 1991, 1994), England (Department for Education and Employment [DEE], 1999), New Zealand (Ministry of Education [ME], 1992) and the United States (National Council of Teachers for Mathematics [NCTM], 2000). Students are asked to engage steps in a multistep process (a) to hypothesize a relationship between two variables, (b) to collect data, (c) to represent the data graphically or analyze them numerically, and (d) to draw conclusions about the relationship in verbal statements. This multistep process reflects professional use in the social and physical sciences, in which covariation is often observed within bivariate data sets, and causal inferences are made. In Australia, representation tasks are suggested for lower secondary students, such as "represent two-variable data in scatter plots and make informal statements about relationships" (AEC, 1994, p. 93), and "represent bivariate time series data in line graphs" (AEC, 1994, p. 109). In England (DEE, 1999), secondary students are expected to draw scatter graphs and line graphs for time-series data, to "look for cause and effect when analyzing data" (p. 40), and to "draw lines of best fit by eye, understanding what these represent" (p. 41). In New Zealand (ME, 1992), time-series data are emphasized at many levels, and for senior secondary school years scatterplots are suggested to assess bivariate association. In the United States (NCTM, 2000), it is recommended that sixth- to eighth-grade students use scatterplots as an important tool in data analysis, and students are encouraged to interpret lines of fit. Causal inference is also considered in these curricula; for example, secondary students in Australia should "investigate and interpret relationships, distinguishing association from cause and effect" (AEC, 1991, p. 178).

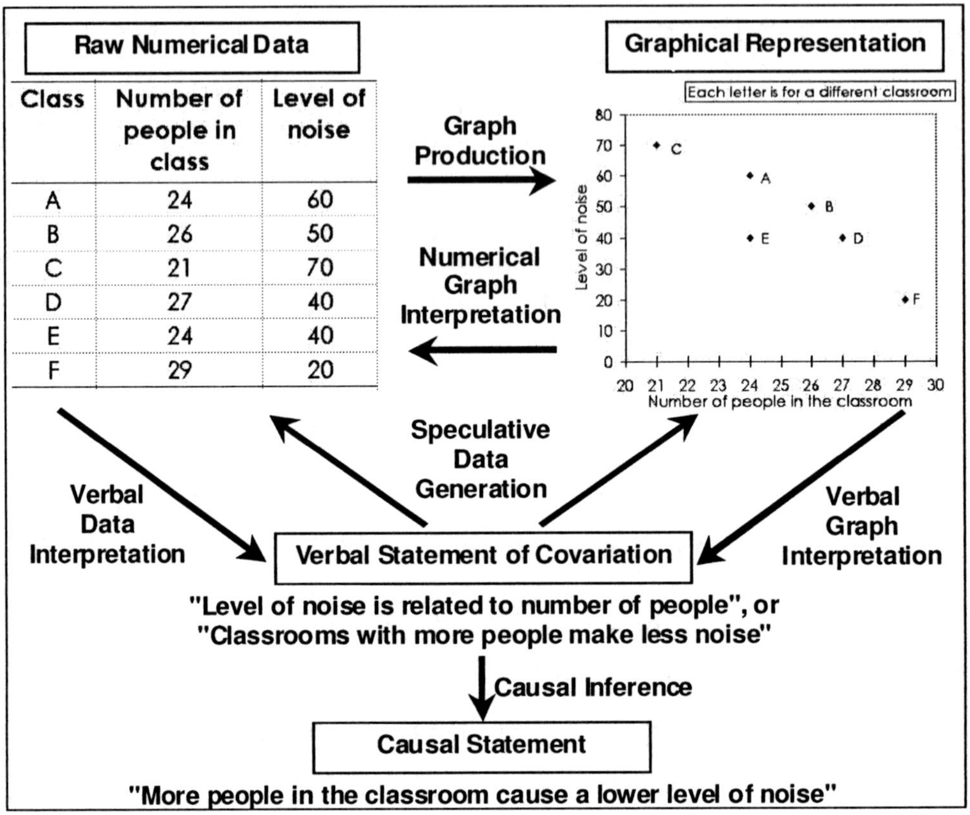

Figure 1. Forms of representing statistical covariation and skills to translate them.

Apart from statistical contexts, curricula (e.g., AEC, 1991; NCTM, 2000) for early algebra courses include covariation relating familiar variables, particularly involving time. Australian primary students are expected to "represent (verbally, graphically, in writing and physically) and interpret relationships between quantities [...] such as variations in hunger through the day" (AEC, 1991, p. 193). Similar suggestions are made for upper-primary students in England (DEE, 1999) and New Zealand (ME, 1992). In the United States (NCTM, 2000), suggestions for activities such as the growth of a plant over time have been proposed for third- to fifth-grade students as part of the algebra standard of "analyze change."

History of Graphing

A brief history of graphing illustrates some of the cognitive difficulties and milestones in reasoning about covariation. Statistical graphs were infrequent before the late 1700s (Tufte, 1983, p. 9), although mapping geographic coordinates was common. From 1663 to 1815, ideas for time-series graphs were developed involving mechanical devices, for example, the invention that could record temperature change over time "on a moving chart by means of pen attached to a float on the surface of a thermometer" (Tilling, 1975, p. 195), although "such automatic graphs were considered useless for analysis and were routinely translated into tabular logs (Beniger & Robyn, 1978, p. 2). More abstract graphs that relate two abstract measures (i.e., not time or position) were a later development. They are still rarely used in daily settings: less than 10% of newspaper and magazine graphs surveyed by Tufte were based on more than one variable (but not a time-series or a map; see p. 83).

These historical developments may be considered to have educational implications for the ordering of curriculum. At a simple level, maps involve the use of coordinates denoting position in which representation is a stylized picture. Some tasks noted in the curriculum documents just cited, for example involving plant growth, exploit the natural mapping of height on the vertical axis, which may assist students because the measure on the graph corresponds to the visual appearance of the actual object. Besides horizontal position, time is a natural covariate; one can read a graph from left to right, just as one reads English language, and a narrative of continuous variation unfolds in temporal order. Graphs of one variable over time permit students to describe in a verbal statement bivariate association as the change of the variable over time, naturally expressed with use of English tense (e.g., "it started to grow faster, then it slowed down again," NCTM, 2000, p. 163). Such verbal statements are bivariate in nature, although time is implicit and only one variable is explicit as changing. Despite the feature of continuous variation in a graph, students may still tend to approach the data pointwise, just as historically graphs were transcribed into tabular form.

Understanding Covariation

Piaget's theory of cognitive development (e.g., Piaget, 1983) highlights some of the key concepts of students' development of reasoning about covariation. *Correspondence* (to confirm identity or a one-one mapping), *classification* (to identify as one of a class or group), and *seriation* (to order a series) were among the logical operations Piaget observed across many studies and considered to be universally fundamental to cognitive development. *Conservation* is perhaps the most renowned indication of the developmental stage called concrete operations. For example, when pouring a given quantity of liquid from a thin glass to a wide glass, most young children attend to only one aspect, such as the height, and proclaim the thin glass has more. The coordination (correspondence of seriations) of height and width is what encourages the learner to rely not on the configurations but rather the transformation or operations (Piaget, 1983, p. 122).

Teaching and reasoning about covariation often focus on either correspondence of bivariate data points, or variation within variables, and aim to build one aspect upon the other. Nemirovsky (1996a) described these two approaches with reference to algebra teaching as (a) a *pointwise approach* of comparing bivariate pairs to identify the functional rule for translating one to the other, and (b) a *variational approach* that considers change in a single variable across a number of cases. These approaches are similar to two Piagetian schema that Wavering (1989) suggested are developed in reasoning to create bivariate graphs: (a) one-to-one correspondence of bivariate data values, and (b) seriation of values of a variable, necessary for scaling of graphs to produce a coordinate system. The two approaches are also similar to two competence models for Cartesian graphing of covariation suggested by Clement (1989): a static model involving translating bivariate data values to points in coordinate space, and a dynamic model involving concepts of variation. Clement noted that a basic qualitative form of the dynamic model involves simply the direction of change with no indication of how the variables are quantitatively measured (e.g., "the more I work, the more tired I'll get," p. 80). Carlson, Jacobs, Coe, Larsen, and Hsu (2002) have proposed a framework for how such qualitative understanding further develops to reasoning about rates of change. The variational approach has been advocated by researchers of early algebra learning (e.g., Nemirovsky, 1996a, 1996b; Yerushalmy, 1997). Nemirovsky (1996b) discussed the importance of time-based mathematical narratives without specific data values, with verbal and graphical language both read left to right to express generalities of how a quantity varies over time. Yerushalmy (1997) used various graphic icons with computer software to provide a graphic language that corresponds to verbal terms *increasing, decreasing,* and *constant,* often with time as the implicit covariate. These studies indicate that verbal phrases and graphs are important forms for understanding covariation.

Representing Covariation in Graphs

Most research into the developing understanding of covariation has come from tasks involving graphs. The broader research literature on graphing has often reported on pointwise tasks of construction and interpretation, such as plotting points or locating values (Leinhardt, Zaslavsky, & Stein, 1990). Tasks involving variation and qualitative graphs—that is, without specific data values—have been considered by some researchers (Leinhardt et al., 1990) to be an underutilized avenue for exploring understanding of general features of graphs, including covariation. Students need to develop skills that flexibly combine local and global views (Ben-Zvi & Arcavi, 2001).

Some researchers have employed tasks to translate verbal descriptions into graphical representations (e.g., Bell, Brekke, & Swan, 1987a, 1987b; Coulombe & Berenson, 2001; Krabbendam, 1982; Mevarech & Kramarsky, 1997; Moritz, 2000; Swan, 1985, 1988). Krabbendam gave 12- to 13-year-olds various graphing tasks, such as involving a newspaper text about the gathering and dispersion of a crowd of people. He concluded, "it appears to be rather difficult for children to keep an eye on two variables" (p. 142), but that "time could play an important part in recording a relation" (p. 142) provided it is seen to pass gradually (i.e., continuously), thus supporting a view of continuous variation rather than a pointwise approach. For a task to represent "how the price of each ticket will vary with the size of the party" on a bus with a fixed total cost, Swan (1988) found that 37% of 192 thirteen- to fourteen-year-olds drew a graph that was decreasing. Mevarech and Kramarsky found that about 55% of 92 eighth-grade students appropriately used labeled two-axis graphs to represent verbal statements of positive association ("the more she studies, the better her grades"), negative association, and no association, whereas only 38% of students correctly represented curvilinear association. Three alternative conceptions were identified: (a) only a single point represented in a graph (25% of responses), (b) only one factor represented in each of a series of graphs (30% of responses), and (c) an increasing function represented irrespective of task requirements (5% of responses). The first two conceptions may reflect students' attempts to reduce the complexity of bivariate data sets. After teaching about Cartesian conventions, distance-time graphs, and graphs found in newspapers, more students included labels and scales, and there was a reduction but not an elimination of these three conceptions. Chazan and Bethell (1994) briefly described a range of dilemmas students encounter in graphing verbal statements of relationships, including identifying the variables, specifying the units of measurements, deciding which variables are independent and dependent, and deciding whether to represent a continuous line or discrete points. Watson (2000; Watson & Moritz, 1997) asked students to represent "an almost perfect relationship between the increase in heart deaths and the increase in the use of motor vehicles" (p. 55) as reported in a newspaper article. Some students' graphs were pictures of the context or basic graphs with no context. Some compared single values of each measure without variation, whereas others showed variation but just for one measure. Successful responses were those that displayed the relationship in a Cartesian coordinate system, or by displaying two data series compared over time on the horizontal axis.

Some researchers of students' skills for graph production have exploited contexts in which students have prior beliefs about covariation, and there is a natural mapping of height on the vertical axis and time on the horizontal axis, by asking students to plot height versus age graphs (Ainley, 1995; Moritz, 2000), a context used by Janvier (1978). Compared to the graphs observed by other researchers, on these tasks students of a young age achieved remarkable success for representing covariation trends in data, possibly because of familiarity with the covariation and with the measurement of the variables. Moritz (2000) observed that students represented a curvilinear relationship demonstrating growth ceasing, but a multivariate task incorporating differences between males and females proved more difficult: Some students represented a *single comparison* of one male and one female to reduce complexity, some a *double comparison* of heights for two specific ages, and some a *series comparison* of two trend lines over a series of ages.

Konold (2002) has suggested that a variety of graph forms are valid alternatives to scatterplots for representing and interpreting covariation, such as *ordered case-value bars*, which involve a bar graph ordered by cases of one variable to examine any pattern in the other variable. For ordered case-value bars, ordering was considered important to assist scanning values to offer a global summary. Similar graphs, which showed two variables measured across a number of cases and described as series comparison graphs by Moritz (2000), were drawn by students in studies by Brasell & Rowe (1993), Moritz (2000), and Cobb, McClain, and Gravemeijer (2003).

Interpreting Covariation

Pinker (1990) has suggested that graph comprehension divides at the most fundamental level into (a) comprehension of the axis framework and scale, and (b) comprehension of the data elements. The scale is necessary for reading numerical values, whereas the data cases without the scale permit trend identification and qualitative comparison of cases. This is the basis for the distinction between skills of verbal graph interpretation and numerical graph interpretation as sustained in this chapter. Curcio (2001) suggested three levels of graph comprehension involving numerical values, described as "reading the data" values, "reading between the data" involving comparison of values, and "reading beyond the data" involving interpolation or extrapolation. Many studies have involved numerical tasks and found that students construct and read graphs as individual numerical points rather than a global whole (e.g., Bell et al., 1987a; Brasell & Rowe, 1993). When a variety of tasks were compared, however, Meyer, Shinar, and Leiser (1997) found trend judgments from line graphs and bar graphs were performed faster and more accurately than tasks to read values, to compare values from the same data series for different *X* values (*X* comparisons), to compare values from different data series with the same *X* value (series comparisons), or to identify the maximum. The terms *X comparison* and *series comparison* match those of Moritz (2000) as qualitative operations on data elements not requiring reference to the numerical scale.

Subjects' judgments of statistical association in a variety of situations have been investigated by researchers in social psychology (e.g., Alloy & Tabachnik, 1984; Crocker, 1981), science education (e.g., Donnelly & Welford, 1989; Ross & Cousins, 1993; Swatton, 1994; Swatton & Taylor, 1994), and statistics education (e.g., Batanero, Estepa, Godino, & Green, 1996; Batanero, Estepa, & Godino, 1997). Many studies have followed Inhelder and Piaget (1958) in considering association of dichotomous variables in contingency tables, whereas few have considered covariation of two numerical variables (Ross & Cousins, 1993). Crocker (1981) outlined six steps for statistically correct judgments of covariation in social settings, as well as some common errors at each step. The six steps included deciding what data are relevant, sampling cases, classifying instances, recalling evidence, integrating the evidence, and using the covariation for predictions.

People often hold prior beliefs about causal associations between the real-world variables that may influence judgments (e.g., Jennings, Amabile, & Ross, 1982). Topic knowledge may result in ignoring the available data (Alloy & Tabachnik, 1984; Batanero et al., 1996), or dismissing an association in the data because there is no apparent causal relationship or because other variables are more plausible causes (Batanero et al., 1997; Crocker, 1981; Estepa & Batanero, 1996).

In using statistical data, some people hold deterministic or unidirectional concepts of association (Batanero et al., 1996, 1997; Crocker, 1981), similar to the alternative conception of an increasing function irrespective of the direction of covariation (Mevarech & Kramarsky, 1997). Some attend to selected data or selected variables as a means of reducing the complexity of the data (Bell et al., 1987a), similar to the alternative conceptions of representing a single point, a single pair of values, or a single variable (Mevarech & Kramarsky, 1997; Moritz, 2000). Attention to selected data points may involve only the extreme points in a scatterplot (Batanero et al., 1997) or the cells with confirming cases in contingency tables (e.g., Batanero et al., 1996; Crocker, 1981; Inhelder & Piaget, 1958). Attention to selected variables has been observed in some studies that have identified levels of response based on the number of variables students have referred to in verbal graph interpretations (e.g., Donnelly & Welford, 1989; Ross & Cousins, 1993; Swatton, 1994; Swatton & Taylor, 1994). Swatton showed sixth-grade students scatter graphs and line graphs and asked, "what do you notice about [*X*] and [*Y*]?" Level 0 responses involved only the context of the data or syntactic/visual patterns in a graph, Level 1 responses described univariate data patterns, Level 2 involved both variables, and Level 3 responses involved both variables with appropriate directionality. Ross and Cousins asked students from grades 5 to 13 to "find out if there was a relationship" between two continuous variables in situations where a third, categorical, variable was involved. Their analysis concerned the numbers of variables students appropriately ordered or described, including 0, 1, 2, or 2 with a control. Thus the complexity of the data cases, the number of variables given, the topic of variables, and possible lurking variables can affect judgments of covariation.

Questioning causal inferences has been considered by some researchers (e.g., McKnight, 1990; Watson, 2000). McKnight considered different levels of data-based tasks including (a) observation of facts, (b) observation of relationships, (c) interpretation of relationships, and (d) critical evaluation of inferential claims. These levels of tasks correspond closely to the three tiers of the statistical literacy hierarchy of Watson (2000), involving basic understanding of terms, understanding concepts in context, and questioning inferential claims. Cobb et al. (2003) noted that seventh-grade students given time to consider the context of the data collection, prior to data analysis, were able to raise issues of sampling procedures and control of extraneous variables that might affect conclusions. Thus questioning claims need not require the level of questioning statistical inference, but may also be addressed at simpler levels related to the context of the data, such as sampling or measurement.

The Current Study

The current study aimed to explore three of the skills of reasoning about covariation shown in Figure 1: speculative data generation (translating a verbal statement into a graph), verbal graph interpretation (translating a scattergraph into a verbal statement), and numerical graph interpretation (reading values and interpolating). Speculative data generation was assessed with respect to demonstration of numerical covariation, not contextual understanding of data elements, and as much as possible without assessing graph production skills.

METHOD

Participants

Participants were from two Tasmanian private schools, one a boys' school and the other a girls' school. Both schools would be expected to draw students of a higher socioeconomic status than the general school population in Tasmania. At each school, one class group from the third, fifth, seventh, and ninth grades was surveyed. Specific classes were selected based on their availability to undertake the survey with minimal interruption to their teaching program. Females described as fifth grade were from a composite class of fourth- or fifth-grade students, with 13 students at each grade level. Ninth-grade students were streamed in basic, intermediate, and advanced mathematics courses; the female class surveyed was undertaking the basic course, and the male class the advanced course.

Tasks

The tasks in this study are shown in Figures 2 and 3. Contexts were chosen such that students would be familiar with the variables. Study time and academic grades are experiential for students, and were used by Mevarech and Kramarsky (1997). Noise level and number of people in a classroom, though rarely measured, are at least intuitively experienced by students in schools. The contexts were also chosen such that students would expect a positive covariation between the variables, but the task described a negative covariation so that students were forced to rely on the data rather than prior beliefs. Task 1 was administered in a positive covariation form instead of the negative form to third- and fifth-grade males. These different forms were designed to explore whether students might respond differently due to their prior beliefs about the covariation. The tasks were worded to support a statistical context for covariation, such as awareness of the data collected and of possible variability from a perfect linear fit. For each task, the data were six cases, and for Task 2, the data included repeated values of each variable.

For the speculative data generation question (Q1), no axes were provided, to permit students to decide the numbers and types of variables to represent and to develop their own form of representation. Verbal graph interpretation was assessed using Q2a and Q2d. Q2a was worded in an open manner to avoid the assumption that an association exists (Donnelly & Welford, 1989). Because students may have avoided comment on covariation in Q2a, Q2d* was included and then revised to Q2d to provide a more specific cue about covariation. Numerical graph interpretation was assessed using Q2b and Q2c. Q2b involved reading a value, and Q2c was designed to identify whether students based interpolation on proximity to one or more of Classes A, C, and E.

Task 1 (Negative association)

Anna and Cara were doing a project on study habits.
They asked some students two questions:

- "What time did you spend studying for the spelling test?"
- "What score did you get on the test?"

Anna asked 6 students. She used the numbers to draw a graph.
She said, "People who studied for more time got lower scores."

Q1. Draw a graph to show what Anna is saying for her 6 students. Label the graph.

Task 1 (Positive association)

She said, "People who studied for more time got higher scores."

Q1*. Draw a graph to show what Anna is saying for her 6 students. Label the graph.

Figure 2. Task 1 to assess speculative data generation.
(Third- and fifth-grade males received Q1* in place of Q1.)

Procedure

The items were among a total of six or seven tasks in a written survey administered to students during class time. The items were Q2 (Task 1) and Q6 (Task 2) on the survey. Q1 on the survey concerned graphing three statements related to height growth with age (Moritz, 2000), Q3 (for secondary students) concerned graphing a verbal statement concerning motor vehicle use and heart death incidence (Watson, 2000), and Q4 concerned graphing a table of raw data about six temperatures recorded with corresponding times at regular intervals. Graphing tasks were placed before interpretation tasks to ensure exposure to the printed graphs did not suggest a graphing method.

The time available was 40–70 minutes, although ninth-grade females had only 25 minutes available; in this case after about 15 minutes, students were instructed to attempt Task 2. Sample sizes vary between questions because those who did not appear to have attempted the item were removed from the analysis, whereas those who appeared to have read and attempted the item but offered no response were included at the lowest response level. Each session began with a brief verbal introduction to the purpose of the survey. The first question and other selected questions were read to students on a class or individual basis as required.

Analysis

Students' representations were scanned into computer graphic files, and their written responses were typed into a spreadsheet. Responses were categorized using iterative techniques (Miles & Huberman, 1994), successively refining categories and subcategories by comparing and contrasting features of graphs or written responses. Frameworks of four levels were developed that described the degree of success students had in generating a data set, in verbally generalizing the required covariation, and in numerically interpreting covariation. The levels—Nonstatistical, Single Aspect, Inadequate Covariation, and Appropriate Covariation—were informed by the frameworks used by others (Moritz, 2000; Ross & Cousins, 1993; Swatton, 1994; Watson & Moritz, 1997) who assigned levels according to the number of aspects, variables, or data elements used, including no use, a single variable, both variables but not related, and all variables successfully related. These levels also relate closely to a theoretical model of cognitive development judged by the structure of the observed learning outcome (Biggs & Collis, 1982), which identifies four levels as prestructural, unistructural (single aspect), multistructural (multiple aspects unrelated), and relational. Further details are provided in the results that follow.

Task 2
Some students were doing a project on noise.
They visited 6 different classrooms.
They measured the level of noise in the class with a sound meter.
They counted the number of people in the class.
They used the numbers to draw this graph.

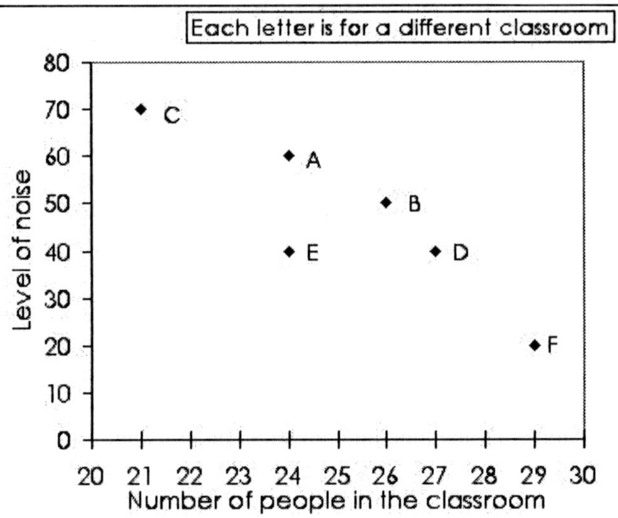

Q2a. Pretend you are talking to someone who cannot see the graph.
Write a sentence to tell them what the graph shows. "The graph shows...

Q2b. How many people are in Class D?

Q2c. If the students went to another class with 23 people, how much noise do you think they would measure? (Even if you are not sure, please estimate or guess.) Please explain your answer.

Q2d. Jill said, "The graph shows that classrooms with more people make less noise". Do you think the graph is a good reason to say this?
❑ YES or ❑ NO Please explain your answer.

Q2d*. Jill said, "The graph shows that the level of noise is related to the number of people in the class". Do you think the graph is a good reason to say this?
❑ YES or ❑ NO Please explain your answer.

Figure 3. Task 2 to assess verbal and numerical graph interpretation. (Third- and fifth-grade males received Q2d* in place of Q2d.)

RESULTS

Results are discussed for three of the skills shown in Figure 1: speculative data generation (Q1), verbal graph interpretation (Q2a and Q2d), and numerical graph interpretation (Q2b and Q2c). For each skill, examples of levels and types of responses are provided. Quoted examples are annotated with labels indicating grade and sex, such as "G3f" for a third-grade female.

Speculative Data Generation

Subsets of students were asked to graph a negative covariation (Q1) or a positive covariation (Q1*). Responses were coded according to the four levels in Table 1. A descriptive analysis of some responses has been reported previously (Moritz, 2002). To be coded at the level of Appropriate Covariation, responses showed the correspondence of variation in two variables, in that (a) the variables were identified with adequate variation and (b) the direction of the correspondence of variation was appropriately specified. Variables were considered adequate if (a) labels were explicit, or units (e.g., hours/minutes) or values (e.g., digital time format) were used that indicated which variable was denoted, using the notion of indicative labeling (Moritz, 2000), and (b) the graph included adequate variation of at least three bivariate values; although the context described six data cases, three were considered sufficient to demonstrate the covariation. The direction of the correspondence of variation was appropriately specified either by values at least ordinal in nature (e.g., "not at all," "not much," "a lot") or by convention of height/sector angle.

Table 1. Characteristics of four levels of speculative data generation

Level	Description
0. Nonstatistical	Responses represent either: (a) context in a narrative but without a data set of more than one value of one variable, or (b) graph axes or values, denoted by number or spatial position, but without a context indicating a data variable
1. Single Aspect	Responses represent either: (a) correspondence in a single bivariate case, or (b) variation of values for a single variable
2. Inadequate Covariation	Responses represent both variables but either: (a) correspondence is shown with inappropriate variation for at least one variable, such as one variable only has two distinct values (often categorical), or (b) variation is shown for each variable with inappropriate correspondence, such as not in the correct direction
3. Appropriate Covariation	Responses represent both variables with appropriate correspondence between the variation of values for each variable

Most students demonstrated at least Inadequate Covariation, and many older students showed Appropriate Covariation, as shown in Table 2. Further details are noted below. Numbers in text are divided into the two forms of the questions (Q1 and Q1*), whereas numbers in Table 2 are combined.

Table 2. Percentage of student responses at four levels of speculative data generation by gender and by grade (N = 167)

Levels of Speculative	**Female Grade**				**Male Grade**				**Total**
Data Generation	3	5	7	9	3[a]	5[a]	7	9	(*N*)
0–Nonstatistical	23	12	5	40	11	11	8	0	18
1–Single Aspect	0	12	14	20	42	11	4	0	18
2–Inadequate Covariation	42	38	18	20	5	0	21	11	35
3–Appropriate Covariation	35	38	64	20	42	78	67	89	96
Total (*N*)	26	26	22	5	19	18	24	27	167[b]

[a] Third- and fifth-grade males were administered Q1* rather than Q1.

[b] Percentages do not always sum to 100 due to rounding.

Graphing a Negative Covariation (Q1)

Level 0: Nonstatistical. Fourteen students responded with no evidence of a data set of covariation for test scores and study time. Two students gave no response. Five students identified the narrative context without a data set, such as a written narrative with names for individuals and a single test score of "10/10" (Figure 4a, G3f). Three students drew graphs that identified each variable but without clear data points, such as labeled axes. Four students drew a basic graph that gave no indication of the data set for the variables being measured and also failed to show six data cases (e.g., Figure 4b, G5m).

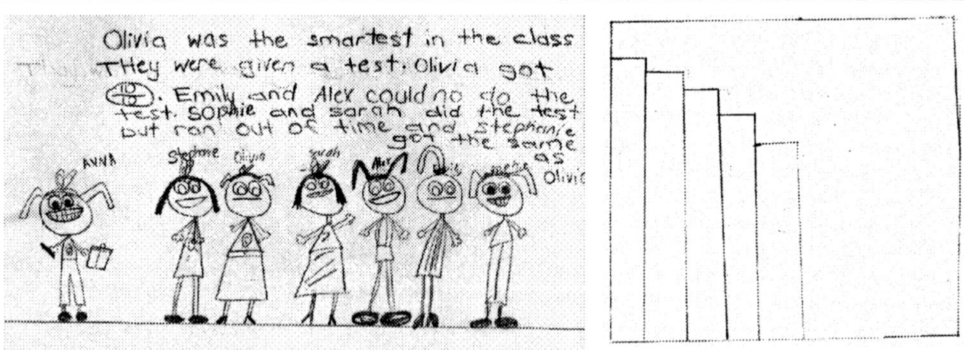

(a) Narrative (b) Basic graph

Figure 4. Student responses to Q1 at Level 0.

Level 1: Single Aspect. Eight students showed a single aspect, either correspondence or variation, in an attempt to show covariation. One student gave a single bivariate data point, presented in a rudimentary table of raw data (Figure 5a, G5f). Seven students represented a single variable: Two showed test scores without indication of study times (e.g., Figure 5b, G7f), and five showed six data cases ordered by values of the single variable, which was not labeled.

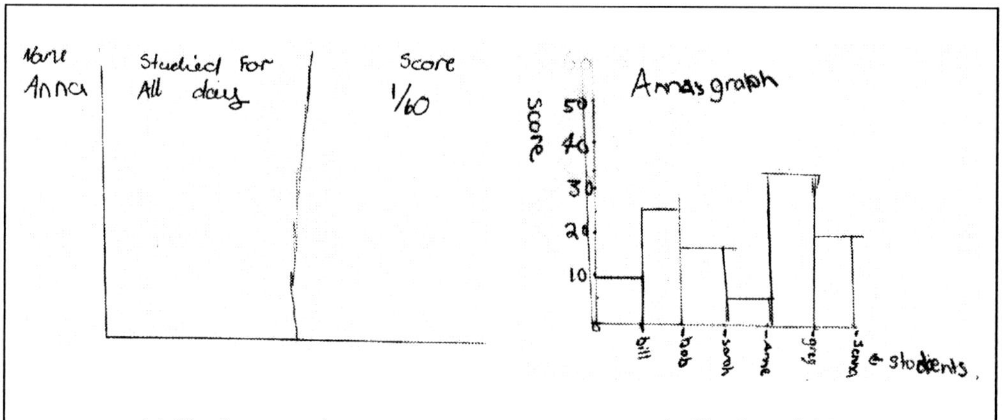

(a) Single comparison (b) Single variable

Figure 5. Student responses to Q1 at Level 1.

Level 2: Inadequate Covariation. Thirty-four students showed some features of the required negative covariation but lacked either appropriate variation or appropriate correspondence. Fifteen students treated study time as a binary variable, five students giving a double comparison involving two bivariate pairs (e.g., Figure 6a, G7m) and 10 representing a group comparison including test scores of six students (e.g., Figure 6b, G3f). Nineteen students did not adequately show the direction of covariation, nine failing to clearly indicate any covariation (e.g., Figure 6c, G7m), seven representing a positive covariation (e.g., Figure 6d, G5f), and three showing a negative trend with some explicit numbers but without labels or units to indicate the variables, such as a pie graph with larger sectors corresponding to labels of smaller percentage values.

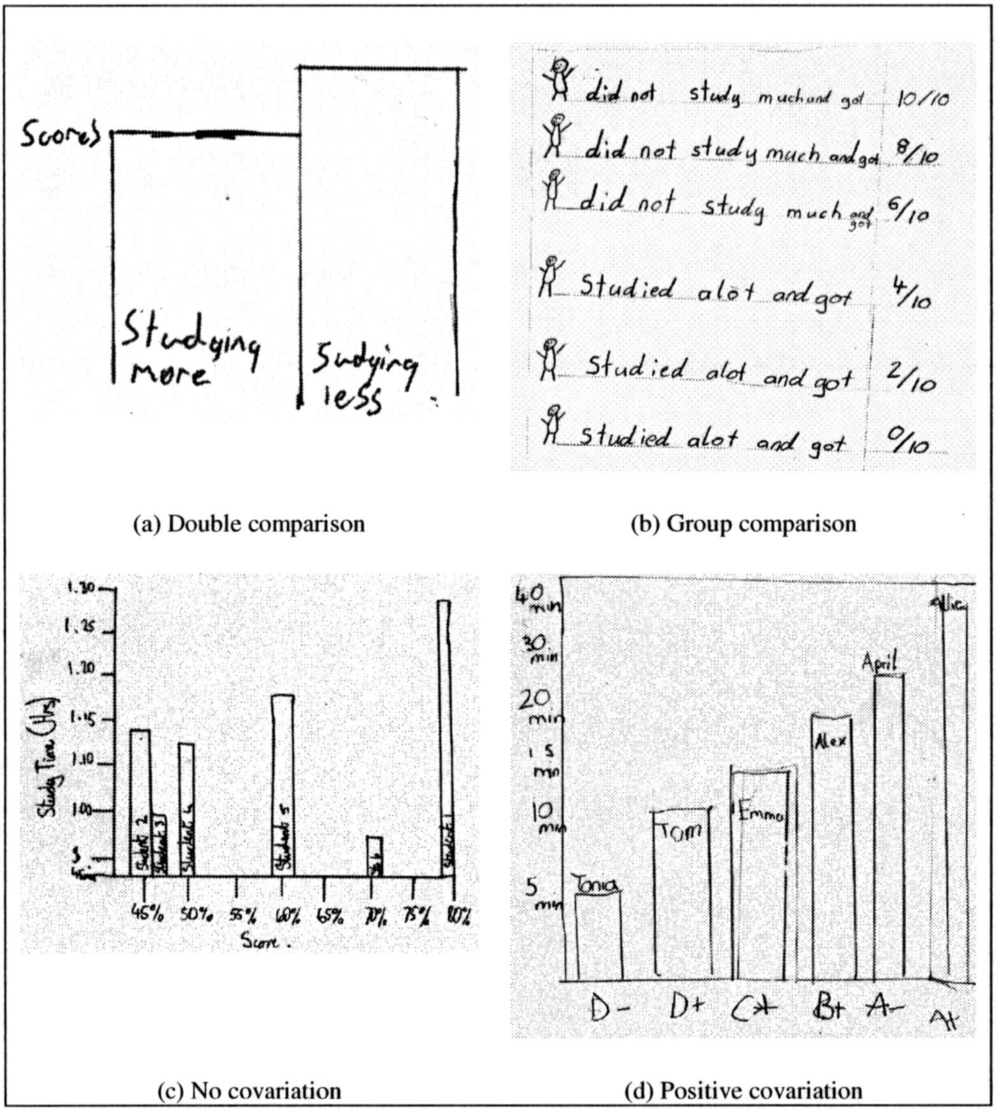

(a) Double comparison

(b) Group comparison

(c) No covariation

(d) Positive covariation

Figure 6. Student responses to Q1 at Level 2

Level 3: Appropriate Covariation. Seventy-four responses provided data for study times for which higher values were associated with lower test scores, with the conditions that at least three bivariate data points were shown and study time was not a binary variable. Thirteen students drew a *table* of raw data, that is, bivariate values were written and spatial position was not used to denote value—for example, including names and repeated values (Figure 7a, G3f), or values placed on a diagonal but without clear use of coordinates (Figure 7b, G5f). Eleven students drew *series comparison* graphs for which the horizontal axis represented the six students that Anna asked, and the vertical axis displayed study times and test scores, either in two graphs or superimposed in one graph, often with two scales (e.g., Figure 7c, G7m). Seven of these were bar graphs, three line graphs, and one was a double pie graph; seven graphs were unordered on the horizontal axis, and four were ordered on one variable. Figure 7c illustrates an unordered horizontal axis, although after the first two cases, the student appears to order the remaining cases. Fifty students represented *orthogonal covariation* with the variables on opposing axes. In some cases axes were unlabeled, but units made clear the variable measured on at least one axis. Thirty represented study time on the horizontal axis and scores on the vertical, whereas 20 interchanged the axes. Forty students used conventional ordering of values on the axes, that is, increasing value as one moves up or right; seven reversed the values on one axis (giving the visual impression of a positive covariation); and three showed values unordered in bar graphs (giving the visual impression of no covariation). Thirty-one responses appeared to indicate a perfect linear fit with values of equal spacing on each variable, and the other 19 showed some variation of a perfect linear fit. Students differed in the form of graph used: bar graphs (25), scattergraphs (7), line graphs (5), and line graphs of connected dots (13). Figure 7d (G9m) shows a line graph of connected dots with conventional axes and linear fit.

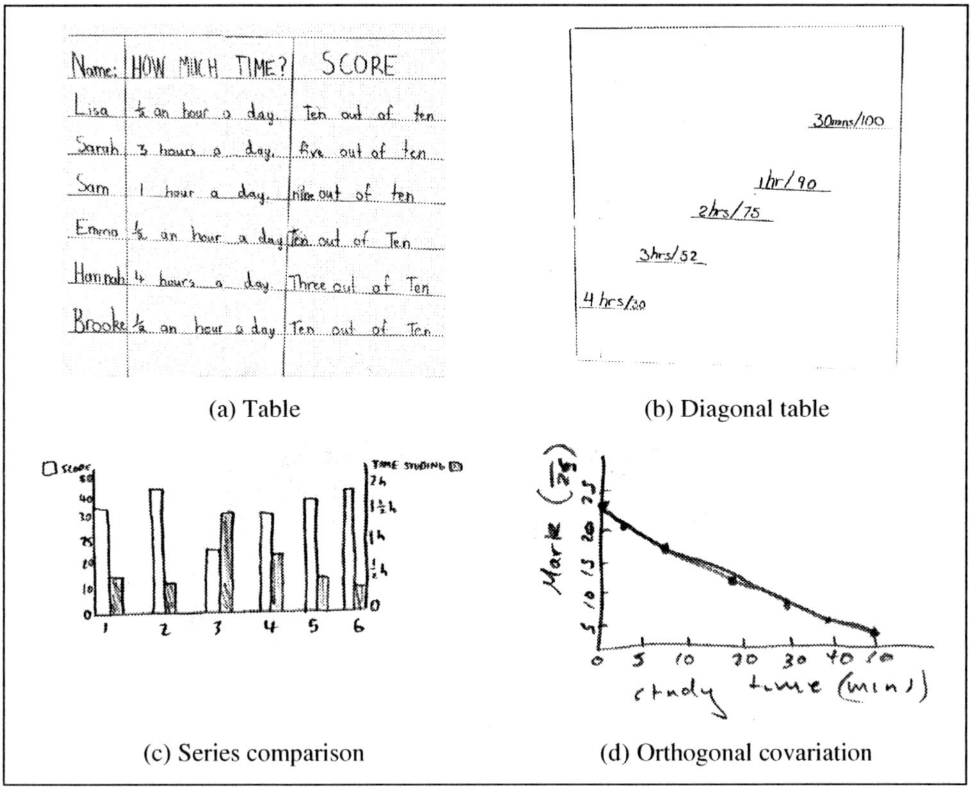

(a) Table

(b) Diagonal table

(c) Series comparison

(d) Orthogonal covariation

Figure 7. Student responses to Q1 at Level 3.

Graphing a Positive Covariation (Q1)*

A positive covariation is consistent with prior beliefs. Compared with the negative covariation task format (see Table 2), fewer students gave graphs with an incorrect direction (Level 2) and more students gave a single-variable (Level 1) graph, as if both variables could be aligned into a single axis of corresponding or identical values. Figure 8 shows examples of student responses from third-graders (Figures 8a, 8b, and 8c) and fifth-graders (Figures 8d and 8e). Many students included names for individual data cases (e.g., Figures 8a, 8b, and 8d), and others denoted cases by numbers (e.g., Figure 8e) or by separate representations (e.g., Figure 8c). Figure 8a was considered to show a single variable of study time, although if the student had indicated that position on the horizontal axis denoted score, the response would have been coded at Level 3.

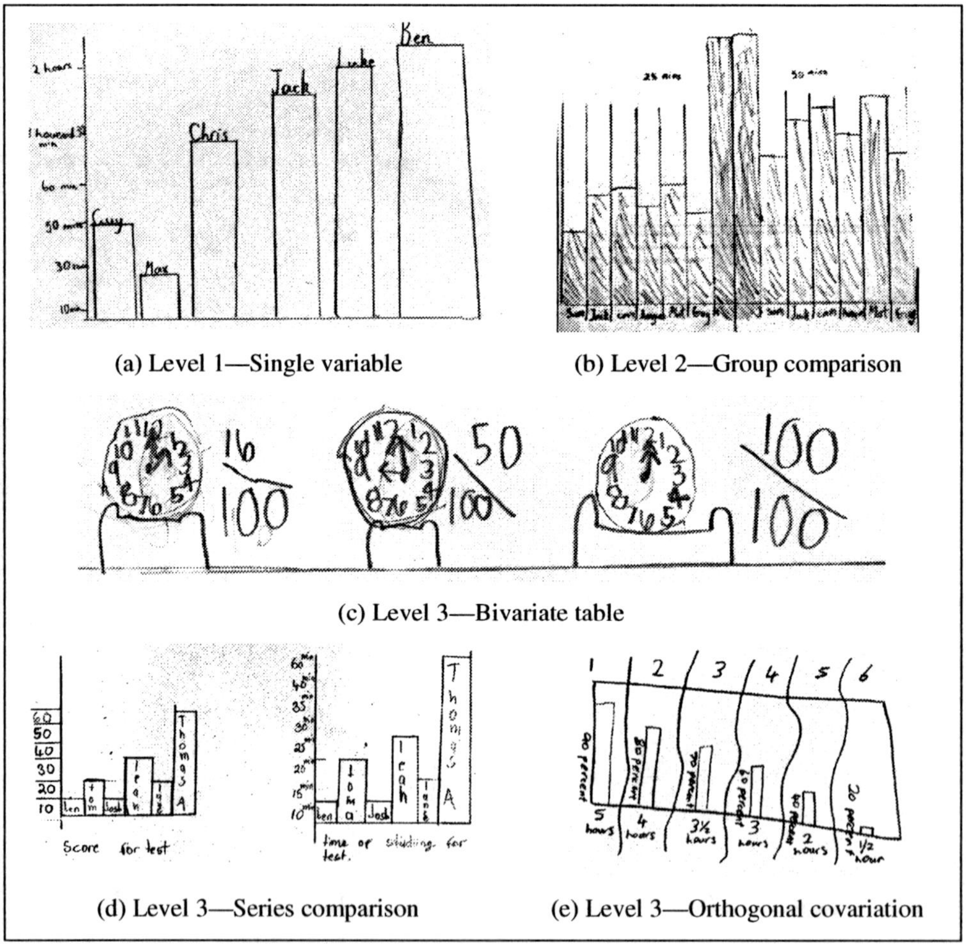

(a) Level 1—Single variable (b) Level 2—Group comparison

(c) Level 3—Bivariate table

(d) Level 3—Series comparison (e) Level 3—Orthogonal covariation

Figure 8. Student responses to Q1* (Positive Covariation task format).

Verbal Graph Interpretation (Q2a and Q2d)

Task 2 asked students to interpret a scattergraph (see Figure 3). Questions Q2a and Q2d (and Q2d*) involved verbal responses. To express the dual notions of appropriate variation and correspondence, responses to Q2a needed (a) to identify "noise" and "number of people" or paraphrases, and (b) make appropriate use of comparative values such as "less" or "more." The characteristics of the four levels of responses are shown in Table 3. In most cases coding was based on response to Q2a; however, in some cases Q2d (and Q2d*) served to demonstrate the student's ability to interpret verbally at a high level than demonstrated in Q2a. Further details are provided later for each level of response. As seen in Table 4, older students tended to respond at higher levels; and in particular, all students in grades 7 and 9 were able to identify at least a single aspect at level 1. Seventh- and ninth-grade males performed better than their female counterparts, although this is likely due to classes sampled rather than the students' gender.

Table 3. Characteristics of four levels of verbal and numerical graph interpretation

Level	Verbal Graph Interpretation	Numerical Graph Interpretation
0. Nonstatistical	Refers to: (a) context but not variables or the association, or (b) visual features, e.g., "dots"	Fails to read data values from axes. May refer to (a) context based "guesses," or (b) visual features, e.g., the maximum on the scale
1. Single Aspect	Refers to either (a) a single data point, or (b) a single variable (dependent)	Reads a value given corresponding bivariate value (Q2b: 27) but fails to use data to interpolate
2. Inadequate Covariation	Refers to both variables but: (a) *correspondence* is noted by comparing two or more points without generalizing to all 6 classes or to classes in general, or (b) *variable*s are described but the correspondence is not mentioned or is not in the correct direction	Reads values (Q2b: 27) and interpolates within local range but without accuracy (Q2c: 39–54 or 71–80)
3. Appropriate Covariation	Refers to both variables and indicates appropriate direction	Reads values (Q2b: 27) and interpolates with accuracy (Q2c: values 55–70)

Table 4. Percentage of student responses at four levels of verbal graph interpretation by gender and by grade (N = 121)

Levels of Verbal Graph Interpretation	Female Grade				Male Grade				Total
	3	5	7	9	3[a]	5[a]	7	9	(*N*)
0–Nonstatistical	31	13	0	0	31	29	0	0	13
1–Single Aspect	38	22	20	8	46	14	8	10	25
2–Inadequate Covariation	23	57	45	67	15	43	33	5	43
3–Appropriate Covariation	8	9	35	25	8	14	58	86	40
Total (*N*)	13	23	20	12	13	7	12	21	121[b]

[a] Third- and fifth-grade males were administered Q2d* rather than Q2d.

[b] Percentages do not always sum to 100 due to rounding.

Level 0: Nonstatistical. Some students offered responses that described no covariation. These included non-responses, responses generically about the topic, such as “that there is 6 classrooms and each dot shows that that is each classroom” (G3f) or “the graph shows class C, class A, class B, class D, class F, class E and numbers” (G5f).

Level 1: Single Aspect. One student commented on a single data point: “it shows that class C had 21 children in there and sound level is 70” (G3m). Many students referred to one variable, the level of noise, without reference to number of people in the classroom, although some mentioned that classrooms were involved. Some of these mentioned no values, with responses such as “noise” (G3m). Some commented that noise values varied, such as “it shows that some classes are noisier” (G3f). Others referred to specific values of noise, such as “80 is the most loud and zero is the most soft” (G3f).

Level 2: Inadequate Covariation. Some students referred to both variables but did not describe any covariation in the data, such as “the number of people in each class and the noise level” (G5f), or “level of noise goes up in 10’s and going across is the number of people in the class room which is going up from 20, 21, 22, to 30” (G9f). Possibly these students read the axis labels but not the data series. Others mentioned both variables and gave some evidence of generalizing covariation between the two variables, such as “that the classroom with the least people is the noisiest and the classroom with the most is the quietest” (G7f), and “that the class with the least people in it is making the most noise” (G5m).

Level 3: Appropriate Covariation. Some students generalized the graphs into a pattern statement, namely a description of the negative covariation. Some responses were simply stated, such as “that less people make more sound” (G7m), and some built up to the idea, for example, “Room C is the noisiest then A followed by B, E and D are each forty, then F brings up the rear, so the more people the less noise” (G7m). Some emphasized both ends of the generalization, similar to those at the previous level but describing “classes” in the plural to generalize either to the set of six classes or to classes in general: “The classes with less people are the loudest. The rooms with more people are the quietest” (G9m). Other students mentioned the imperfect nature of the covariation: “In most cases the higher the amount of noise the lower the amount of people with the exception of E" (G9m). Responses included statements that emphasized variation by comparison across cases such as "the more *X*, the less *Y*," "cases with more *X* have less *Y*," and "as *X* increases, *Y* decreases." No students gave responses that objectified the correspondence or relationship at the expense of variation, such as "*X* and *Y* are negatively/inversely related."

Numerical Graph Interpretation (Q2b and Q2c)

Numerical graph interpretation was assessed by two questions, one involving reading a value (Q2b), and the other involving interpolation (Q2c). The coding of the levels of response is shown in Table 3. There was a high degree of consistency between responses to Q2b and Q2c, in that of 83 responses showing some evidence of interpolation at Levels 2 and 3, only six students were unable to read values, and three of these responded "40" by reading point E. Of 12 nonstatistical (Level 0) responses, four did not respond to Q2b and five responded "23"—probably because of it appearing on the next line for Q2c. Nonstatistical responses to Q2c were idiosyncratic, such as "50, because some talk and some don't" (G3f). Single Aspect responses read a single value from the graph, but for Q2c, either acknowledged they did not know or gave responses that used single points in an idiosyncratic argument such as "30, under E" (G7m) and "80, because there would have been 50 people in the room" (G5f). Responses interpolating at Level 2 offered values in the ranges 39–54 or 71–80, and/or provided reasons related to adjacent data points, such as "If 23 people were in the class I would estimate 50 because in the classes of 24 they're 40 and 60 and 50 was in the middle" (G9f). Responses coded at Level 3 showed evidence of interpolation using the trend of the data to predict a value in the range 55–70. Many predicted a value of 65, often with reasoning such as, "about 65 because in the class of 24 it is 60 and in the class of 21 it is 70" (G7f); and some predicted other values such as, "60 because that is the trend of the graph" (G9m). The percentages of students who responded at each level are shown in Table 5. Notably, no third- or fifth-grade students responded at Level 3, whereas no seventh- or ninth-grade students responded at Level 0.

Associations among Skills

Associations among the skills of speculative data generation, verbal graph interpretation, and numerical graph interpretation are shown in Table 6. Using the scores of the levels on an interval scale from 0 to 3, numerical graph interpretation was highly correlated with verbal graph interpretation ($r_{119} = 0.54$) and with speculative data generation ($r_{109} = 0.47$), whereas the correlation of verbal graph interpretation with speculative data generation was weaker ($r_{109} = 0.30$).

Table 5. Percentage of student responses at four levels of numerical graph interpretation by gender and by grade (N = 121)

Levels of Numerical Graph Interpretation	Female Grade				Male Grade				Total
	3	5	7	9	3	5	7	9	(*N*)
0 – Nonstatistical	46	13	0	0	15	14	0	0	12
1 – Single Aspect	23	52	25	33	62	29	25	10	39
2 – Inadequate Covariation	31	35	35	50	23	57	50	10	40
3 – Appropriate Covariation	0	0	40	17	0	0	25	81	30
Total (*N*)	13	23	20	12	13	7	12	21	121[a]

[a] Percentages do not always sum to 100 due to rounding.

Table 6. Percentage of student responses at four levels of one skill by level of another skill

Response	Speculative Data Generation				Total	Verbal Graph Interpretation				Total
Level	0	1	2	3	(*N*)	0	1	2	3	(*N*)
Verbal Graph Interpretation										
0	0	23	19	9	13	—	—	—	—	—
1	42	46	24	12	24	—	—	—	—	—
2	42	23	48	26	35	—	—	—	—	—
3	17	8	10	52	39	—	—	—	—	—
Numerical Graph Interpretation										
0	33	23	24	0	12	15	24	9	0	12
1	17	62	48	25	36	54	40	44	8	39
2	50	8	29	35	36	31	28	35	35	40
3	0	8	0	40	27	0	8	12	58	30
Total (*N*)	12	13	21	65	111	13	25	43	40	121[a]

[a] Percentages do not always sum to 100 due to rounding.

DISCUSSION

Four levels of response were detailed for tasks concerning speculative data generation, verbal graph interpretation, and numerical graph interpretation. These levels relate closely to levels described in previous research of correlational reasoning (Ross & Cousins, 1993; Swatton, 1994) and graph comprehension (Curcio, 2001). Most students, even third-graders, offered responses that identified at least a single aspect related to the data, such as reading a value from a scatterplot, which demonstrated they could engage the task. Levels of verbal and numerical graph interpretation were highly correlated, possibly in part because the coding of both required reading from the axes, whether the variable label or the value on the scale.

Many students, even third-graders, demonstrated a negative covariation by speculative data generation. This finding extends to a younger age the findings of Swan, 1988 (success 37% of 13- to 14-year-olds) and of Mevarech & Kramarsky, 1997 (55% of eighth-graders). Reasons for this success rate may include the above-average capabilities of the sample of students, or the context of the task involving six discrete data cases in a familiar setting. A notable difference of the current study from previous research was the open-ended response format and the coding, which did not insist students represent the data in a certain form, such as with Cartesian axes.

This study set out to assess the skill of speculative data generation, irrespective of representational form. The highest level of response illustrates this skill with different forms of representation—tables of raw data, series comparison graphs, and orthogonal covariation graphs as well as bar graphs, line graphs, and scattergraphs—each with potential to be ordered or unordered. That some students drew tables rather than graphs may reflect the historical tradition noted to reproduce accurately all aspects of the data in a table rather than a graph (Beniger & Robyn, 1978), and raises the question of the graph constructor being aware of audience and of the purpose for the representation. In this study series comparison graphs were considered to demonstrate the highest level of speculative data generation (Konold, 2002; Watson & Moritz, 1997), whereas for the purposes of assessing graph production skills, Brasell and Rowe (1993) considered such graphs were Cartesian failures. Further, the principle of indicative labeling (Moritz, 2000) was used to assist assessing poorly labeled graphs. Figure 8c, for example, has only three bivariate data points; time is unlabeled, but can be inferred by the clock representation, and score is only inferred by the notation of "/100." The student did not use labels as requested, nor show six data points, but the representation illustrates the student expressed the two aspects of covariation, namely correspondence and variation. Clearly there are many aspects of student understanding we may seek to assess, including graph production skills to conform to various conventions. If we want to encourage the view of graphs as tools for analysis rather than ends in themselves (e.g., NCTM, 2000), then we need to permit and even encourage a variety of representations to achieve the purposes of engaging the data and reasoning about covariation. In short, there is a place for assessing the skill of speculative data generation, and this study indicates this assessment is appropriate by third grade.

Many of the different approaches to graphing observed—difficulties with labels or units, inversion of axes, reversal or uneven metric scales, and continuity versus discrete data points—have been observed previously (e.g., Chazan & Bethell, 1994). Selecting familiar and distinct variables for tasks may be important for students' reasoning and in particular for labeling and use of units. In light of the interpretation difficulties of some students in this study, such as reading values from the wrong axis, it may also be helpful to use distinctively different values for each variable—such as 1, 2, 3 versus 10, 20, 30—so that students can be clear which variable is referred to by a value. Use of discrete data appeared to encourage many students to consider six different cases in tables or bar graphs. Other students connected these data points by lines, or showed a line without data points. In algebra classes, covariation is often represented in a graph by a line or a line connecting points, whereas statistics is typified by the notion of data sets, which tend to be classified according to the number and type of variables and the number of cases of discrete values. What a line segment in a graph denotes should be clarified with respect to the variable. In some situations, such as measuring temperature, it may be a valid interpolation between known data points, and in other cases with discrete data, a connecting line may confuse what is measured. A slightly more sophisticated notion, acceptable in both statistics and algebra classes, is a straight line of best fit of the points, which may be formalized into an algebraic expression of the function.

IMPLICATIONS

Three difficulties students encountered, also observed by Mevarech and Kramarsky (1997), included (a) focusing on isolated bivariate points only, such as reducing study time from a numerical variable to a measure with only two categorical values; (b) focusing on a single variable rather than bivariate data; and (c) handling a negative covariation that was counter to prior belief in a positive association. These difficulties are discussed in the next three sections with suggestions for how teachers may build student understanding.

From Single Data Points to Global Trends

Many students described the scattergraph by reference to one or two bivariate data points, and several students drew single or double comparison graphs, that is, comparing one or two bivariate data points (e.g., Figures 5a and 6a). Pointwise approaches may provide an important way into many statistical issues—such as repeated values in either variable and the contextual understanding of data elements involving measurement and sampling issues—that do not occur in algebraic studies of continuous functions. In this respect, tables and series comparison graphs (e.g., Figures 7a, 7c, and 8d) may be significant representations for reasoning about covariation, since they devote a feature (column or axis) to retain case information, such as the name of a person, and can represent two cases with identical bivariate values, which are slightly problematic to display in Cartesian coordinates.

Students' reasoning about isolated data points emphasized correspondence of two measures but did not describe variation to indicate covariation adequately. Development of the pointwise approach in verbal interpretations may be considered as a progression of comparisons within variables, from single-point values ("class C had 21 children …") to comparison of points ("the classroom with the least people is the noisiest …") to generalizing beyond the available points ("the more people the less noise …"). This follows the levels of "reading the data," "reading between the data," and "reading beyond the data" described by Curcio (2001). For speculative data generation, a pointwise approach was the building block used by some young students who added more data points; for example, the student who drew Figure 7a probably began with a representation much like Figure 5a. In generating more points, students appeared to find it easy to maintain the appropriate correspondence between the measures: Students who drew double or group comparisons conceived of study times as two high and low extremes, and generated scores that were corresponding low and high extremes (e.g., Figures 6a and 6b). Even the student who drew the table in Figure 7a appears to have clustered times into high (3 and 4) and low (½ and 1) values and corresponding scores into low (3 and 5) and high (9 and 10). The difficulty in generating more data points appeared to be generating appropriate variation that ensured both numerical variables do vary.

An important idea for development of reasoning beyond isolated points or dichotomous extremes may be the ordering of cases on a single variable (Ross & Cousins, 1993; Wavering, 1989). For speculative data generation, one can generate new cases that have incrementally more or less of one measure, often at fixed differences, and then simply increment the other variable appropriately. Such fixed differences move a student away from considering isolated cases that may include repeated values, to a generation of patterns within a variable that is frequent in algebra. For real-world data variables, generating new values may be restricted by the minimum or maximum possible values. Figure 7c illustrates the impact of ordering and extremes values, where, after generating two cases, the student reached the maximum score on the scale of 50, and thus broke the pattern to generate the rightmost four cases in order. Extremes of possible values may also explain why Figures 7b and 8c did not include six data cases: Having reached a score of 100, the students could not generate another score in the order. For verbal interpretation, ordering of one variable allows variation of the other variable to be observed as an increasing or decreasing feature of the data series (a trend) verbally summed up as a single phrase, thus corresponding to the graphic language of the data series with the verbal language of change (Yerushalmy, 1997).

From Single Variables to Bivariate Data

Sixteen students drew graphs of single variables, and many described only the variable noise in verbal descriptions of a scattergraph. These students emphasized variation but did not describe correspondence of two measures to indicate covariation adequately. Those who had success in verbally describing the covariation all used the language of incremental change across cases, implied by ordering each variable, rather than objectifying the correspondence as "*X* is related to *Y*." Interpolation tasks, though numerical and often involving reference to specific points, may in fact encourage students to discuss differences between points and lead to discussion of increments more globally.

A change-over-time approach to covariation has been recommended by algebra curricula (e.g., NCTM, 2000) and researchers (e.g., Nemirovsky, 1996b). Such an approach carries with it implicitly the understanding that time is ordered, and thus verbal phrases such as "it started to grow faster, then it slowed down again" (NCTM, 2000, p. 163) allow students to focus on change of one variable without attending to the correspondence of the variables, as is required if the independent variable is not time. Tables and series comparison graphs may be significant representations not just for developing reasoning to include more cases as noted earlier but also for emphasizing both variables and the correspondence of individual data values. Both of these representations treat each variable as a measured variable (often termed dependent and, if graphed, represented on the vertical axis) across a number of cases, whereas Cartesian graphs have axes conventionally considered independent (horizontal) and dependent (vertical). Aside from the implication of dependency and possibly causation (difficulties discussed in the next section), some students do not attend to the variable on the horizontal axis, such as the many interpretations involving only the variable noise. Tables and series comparison graphs (e.g., Figures 7c and 8d) may be considered as natural progressions composed of two univariate tables or graphs (e.g., Figures 5b and 8a). As already noted, ordering of values is a key concept that allows not only handling of variation, but also establishing correspondence case-wise. Once cases are ordered by one variable, such as in the horizontal dimension, the foundation is set for coordinating the correspondence of two variables in Cartesian coordinates. The transformation from an ordered table (no use of dimension), or from an ordered-series comparison graph with both data series in an axis framework (both variables denoted by vertical dimension), to the orthogonal covariation of Cartesian coordinates can be seen in Figures 7b and 8e, where bivariate cases have been (reverse) ordered by study times in the horizontal dimension, and vertical height incorporated to denote variation in test scores. In these representations, moving the written value labels from the data elements to the axes results in Cartesian coordinates, as in Figure 7d.

From Prior Beliefs to Data-Based Judgments

Some students generated or interpreted a data set as a positive covariation based on prior beliefs when a negative covariation existed in the data. Others wrote the values on one axis in reverse order, thus displaying a negative covariation but appearing visually as an increasing function, in accord with an alternative conception that all covariation graphs should appear in a positive direction (Mevarech & Kramarsky, 1997). The counterintuitive nature of the tasks was important for assessment purposes in eliciting these responses. An important level for these students to achieve was appreciating covariation in context, similar to Tier 2 of Watson's (2000; Watson & Moritz, 1997) statistical literacy hierarchy, evident by representing a verbal claim in a graph or by interpreting a graph in a verbal statement. To do this, students must be encouraged to suspend prior beliefs temporarily to look at the data and examine what covariation might be indicated.

Once the claim of covariation is understood in context, students must question the process of inference from statistical data to causal claim—Tier 3 of Watson's hierarchy. At this level, awareness of prior beliefs should be encouraged, as well as its balanced integration with available data. An important feature of using tasks involving counterintuitive covariation is that they should naturally raise questions about reliability of the data set, and about generalizability to a causal inference. The tasks involving only six data points were designed to be easy for students to break down the tasks to represent covariation as a series of corresponding cases and draw it quickly, but also importantly introduced the issue of sample size. Other questions used as part of this wider study have elicited student responses noting that small sample size made generalization difficult. These responses will be discussed in future research reports.

Future Teaching and Research

This study has shown that graphing and verbalizing covariation, using familiar contexts, can occur before the standardization of graphing conventions. Teaching of standard graphs forms, such as Cartesian coordinates, might not eliminate alternative conceptions (Mevarech & Kramarsky, 1997), and might even inhibit reasoning about covariation, if students are able to interpret only their own representation. Instruction may be more effective if it builds on students' existing reasoning and challenges further development of this reasoning. Employing the Piagetian principle of cognitive conflict, Watson and Moritz (2001) asked students to construct a pictograph, and then showed students different representations and asked them to comment; many students could acknowledge the merits of more structured graphs. For research this procedure has potential for students to have moments of learning during observation, as they recognize the merits of another way of reasoning. The new ideas can be selectively shown in order to build on a students' existing idea. For teaching situations, it may prove helpful to use graphs hand-drawn by anonymous students, similar to the student's own, since this removes the emotional personal threat of one's own work being critiqued unfavorably. Once students have begun to engage the context of the variables, they can begin to investigate covariation among variables, discuss ways of reasoning about covariation, and only slowly be introduced to conventions for expressing their reasoning in graphs, words, and numerical methods.

REFERENCES

Ainley, J. (1995). Re-viewing graphing: Traditional and intuitive approaches. *For the Learning of Mathematics, 15*(2), 10–16.

Alloy, L. B., & Tabachnik, N. (1984). Assessment of covariation by humans and animals: The joint influence of prior expectations and current situational information. *Psychological Review, 91*(1), 112–149.

Australian Education Council. (1991). *A national statement on mathematics for Australian schools.* Carlton, Vic.: Author.

Australian Education Council. (1994). *Mathematics—A curriculum profile for Australian schools.* Carlton, Vic.: Curriculum Corporation.

Batanero, C., Estepa, A., & Godino, J. D. (1997). Evolution of students' understanding of statistical association in a computer based teaching environment. In J. Garfield & G. Burrill (Eds.), *Research on the role of technology in teaching and learning statistics* (pp. 191–205). Voorburg, The Netherlands: International Statistical Institute.

Batanero, C., Estepa, A., Godino, J. D., & Green, D. R. (1996). Intuitive strategies and preconceptions about association in contingency tables. *Journal for Research in Mathematics Education, 27*, 151–169.

Bell, A., Brekke, G., & Swan, M. (1987a). Diagnostic teaching: 4 Graphical interpretations. *Mathematics Teaching, 119*, 56–59.

Bell, A., Brekke, G., & Swan, M. (1987b). Diagnostic teaching: 5 Graphical interpretation teaching styles and their effects. *Mathematics Teaching, 120*, 50–57.

Bell, A., & Janvier, C. (1981). The interpretation of graphs representing situations. *For the Learning of Mathematics, 2*(1), 34–42.

Beniger, J. R., & Robyn D. L. (1978). Quantitative graphics in statistics: A brief history, *American Statistician, 32*, 1–10.

Ben-Zvi, D., & Arcavi, A. (2001). Junior high school students' construction of global views of data and data representations. *Educational Studies in Mathematics, 45*, 35–65.

Biggs, J. B., & Collis, K. F. (1982). *Evaluating the quality of learning: The SOLO taxonomy*. New York: Academic Press.

Brasell, H. M., & Rowe, M. B. (1993). Graphing skills among high school physics students. *School Science and Mathematics, 93*(2), 63–70.

Carlson, M., Jacobs, S., Coe, E., Larsen, S., & Hsu, E. (2002). Applying covariational reasoning while modeling dynamics events: A framework and a study. *Journal for Research in Mathematics Education, 33*, 352–378.

Chazan, D., & Bethell, S. C. (1994). Sketching graphs of an independent and a dependent quantity: Difficulties in learning to make stylized, conventional "pictures." In J. P. da Ponte & J. F. Matos (Eds.), *Proceedings of the 18th Annual Conference of the International Group for the Psychology of Mathematics Education, 2*, 176–184. Lisbon: University of Lisbon.

Clement, J. (1989). The concept of variation and misconceptions in Cartesian graphing. *Focus on Learning Problems in Mathematics, 11*, 77–87.

Cobb, P., McClain, K., & Gravemeijer, K. (2003). Learning about statistical covariation. *Cognition and Instruction, 21*(1), 1–78.

Coulombe, W. N., & Berenson, S. B. (2001). Representations of patterns and functions: Tools for learning. In A. A. Cuoco & F. R. Curcio (Eds.), *The roles of representation in school mathematics (2001 Yearbook)* (pp. 166–172). Reston, VA: National Council of Teachers of Mathematics.

Crocker, J. (1981). Judgment of covariation by social perceivers. *Psychological Bulletin, 90*, 272–292.

Curcio, F. R. (2001). *Developing data-graph comprehension in grades K through 8* (2d ed.). Reston, VA: National Council of Teachers of Mathematics.

Department for Education and Employment. (1999). *Mathematics: The national curriculum for England*. London: Author and Qualifications and Curriculum Authority.

Donnelly, J. F., & Welford, A. G. (1989). Assessing pupils' ability to generalize. International *Journal of Science Education, 11*, 161–171.

Estepa, A., & Batanero, C. (1996). Judgments of correlation in scatterplots: Students' intuitive strategies and preconceptions. *Hiroshima Journal of Mathematics Education, 4*, 21–41.

Inhelder, B., & Piaget, J. (1958). Random variations and correlations. In B. Inhelder & J. Piaget, *The growth of logical thinking from childhood to adolescence* (A. Parsons & S. Milgram, Trans.) (pp. 224–242). London: Routledge & Kegan Paul.

Janvier, C. (1978). The interpretation of complex Cartesian graphs representing situations: Studies and teaching experiments. Unpublished doctoral dissertation, University of Nottingham.

Jennings, D. L., Amabile, T. M., & Ross, L. (1982). Informal covariation assessment: Data-based versus theory-based judgments. In D. Kahneman, P. Slovic, & A. Tversky (Eds.), *Judgment under uncertainty: Heuristics and biases* (pp. 211–230). Cambridge, UK: Cambridge University Press.

Konold, C. (2002). Alternatives to scatterplots. In B. Phillips (Ed.), *Proceedings of the Sixth International Conference on Teaching Statistics*. Cape Town, South Africa.

Krabbendam, H. (1982). The non-qualitative way of describing relations and the role of graphs: Some experiments. In G. Van Barnveld & H. Krabbendam (Eds.), *Conference on functions* (Report 1, pp. 125–146). Enschede, The Netherlands: Foundation for Curriculum Development.

Leinhardt, G., Zaslavsky, O., & Stein, M. K. (1990). Functions, graphs and graphing: Tasks, learning and teaching. *Review of Educational Research, 60*(1), 1–64.

McKnight, C. C. (1990). Critical evaluation of quantitative arguments. In G. Kulm (Ed.), *Assessing higher order thinking in mathematics* (pp. 169–185). Washington, DC: American Association for the Advancement of Science.

Mevarech, Z. R., & Kramarsky, B. (1997). From verbal descriptions to graphic representations: Stability and change in students' alternative conceptions. *Educational Studies in Mathematics, 32*, 229–263.

Meyer, J., Shinar, D., & Leiser D. (1997). Multiple factors that determine performance with tables and graphs. *Human Factors, 39*(2), 268–286.

Miles, M. B., & Huberman, A. M. (1994). *Qualitative data analysis: An expanded sourcebook* (2nd ed.). Thousand Oaks, CA: Sage.

Ministry of Education. (1992). *Mathematics in the New Zealand curriculum*. Wellington, NZ: Author.

Moore, D. S. (1990). Uncertainty. In L. A. Steen (Ed.), *On the shoulders of giants: New approaches to numeracy* (pp. 95–137). Washington, DC: National Academy Press.

Moritz, J. B. (2000). Graphical representations of statistical associations by upper primary students. In J. Bana & A. Chapman (Eds.), *Mathematics education beyond 2000* (Proceedings of the 23rd Annual Conference of the Mathematics Education Research Group of Australasia, 2, 440–447. Perth: Mathematics Education Research Group of Australasia.

Moritz, J. B. (2002). Study times and test scores: What students' graphs show. *Australian Primary Mathematics Classroom, 7*(1), 24–31.

National Council of Teachers of Mathematics. (2000). *Principles and standards for school mathematics*. Reston, VA: Author.

Nemirovsky, R. (1996a). A functional approach to algebra: Two issues that emerge. In N. Bednarz, C. Kieran, & L. Lee (Eds.), *Approaches to algebra: Perspectives for research and teaching* (pp. 295–313). Dordrecht, The Netherlands: Kluwer Academic Publishers.

Nemirovsky, R. (1996b). Mathematical narratives, modeling, and algebra. In N. Bednarz, C. Kieran, & L. Lee (Eds.), *Approaches to algebra: Perspectives for research and teaching* (pp. 197–220). Dordrecht, The Netherlands: Kluwer Academic Publishers.

Piaget, J. (1983). Piaget's theory. (G. Cellerier & J. Langer, Trans.) In P. Mussen (Ed.), *Handbook of child psychology* (4th ed., Vol. 1, pp. 103–128). New York: Wiley.

Pinker, S. (1990). A theory of graph comprehension. In R. Freedle (Ed.), *Artificial intelligence and the future of testing* (pp. 73–126). Hillsdale, NJ: Erlbaum.

Ross, J. A., & Cousins, J. B. (1993). Patterns of student growth in reasoning about correlational problems. *Journal of Educational Psychology, 85*(1), 49–65.

Swan, M. (1985). *The language of functions and graphs*. University of Nottingham: Shell Center.

Swan, M. (1988). Learning the language of functions and graphs. In J. Pegg (Ed.), *Mathematics Interfaces: Proceedings of the 12th Biennial Conference of the Australian Association of Mathematics Teachers* (pp. 76–80). Newcastle, NSW: The New England Mathematical Association.

Swatton, P. (1994). Pupils' performance within the domain of data interpretation, with particular reference to pattern recognition. *Research in Science and Technological Education, 12*(2), 129–144.

Swatton, P., & Taylor, R. M. (1994), Pupil performance in graphical tasks and its relationship to the ability to handle variables. *British Educational Research Journal, 20*, 227–243.

Tilling, L. (1975). Early experimental graphs. *British Journal for the History of Science, 8*, 193–213.

Tufte, E. R. (1983). *The visual display of quantitative information*. Cheshire, CT: Graphics Press.

Watson, J. M. (2000). Statistics in context. *Mathematics Teacher, 93*(1), 54–58.

Watson, J. M., & Moritz, J. B. (1997). Student analysis of variables in a media context. In B. Phillips (Ed.), *Papers on Statistical Education Presented at ICME-8* (pp. 129–147). Hawthorn, Australia: Swinburne Press.

Watson, J. M., & Moritz, J. B. (2001). Development of reasoning associated with pictographs: Representing, interpreting, and predicting. *Educational Studies in Mathematics, 48*(1), 47–81.

Wavering, M. J. (1989). Logical reasoning necessary to make line graphs. *Journal of Research in Science Teaching, 26*(5), 373–379.

Yerushalmy, M. (1997). Mathematizing verbal descriptions of situations: A language to support modeling. *Cognition and Instruction, 15*(2), 207–264.

Chapter 11

STUDENTS' REASONING ABOUT THE NORMAL DISTRIBUTION[1]

Carmen Batanero[1], Liliana Mabel Tauber[2], and Victoria Sánchez[3]
Universidad de Granada, Spain[1], Universidad Nacional del Litoral, Santa Fe, Argentina[2], and Universidad de Sevilla, Spain[3].

OVERVIEW

In this paper we present results from research on students' reasoning about the normal distribution in a university-level introductory course. One hundred and seventeen students took part in a teaching experiment based on the use of computers for nine hours, as part of a 90-hour course. The teaching experiment took place during six class sessions. Three sessions were carried out in a traditional classroom, and in another three sessions students worked on the computer using activities involving the analysis of real data. At the end of the course students were asked to solve three open-ended tasks that involved the use of computers. Semiotic analysis of the students' written protocols as well as interviews with a small number of students were used to classify different aspects of correct and incorrect reasoning about the normal distribution used by students when solving the tasks. Examples of students' reasoning in the different categories are presented.

THE PROBLEM

One problem encountered by students in the introductory statistics course at university level is making the transition from data analysis to statistical inference. To make this transition, students are introduced to probability distributions, with most of the emphasis placed on the normal distribution. The normal distribution is an important model for students to learn about and use for many reasons, such as:

[1] This research has been supported by DGES grant BS02000-1507 (M.E.C., Madrid).

D. Ben-Zvi and J. Garfield (eds.),
The Challenge of Developing Statistical Literacy, Reasoning and Thinking, 257–276.

- Many physical, biological, and psychological phenomena can be reasonably modeled by this distribution such as physical measures, test scores and measurement errors.
- The normal distribution is a good approximation for other distributions—such as the binomial, Poisson, and *t* distributions—under certain conditions.
- The Central Limit Theorem assures that in sufficiently large samples the sample mean has an approximately normal distribution, even when samples are taken from nonnormal populations.
- Many statistical methods require the condition of random samples from normal distributions.

We begin by briefly describing the foundations and methodology of our study. We then present results from the students' assessment and suggest implications for the teaching of normal distributions. For additional analyses based on this study see Batanero, Tauber, and Meyer (1999) and Batanero, Tauber, and Sánchez (2001).

THE LITERATURE AND BACKGROUND

Previous Research

There is little research investigating students' understanding of the normal distribution, and most of these studies examine isolated aspects in the understanding of this concept. The first pioneering work was carried out by Piaget and Inhelder (1951), who studied children's spontaneous development of the idea of stochastic convergence. The authors analyzed children's perception of the progressive regularity in the pattern of sand falling through a small hole (in the Galton apparatus or in a sand clock). They considered that children need to grasp the symmetry of all the possible sand paths falling through the hole, the probability equivalence between the symmetrical trajectory, the spread and the role of replication, before they are able to predict the final regularity that produces a bell-shaped (normal) distribution. This understanding takes place in the formal operations stage (13- to 14-year-olds).

Regarding university students, Huck, Cross, and Clark (1986) identified two erroneous conceptions about normal standard scores: On the one hand, some students believe that all standard scores will always range between –3 and +3, while other students think there is no restriction on the maximum and minimum values in these scores. Each of those beliefs is linked to a misconception about the normal distribution. The students who think that *z*-scores always vary from –3 to + 3 have frequently used either a picture or a table of the standard normal curve, with this range of variation. In a similar way, the students who believe that *z*-scores have no upper or lower limits have learned that the tails of the normal curve are asymptotic to the abscissa; thus they make an incorrect generalization, because they do not notice that no finite distribution is exactly normal.

For example, if we consider the number of girls born out of 10 newborn babies, this is a random variable X, which follows the binomial distribution with $n = 10$ and $p = 0.5$. The mean of this variable is $np = 5$ and the variance is $npq = 2.5$. So the maximum z-score that could be obtained from this variable is $z_{max} = (10 - 5)/\sqrt{2.5} = 3.16$. Thus we have a finite limit, but it is greater than 3.

In related studies, researchers have explored students' understanding of the Central Limit Theorem and have found misconceptions regarding the normality of sampling distributions (e.g., Vallecillos, 1996, 1999; Méndez, 1991; delMas, Garfield, & Chance, 1999). Wilensky (1995, 1997) examined student behavior when solving problems involving the normal distribution. He defined *epistemological anxiety* as the feeling of confusion and indecision that students experience when faced with the different paths for solving a problem. In interviews with students and professionals with statistical knowledge, Wilensky asked them to solve a problem by using computer simulation. Although most subjects in his research could solve problems related to the normal distribution, they were unable to justify the use of the normal distribution instead of another concept or distribution, and showed a high epistemological anxiety.

Meaning and Understanding of Normal Distributions in a Computer-Based Course

Our research is based on a theoretical framework about the meaning and understanding of mathematical and statistical concepts (Godino, 1996; Godino & Batanero, 1998). This model assumes that the understanding of normal distributions (or any other concept) emerges when students solve problems related to that concept. The meaning (understanding) of the normal distribution is conceived as a complex system, which contains five different types of elements:

1. *Problems and situations from which the object emerges.* In our teaching experiments, students solved the following types of problems: (a) fitting a curve to a histogram or frequency polygon for empirical data distributions, (b) approximating the binomial or Poisson distributions, and (c) finding the approximate sampling distribution of the sample mean and sample proportion for large samples (asymptotic distributions).
2. *Symbols, words, and graphs used to represent or to manipulate the data and concepts involved.* In our teaching, we considered three different types of representations:
   a) *Static paper-and-pencil graphs and numerical values of statistical measures,* such as histograms, density curves, box plots, stem-leaf plots, numerical values of averages, spread, skewness, and kurtosis. These might appear in the written material given to the students, or be obtained by the students or teacher.
   b) *Verbal and algebraic representations of the normal distribution; its properties or concepts related to normal distribution,* such as the words *normal* and *distribution;* the expressions *density curve, parameters of the*

normal distribution, the symbol N (μ, σ), equation of density function, and so forth.

c) *Dynamic graphical representations on the computer.* The Statgraphics software program was used in the teaching. This program offers a variety of simultaneous representations on the same screen which are easily manipulated and modified. These representations include histograms, frequency polygons, density curves, box plots, stem-leaf plots, and symmetry and normal probability plots. The software also allows simulation of different distribution, including the normal distribution.

3. *Procedures and strategies to solve the problem.* Beyond the descriptive analyses of the variables studied in the experiment, the students were introduced to computing probabilities under the curve, finding standard scores, and critical values (computed by the computer or by hand).
4. *Definitions and properties.* Symmetry and kurtosis: relative position of the mean, median and mode, areas above and below the mean, probabilities within one, two and three standard deviations, meanings of parameters, sampling distributions for means and proportions, and random variables.
5. *Arguments and proofs.* Informal arguments and proofs made using graphical representation, computer simulations, generalization, analysis, and synthesis.

SUBJECTS AND METHOD

Sample and Teaching Context

The setting of this study was an elective, introductory statistics course offered by the Faculty of Education, University of Granada. The instruction for the topic of normal distributions was designed to take into account the different elements of meaning as just described. Taking the course were 117 students (divided into 4 groups), most of whom were majoring in Pedagogy or Business Studies. Some students were from the School of Teachers Training, Psychology, or Economics.

At the beginning of the course students were given a test of statistical reasoning (Garfield, 1991) to assess their reasoning about simple statistical concepts such as averages or sampling, as well as to determine the possible existence of misconceptions. An examination of students' responses on the statistical reasoning test revealed some errors related to sampling variability (representativeness heuristics), sample bias, interpretation of association, and lack of awareness of the effect of atypical values on averages. There was a good understanding of probability, although some students showed incorrect conceptions about random sequences.

Before starting the teaching of the normal distribution, the students were taught the foundations of descriptive statistics and some probability, with particular emphasis on helping them to overcome the biases and errors mentioned. Six 1.5-

hour sessions were spent teaching the normal distribution, and another 4 hours were spent studying sampling and confidence intervals. Students received written material specifically prepared for the experiment and asked to read it beforehand. Half of these sessions were carried out in a traditional classroom, where the lecturer introduced the normal distribution as a model to describe empirical data, using a computer with projection facility. Three samples (n = 100, 1,000, and 10,000 observations) of intelligence quotient (IQ) scores were used to progressively show the increasing regularity of the frequency histogram and polygon, when increasing the sample size. The lecturer also presented the students with written material, posed some problems to encourage the students to discover for themselves all the elements of meaning described in section 3.2, and guided student discussion as they solved these problems.

The remaining sessions were carried out in a computer lab, where pairs of students worked on a computer to solve data analysis activities, using examples of real data sets from students' physical measures, test scores, and temperatures, which included variables that could be fitted to the normal distribution and other variables where this was not possible. Activities included checking properties such as unimodality or skewness; deciding whether the normal curve provided a good fit for some of the variables; computing probabilities under the normal curve; finding critical values; comparing different normal distributions by using standardization; changing the parameters in a normal curve to assess the effect on the density curve and on the probabilities in a given interval; and solving application problems. Students received support from their partner or the lecturer if they were unable to perform the tasks, and there was also collective discussion of results.

Assessing Students' Reasoning about the Normal Distribution

At the end of the course students were given three open-ended tasks, to assess their reasoning about the normal distribution as part of a final exam that included additional content beyond this unit. These questions referred to a data file students had not seen before, which included qualitative and quantitative (discrete and continuous) variables (See Table 1). The students worked alone with the Statgraphics program, and they were free to solve the problem using the different tools they were familiar with.

Each problem asked students to complete a task and to explain and justify their responses in detail, following guidelines by Gal (1997), who distinguished two types of questions to use when asking students to interpret statistical information. Literal reading questions ask students for unambiguous answers—they are either right or wrong. In contrast, to evaluate questions aimed at eliciting students' ideas about overall patterns of data, we need information about the evidential basis for the students' judgments, their reasoning process, and the strategy they used to relate data elements to each other. The first type of question was taken into account in a questionnaire with 21 items, which was also given to the students in order to assess literal understanding for a wide number of elements of the normal distribution

(Batanero et al., 2001). The second type of question considered by Gal (1997) was considered in the following open tasks given to students.

> *Task 1*
> In this data file, find a variable that could be fitted by a normal distribution. Explain your reasons for selecting that variable and the procedure you have used.

In this task the student is asked to discriminate between variables that can be well fitted to a normal distribution and others for which this is not possible. In addition to determining the student's criteria when performing the selection (the properties they attribute to normal distributions), we expected students to analyze several variables and use different approaches to check the properties of the different variables to determine which would best approximate a normal distribution. We also expected students to synthesize the results to obtain a conclusion from all their analyses. We hoped that student responses to this task would reveal their reasoning.

> *Task 2*
> Compute the appropriate values of parameters for the normal distribution to which you have fitted a variable chosen in question 1.

In this question the students have to remember what the parameters in a normal distribution (mean and variance) are. We also expected them to remember how to estimate the population mean from the sample mean and to use the appropriate Statgraphic program to do this estimation. Finally, we expected the students to discriminate between the ideas of statistics (e.g., measures based on sample data) and parameters (e.g., measures for atheoretical population model).

> *Task 3*
> Compute the median and quartiles for the theoretical distribution you have constructed in Task 2.

The aim is to evaluate the students' reasoning about the ideas of median and quartiles for a normal distribution. Again, discrimination between empirical data distribution and the theoretical model used to fit these data is needed. We expect the student to use the critical value facility of Statgraphics to find the median and quartiles in the theoretical distribution. Those students who do not discriminate will probably compute the median and quartile from the raw empirical data with the summary statistics program.

The three tasks just described were also used to evaluate the students' ability to operate the statistical software and to interpret its results. Since the students were free to solve the tasks using any previous knowledge to support their reasoning, we could evaluate the correct or incorrect use of the different meaning elements (representations, actions, definitions, properties, and arguments) that we defined earlier and examine how these different elements were interrelated.

Each student worked individually with the Statgraphics and produced a written report using the word processor, in which they included all the tables and graphs needed to support their responses. Students were encouraged to give detailed reasoning. Once the data were collected, the reports were printed and we made a content analysis. We identified which elements of meaning each student used correctly and incorrectly to solve the tasks.

In the next section we provide a global analysis for each question and then describe the elements of meaning used by the students.

RESULTS AND ANALYSIS

Students' Perception of Normality

In Table 1, we include the features of variables in the file and the frequency and percentage of students who selected each variable in responding to the first question. The normal distribution provided a good fit for two of these variables: *Time to run 30 m (December)* and *Heartbeats after 30 press-ups*. The first variable, *Time to run 30 m*, was constructed by simulating a normal continuous distribution. Normality can be checked easily in this variable from its graphical representation; the skewness and kurtosis coefficient were very close to zero, although the mean, median, and mode did not exactly coincide. *Heartbeats after 30 press-ups* was a discrete variable; however, its many different values, its shape, and the values of its different parameters suggested that the normal distribution could provide an acceptable fit.

Table 1. Description of the variables that students considered to fit a normal distribution well

Variable	**Variable Features**				**Students choosing this variable (%)**
	Variable type	**Skewness**	**Kurtosis**	**Mean, median, and mode**	
Age	Discrete; three different values	0	–0.56	13, 13, 13	27 (23.1)
Height	Continuous Multimodal	0.85	2.23	156.1, 155.5, †	26 (22.2)
Heartbeats after 30 press-ups*	Discrete; many different values	0.01	–0.19	123.4, 122, 122	37 (31.6)
Time spent to run 30 m.(Dec.)*	Continuous	0.23	–0.42	4.4, 4.4, 5.5	12 (10.3)
Weight	Continuous Atypical values	2.38	9.76	48.6, 46, 45	4 (3.4)
Heartbeats at rest	Discrete; many different values	0.2	–0.48	71.4, 72, 72	6 (5.2)
Time spent to run 30 m. (Sep.)	Continuous	2.4	12.2	5.3, 5.2, 5	4 (3.4)
No answer					9 (7.2)

* Correct answer. The normal distribution is a good approximation for these variables

† Although *Height* had in fact three modes: 150, 155, 157, that were visible from the stem plot, this was noticeable only from the histogram with specific interval widths.

The variable *Height*, despite being symmetric, had kurtosis higher than expected and was multimodal, though this was noticeable only by examining a stem-and-leaf plot or histogram of the data.

Some of these students confused the empirical data distribution for *Age* (Fig. 1a) with the theoretical distribution they fitted to the data. In Figure 1b the data frequency histogram for *Age* and a superimposed theoretical normal curve are plotted. Some students just checked the shape of the theoretical density curve (the normal curve with the data mean and standard deviation) without taking into account whether the empirical histogram approached this theoretical curve or not.

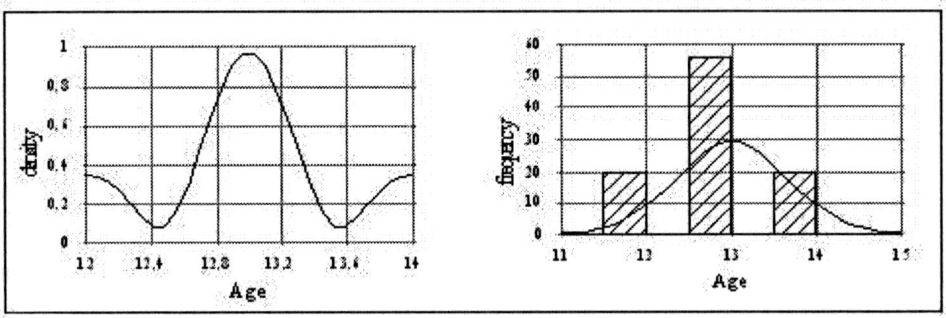

Figure 1. (a) Empirical density curve for *Age* (b) Theoretical normal curve fitted to *Age*.

Twenty-two percent of students selected a variable with high kurtosis (*Height*). In the following example, while the student could perceive the symmetry from the graphical representation of data, this graph was however unproductive as regards the interpretation of the standard kurtosis coefficient (4.46) that was computed by the student. The student did not compute the median and mode. We assume he visually perceived the curve symmetry and from this property he assumed the equality of mean, median, and mode.

Example 2

"I computed the mean (156.1) and standard deviation (8, 93) and they approach those from the normal distribution. Then I represented the data (Figure 2) and it looks very similar to the normal curve. The values of mean, median and mode also coincide. Std Kurtosis = 4.46" (Student 2).

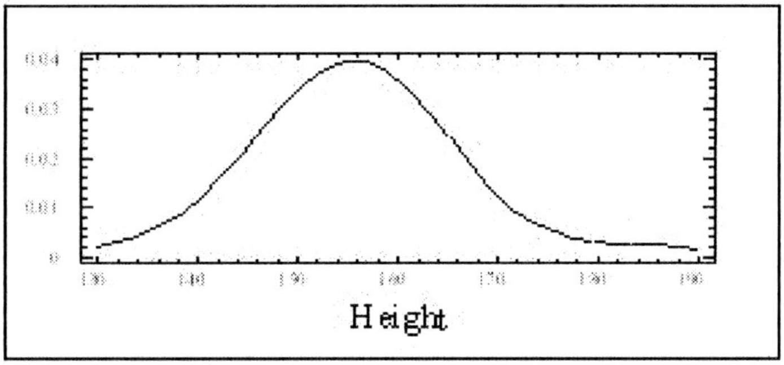

Figure 2. Density trace for *Height*.

Finding the Parameters

Table 2 displays the students' solutions to question 2. Some students provided incorrect parameters or additional parameters such as the median that are not needed to define the normal distribution. In Example 3, the student confuses the tail areas with the distribution parameters. In Example 4, the student has no clear idea of what the parameters are and he provides all the summary statistics for the empirical distribution.

Example 3
"These are the distribution parameters for the theoretical distribution I fitted to the variable pulsation at rest:
area below 98.7667 = 0.08953
area below 111.113 = 0.25086
area below 123.458 = 0.5" (Student 3)

Example 4
"Count=96, Average = 123.458, Median = 122.0, Mode = 120.0, Variance = 337.682, Standard deviation = 18.3761, Minimum = 78.0, Maximum = 162.0, Range = 84.0, Skewness = 0.0109784, Stnd. Skewness = 0.043913, Kurtosis = –0.197793, Stnd. Kurtosis = –0.395585, Coeff. of variation = 14.8845%, Sum = 11852.0" (Student 4).

These results suggest difficulties in understanding the idea of parameter and the difference between theoretical and empirical distributions.

Table 2. Frequency and percentage of responses in computing the parameters

Response	**Number and Percentage**
Correct parameters	60 (51)
Incorrect or additional parameters	18 (15)
No answer	39 (33)

Computing Percentiles in the Theoretical Distribution

Table 3 presents a summary of students' solutions to question 3. About 65% of the students provided correct or partly correct solutions in computing the median and quartiles. However, few of them started from the theoretical distribution of critical values to compute these values. Most of the students computed the quartiles in the empirical data, through different options such as frequency tables or statistical summaries; and a large proportion of students found no solution. In the following example the student is using the percentiles option in the software, which is appropriate only for computing median and quartiles in the empirical distribution. He is able to relate the idea of median to the 50th percentile, although he is unable to relate the ideas of quartiles and percentiles. Again, difficulties in discriminating between the theoretical and the empirical distribution are noticed.

Example 5
"These are the median and quartiles of the theoretical normal distribution for *Age*. The median is 13. Percentiles for *Age*: 1.0% = 12.0, 5.0% = 12.0, 10.0% = 12.0, 25.0% = 13.0, 50.0% = 13.0, 75.0% = 13.0, 90.0% = 14.0, 95.0% = 14.0, 99.0% = 14.0" (Student 1)

Table 3. Frequency and percentages of students' solutions classified by type of distribution

	Type of distribution used		
	Theoretical	**Empirical**	**None**
Correct	21 (17.9)	29 (24.8)	
Partly correct	9 (7.7)	14 (12.0)	1 (0.9)
Incorrect	2 (1.7)	17 (14.5)	4 (3.4)
No solution			20 (17.1)

Students' Reasoning and Understanding of Normal Distribution

Besides the percentage of correct responses to each question, we were interested in assessing the types of knowledge the students explicitly used in their solutions. Using the categorization in the theoretical framework we described in Section 2, we analyzed the students' protocols to provide a deeper picture of the students' reasoning and their understanding of normal distributions. Four students were also interviewed after they completed the tasks. They were asked to explain their procedures in detail and, when needed, the researcher added additional questions to clarify the students' reasoning in solving the tasks. In this section we analyze the results, which are summarized in Table 4 and present examples of the students' reasoning in the different categories.

Symbols and Representations

Many students in both groups correctly applied different representations, with a predominance of density curves, and a density curve superimposed onto a histogram. Their success suggests that students were able to correctly interpret these graphs, and could find different properties of data such as symmetry or unimodality from them as in Example 6, where there is a correct use of two graphs to assess symmetry.

> *Example 6*
> "You can see that the distribution of the variable weight is not symmetrical, since the average is not in the centere of the variable range (Figure 3). The areas over and below the centre are very different. When comparing the histogram with the normal density curve, this skews to the left" (Student 5).

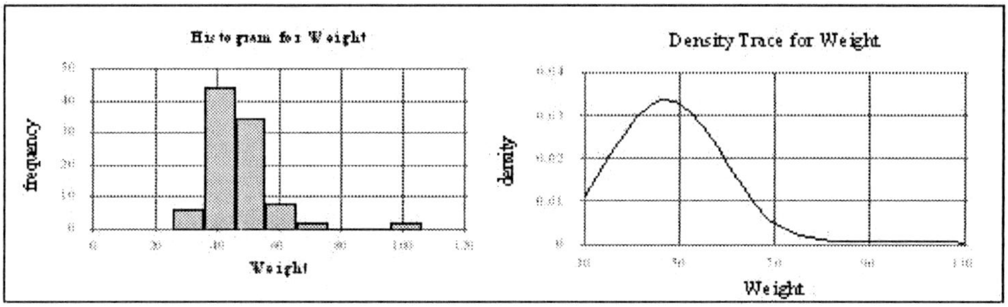

Figure 3. Histogram and density trace for Weight.

Among numerical representations, the use of parameters (mean and standard deviation) was prominent, in particular to solve task 2. Statistical summaries were correctly applied when students computed the asymmetry and kurtosis coefficients, and incorrectly applied when they computed the median and quartiles, since in that question the students used the empirical distribution instead of the theoretical curve (e.g., in Example 5). Few students used frequency tables and critical values. We conclude that graphical representations were more intuitive than numeric values, since a graph provides much more information about the distribution, and the interpretation of numerical summaries requires a higher level of abstraction.

Actions

The most frequent action was visual comparison (e.g., Examples 2, 6), although it was not always correctly performed (such as in Example 2, where the student was unable to use the graph to assess the kurtosis). A high percentage of students correctly compared the empirical density correctly with the theoretical normal density (e.g., Example 6). However, 40% of the students confused these two curves.

Table 4. Frequency of main elements of meaning used by the students in solving the tasks

Elements of Meaning	Correct Use	Incorrect Use
Symbols and Representations		
Graphical representations		
Normal density curve	45 (38.5)	1 (0.9)
Over imposed density curve and histogram	30 (25.6)	
Normal probability plot	6 (5.1)	
Cumulative density curve	2 (1.7)	
Histogram	37 (31.6)	
Frequency polygon	12 (10.3)	
Box plot	2 (1.7)	
Symmetry plot	1 (0.9)	
Numerical summaries		
Critical values	29 (24.8)	4 (3.4)
Tail areas	3 (2.6)	5 (4.3)
Mean and standard deviation (as parameters in the distribution	48 (41.0)	3 (2.6)
Goodness of fit test	2 (1.7)	2 (1.7)
Steam-leaf	5 (4.3)	
Summaries statistics	59 (50.4)	47 (40.2)
Frequency tables	26 (22.2)	
Percentiles	9 (7.7)	9 (7.7)
Actions		
Computing the normal distribution parameters	50 (42.7)	18 (15.4)
Changing the parameters	10 (8.5)	2 (1.7)
Visual comparison	56 (47.9)	49 (41.9)
Computing normal probabilities	13 (11.1)	1 (0.9)
Finding critical values	28 (23.9)	68 (58.1)
Descriptive study of the empirical distribution	39 (33.3)	8 (6.8)
Finding central interval limits	14 (12)	
Concepts and properties		
Symmetry of the normal curve	40 (34.2)	13 (11.1)
Mode, Unimodality in the normal distribution	32 (27.4)	16 (13.7)
Parameters of the normal distribution	51 (46.3)	16 (13.7)
Statistical properties of the normal curve	27 (26.1)	3 (2.6)
Proportion of values in central intervals	13 (11.1)	1 (0.9)
Theoretical distribution	48 (41.0)	50 (42.7)
Kurtosis in the normal distribution; kurtosis coefficients	27 (26.1)	1 (0.9)
Variable: qualitative, discreet, continuous	50 (42.7)	65 (55.6)
Relative position of mean, median, mode in a normal distribution	35 (29.9)	5 (4.3)
Skewness and standard skewness coefficients	34 (29.1)	1 (0.9)
Atypical value	5 (4.3)	
Order statistics: quartiles, percentiles	32 (27.4)	63 (53.8)
Frequencies: absolute, relative, cumulative	13 (11.1)	
Arguments		
Checking properties in isolated cases	18 (15.4)	3 (2.6)
Applying properties	58 (49.6)	7 (6.0)
Analysis	32 (27.4)	5 (4.3)
Graphical representation	58 (49.6)	36 (30.8)
Synthesis	26 (22.2)	4 (3.4)

For example, regarding the variable of *Age* (Figure 1a), the empirical density curve is clearly nonnormal (since there is no horizontal asymptote). The students who, instead of using this empirical density, compared the histogram with the normal theoretical distribution (Figure 1b) did not perceive that the histogram was not well fitted to the same, even when this was clearly visible in the graph.

A fair number of students correctly computed the parameters, although a large percentage made errors in computing the critical values for the normal distribution (quartiles and median, as in Example 5). Even when the computer replaces use of the normal distribution tables, it does not solve all the computing problems, since the students had difficulties in understanding the idea of critical values and in operating the software options. Finally, some students performed a descriptive study of data before fitting the curve.

Concepts and Properties

Students correctly used the different specific properties of the normal distribution as well as the definition of many related concepts. The most common confusion was thinking that a discrete variable with only three different values was normal (e.g., Examples 1, 5). This was usually because students were unable to distinguish between the empirical and the theoretical distribution. Other authors have pointed out the high level of abstraction required to distinguish between model and reality, as well as the difficulties posed by the different levels in which the same concept is used in statistics (Schuyten, 1991; Vallecillos, 1994).

An interesting finding is that very few students used the fact that the proportion of cases within one, two, and three standard deviations is 68%, 95%, and 99%, even when we emphasized this property throughout the teaching. This suggests the high semiotic complexity required in applying this property where different graphical and symbolic representations, numerical values of parameters and statistics, concepts and properties, and actions and arguments need to be related, as shown later in Example 7.

The scant number of students who interpreted the kurtosis coefficient, as compared with the application of symmetry and unimodality, is also revealing. Regarding the parameters, although most students used this idea correctly, errors still remain. Some students correctly compared the relative position of the measures of central position in symmetrical and asymmetrical distributions, although some of them just based their selection on this property and argued it was enough to assure normality.

Arguments

The use of graphical representations was predominant in producing arguments. In addition to leading to many errors, this also suggests the students' difficulty in producing high-level arguments such as analysis and synthesis. Most students just applied or checked a single property, generally symmetry. They assumed that one necessary condition was enough to assure normality. This is the case in Example 7, where the student correctly interprets symmetry from the symmetry plot and then assumes this is enough to prove normality.

> *Example 6*
> "We can graphically check the symmetry of *Time spent to run 30 Mts. in December* with the symmetry plot (Figure 4), as we see the points approximately fit the line; therefore the normal distribution will fit these data" (Student 6).

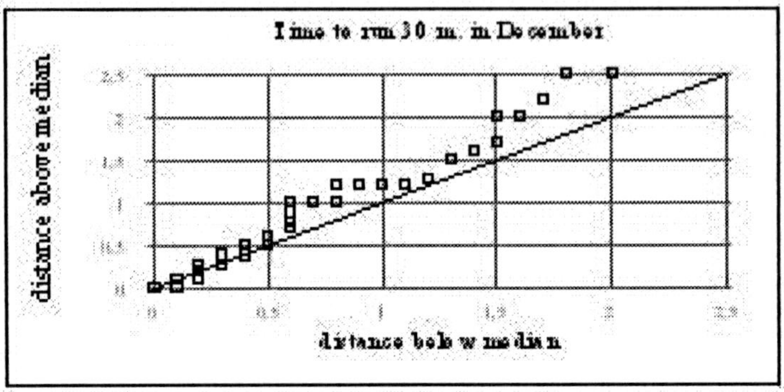

Figure 4. Symmetry plot.

In other cases the students checked several properties, although they forgot to check one of the conditions that is essential for normality, such as in the following interview, where the student studied the type of variable (discrete, continuous), unimodality, and relative position of mean, median and mode. However, he forgot to assess the value of the kurtosis coefficient, which is too high for a normal distribution (Student 7):

> *Teacher*: In the exam you selected *Time to run 30 Mts. in December* as a normal distribution. Why did you choose that variable?
> *Student*: I first rejected all the discrete variables since you need many different values for a discrete variable to be well fitted to a normal distribution. Since the two variables *Time to run 30 Mts. in December* and *Time to run 30 Mts. in September* are continuous I took one of them at random. I just might also have taken *Time to run 30 Mts. in September.* Then I realized the variable has only one mode, the shape was very similar to the normal distribution, mean and median were similar.
> *Teacher*: Did you do any more analyses?
> *Student*: No, I just did those.

A small number of students applied different elements of meaning, and carried out an analysis of each property. Seven percent of them produced a final synthesis, such as the following student.

Example 8

"The variable *Heartbeats after 30 press-ups* is what I consider best fits a normal distribution. It is a numerical variable. The variable is symmetrical, since both the histogram and the frequency polygon (Figure 5) are approximately symmetrical. On the other hand the skewness coefficient is close to zero (0.0109) and standard skewness coefficient falls into the interval (–2, +2). We also observe that the kurtosis coefficient is close to zero (–0.1977) which suggests the variable can fit a normal distribution.

Furthermore, we know that in normal distributions, mean median and mode coincide and in this case the three values are very close (Mean = 123.4; Mode = 120; Median = 122). Moreover there is only one mode. As for the rule 68,95,99.7 in the interval $(\mu - \sigma, \mu + \sigma) \Rightarrow (105.08.141.82)$ there are 68.75% of the observations, in the interval $(\mu - 2\sigma, \mu + 2\sigma) \Rightarrow (86.81,160.19)$ there is 95.84% and in the interval $(\mu - 3\sigma, \mu + 3\sigma) \Rightarrow (68.34,178.56)$ we found 100% of the data. These data are very close. Therefore you can fit a normal distribution to these data" (Student 8).

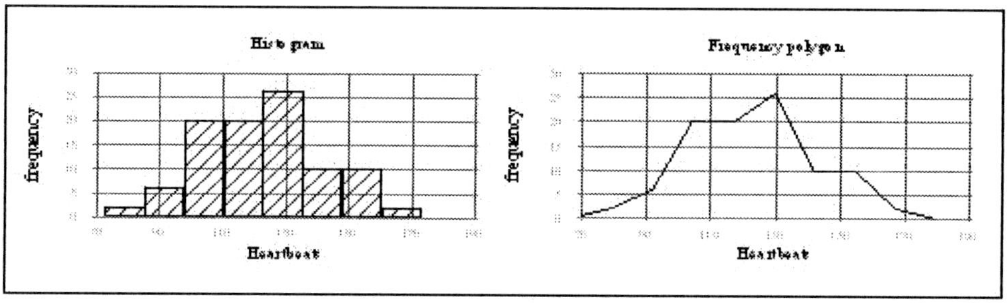

Figure 5. Histogram and frequency polygon for *Heartbeats after 30 press-ups.*

In this answer, the student relates the property of symmetry (concept) to the histogram and frequency polygon (representations). He is able to compute (action) the skewness and kurtosis coefficients (numerical summaries) and compares their values with those expected in normal distributions (properties and concepts). He also applies and relates the property of relative positions of central tendency measure and central intervals in a normal distribution, being able to operate the software (action) in order to produce the required graphs and summaries, which are correctly related and interpreted. This type of reasoning requires the integration of many different ideas and actions by the student.

Other students provided incorrect variables, even when they were able to use the software and to correctly produce a great number of different graphs. In Example 9 the student is able to plot different graphs and compute the quartiles. However, he is neither able to extract the information needed to assess normality from these graphs nor capable of relating the different results with the concepts behind them. No arguments linking these different representations or supporting his election are given. Moreover, he did not relate the high kurtosis coefficient to a lack of normality. The graphs and statistics produced are presented in Figure 6.

Example 9
"I selected *Height* since the normal distribution is used to describe real data. And describing the students' height is a real biological problem. This is also a quantitative variable and normal distribution describes quantitative variables" (Student 9).

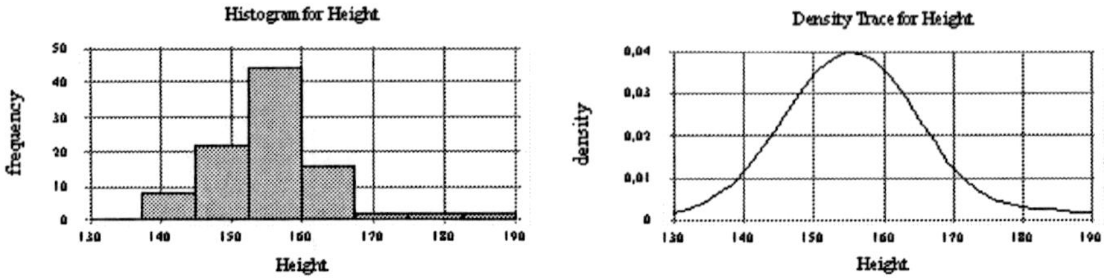

Stem-and-Leaf Display for HEIGHT: unit = 1.0 1|2 represents 12.0

```
 2     13|88
 6     14|0000
16     14|5566777799
40     15|000000002222223333444444
(28)   15|5555555566667777777788889999
28     16|0000001122222244
12     16|555577
 6     17|11

       HI|182,0 182,0 185,0 185,0
```

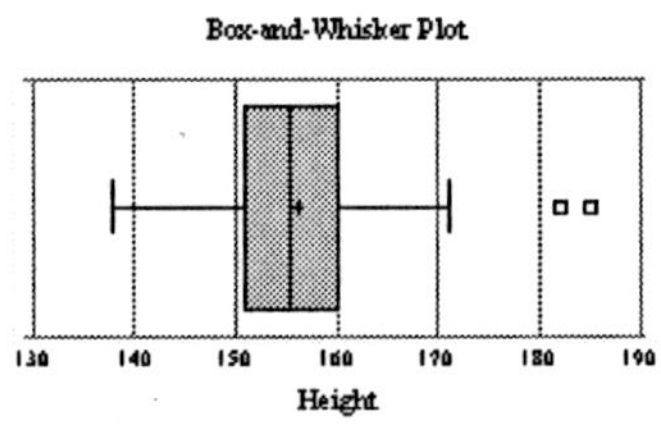

Summary Statistics for Height: Count = 96, Median = 155.5, Lower quartile = 151.0
Upper quartile = 160.0, Stnd. skewness = 3.4341, Stnd. kurtosis = 4.46366

Figure 6. Graphical representations and statistical summaries for *Height.*

Discussion

Many students grasped the idea of model, and showed a good understanding of the usefulness of models, density curves, and areas under the normal curve. Our analysis of the various actions, representations, concepts, properties, and arguments used by the students in solving the tasks suggests that many students were able to correctly identify many elements in the meaning of normal distribution and to relate one to another. Some examples are as follows:

- Relating concepts and properties. For example, relating the idea of symmetry to skewness coefficient or to relative position of mean, median, and mode in Examples 6, 7, and 8.
- Relating graphical representations to concepts. For example, relating the empirical histogram and density curve shapes to the theoretical pattern in a normal curve (e.g., in Example 8).

- Relating the various graphic representations and data summaries to the software options and menus they need to produce them (relating representations and actions in all the examples).
- Relating the definition and properties of normal distribution to the actions needed to check the properties in an empirical data set (e.g., in Example 8).
- There was a good understanding of the idea of mean and standard deviation and its relationship to the geometrical properties of the normal curve (e.g., Example 2).

There was also a clear disagreement between the personal meaning of normal distribution acquired by the students and the meaning we tried to teach them. Here we describe the main difficulties observed:

1. Perceiving the usefulness of theoretical models to describe empirical data. This is shown in the following transcript (Student 10):

   Teacher: Now that you know what the normal distribution is, can you tell me what it is useful for or in which way you can apply the normal distribution?
   Student: For comparing, isn't it? For example to compare data and tables, it is difficult to explain. ... You have some data and you can repeat with the computer what we did in the classroom.

2. Interpreting areas in frequency histograms and computing areas in the cases when a change in the extremes of intervals is needed. This point is not specific to the normal distribution or to the use of computers, and the student should have learned it at the secondary school level. However, in the following interview transcript, the student is not aware of the effect of interval widths on the frequency represented, which is given by the area under the histogram (Student 10):

   Teacher: How would you find the frequency in the interval 0–10 in this histogram?
   Student: The frequency is 5, this is the rectangle height.
   Teacher: What about the frequency for the interval 10–30?
   Student: It is 10, that is the height of this rectangle.

3. Interpreting probabilities under the normal curve. The graphical representation of the areas under the normal curve is the main didactic tool for students to understand the computation of probabilities under the curve and, at the same time to solve different problems involving the normal distribution. However, for some students with no previous instruction, this computation was not easily understood and performed.

4. We also observed difficulties in discriminating between empirical data and mathematical models, interpreting some statistical summaries and graphs,

and a lack of analysis and synthesis ability to relate all these properties when making a decision (Student 11).

Teacher: When you computed the median and quartiles in question 3, which data did you use: the theoretical normal distribution you fit to the data or the real data?
Student: I ... I am not very sure. Well, I used the data file ...

5. There was a great deal of difficulty in discriminating between the cases where a discrete quantitative variable can and cannot be fitted by a normal distribution (e.g., in Example 5) and even in distinguishing between the different types of variables.
6. Other students misinterpreted the skewness coefficient or assumed that the equality of mean, median and mode was enough to show the symmetry of the distribution, accepted as normal a distribution with no horizontal asymptote, made a rough approximation when formally or informally checking the rule ($\mu - k\sigma$, $\mu + k\sigma$), accepted too many outliers in a normal distribution, or misinterpreted the values of kurtosis.

Even when most of the students were able to change from the local to the global view of data (Ben-Zvi & Arcavi, 2001) in taking into account the shape of graphs as a whole, the idea of distribution as a property of a collective, and the variability of data, there is still a third level of statistical reasoning many of these students did not reach. This is the *modeling* viewpoint of data, where students need to deal at the same time with an empirical distribution as a whole (therefore, they need to adopt a global viewpoint of their data) and the mathematical model (the normal distribution in our research). In this modeling perspective, students need to concentrate on the different features of the data set as a whole and on the different features of the model (type of variable, unimodality, skewness, percentage of central cases, horizontal asymptote, etc., in our case). In addition to understanding the model as a complex entity with different components, they should be able to distinguish the model from the real data, to compare the real data to the model, and to make an accurate judgment about how well the model fits the data.

There was also difficulty in using secondary menu options in the software—which, however, are frequently essential in the analysis. Finally, the students showed scant argumentative capacity, in particular regarding analysis and synthesis (e.g., in Example 9).

IMPLICATIONS FOR TEACHING NORMAL DISTRIBUTIONS

The main conclusion in this study is that the normal distribution is a very complex idea that requires the integration and relation of many different statistical concepts and ideas. Recognizing this complexity, our work also suggests that it is possible to design teaching activities that facilitate the learning of basic notions about normal distribution. Since the learning of computational abilities is no longer an important objective, an intuitive understanding about basic concepts is possible for students with moderate mathematical knowledge, whenever we choose appropriate tasks.

Working with computer tools seemed to promote graphical understanding, as students in our experiment easily recognized and used many different plots (such as density curves, histograms, etc.) to solve the problems proposed. Moreover, they also showed a good understanding of many abstract properties, such as the effect of parameters on the density curve shape, and made extensive use of graphs as part of their argumentation. This suggests the essential role of computers to facilitate students' exploration of these properties and representations.

It is important that students understand basic concepts such as probability, density curve, spread and skewness, and histograms before they start the study of normal distribution; its understanding is based on these ideas. They should also be confident in the use of software before trying to solve problems related to the normal distribution, since they often misinterpret or confuse results from different software options.

The student's difficulties in discriminating between theoretical models and empirical data suggest that more activities linking real data with the normal model are needed. Simulating data from normal distributions and comparing them with real data sets might also be used as an intermediate step between mathematical model and reality. As a didactic tool it can serve to improve students' probabilistic intuition, to teach them the different steps in the work of modeling (Dantal, 1997), and to help them discriminate between model and reality. Simulation experiences and dynamic visualization can contribute, as analyzed by Biehler (1991), to provide students with a stochastic experience difficult to reach in the real world.

Finally, it is important to take into account the different components of meaning and understanding when assessing students' learning. Computer-based assessment tasks in which students are asked to analyze simple data sets and provide a sound argument for their responses—such as those presented in this paper—are a good tool to provide a complete picture of students' understanding and ways of reasoning.

REFERENCES

Batanero, C., Tauber, L., & Meyer, R. (1999). From data analysis to inference: A research project on the teaching of normal distributions. *Bulletin of the International Statistical Institute: Proceedings of the Fifty-Second Session of the International Statistical Institute* (Tome LVIII, Book 1, pp. 57–58). Helsinki, Finland: International Statistical Institute.

Batanero, C., Tauber, L., & Sánchez, V. (2001). *Significado y comprensión de la distribución normal en un curso introductorio de análisis de datos* (Meaning and understanding of normal distributions in an introductory data analysis course). *Quadrante*, 10(1), 59–92.

Ben-Zvi, D. (2000). Towards understanding the role of technological tools in statistical learning. *Mathematics Thinking and Learning*, 2(1&2), 127–155.

Ben-Zvi, D., & Arcavi, A. (2001). Junior high school student's construction of global views of data and data representations. *Educational Studies in Mathematics*, 43, 35–65.

Biehler, R. (1991). Computers in probability education. In R. Kapadia & M. Borovcnick (Eds.), *Chance encounters: Probability in education* (pp. 169–211). Dordrecht, The Netherlands: Kluwer.

Dantal, B. (1997). Les enjeux de la modélisation en probabilité (The challenges of modeling in probability). In *Enseigner les probabilités au lycée* (pp. 57–59). Reims: Commission Inter-IREM Statistique et Probabilités.

delMas, R. C., Garfield, J. B., & Chance, B. (1999). *Exploring the role of computer simulations in developing understanding of sampling distributions.* Paper presented at the *Annual Meeting of the American Educational Research Association,* Montreal, Canada.

Gal, I. (1997). Assessing students' interpretations of data: Conceptual and pragmatic issues. In B. Phillips (Ed.), *Papers on Statistical Education presented at ICME-8* (pp. 49–58). Swinburne, Australia: Swinburne University of Technology.

Garfield, J. B. (1991). Evaluating students' understanding of statistics: Developing the statistical reasoning assessment. In R. G. Underhill (Ed.), *Proceedings of the 13th Annual Meeting of the North American Chapter of the International Group for the Psychology of Mathematics Education* (Vol. 2, pp. 1–7). Blacksburg, VA: Comité organizador.

Godino, J. D. (1996). Mathematical concepts, their meaning and understanding. In L. Puig & A. Gutiérrez (Eds.), *Proceedings of the 20th Conference of the International Group for the Psychology of Mathematics Education* (Vol. 2, pp. 417–424). Valencia: Comité organizador.

Godino, J. D., & Batanero, C. (1998). Clarifying the meaning of mathematical objects as a priority area of research in mathematics education. In A. Sierpinska & J. Kilpatrick (Eds.), *Mathematics Education as a research domain: A search for identity* (pp. 177–195). Dordrecht: Kluwer.

Huck, S., Cross, T. L., & Clark, S. B. (1986). Overcoming misconceptions about *z*-scores. *Teaching Statistics,* 8(2), 38–40.

Méndez, H. (1991). *Understanding the central limit theorem.* Ph.D. diss., University of California, University Microfilm International number 1-800-521-0600.

Piaget, J., & Inhelder, B. (1951). *La génese de l'idée de hasard chez l'enfant.* Paris: Presses Universitaires de France.

Rubin, A., Bruce, B., & Tenney, Y. (1991). Learning about sampling: Trouble at the core of statistics. In D. Vere-Jones (Ed.), *Proceedings of the Third International Conference on Teaching Statistics* (pp. 314–319). Voorburg, The Netherlands: International Statistical Institute.

Schuyten, G. (1991). Statistical thinking in psychology and education. In D. Vere-Jones (Ed.), *Proceedings of the III International Conference on Teaching Statistics* (Vol. 2, pp. 486–489). Dunedin, Australia: University of Otago.

Vallecillos, A. (1996). *Inferencia estadística y enseñanza: Un análisis didáctico del contraste de hipótesis estadísticas* (Statistical inference and teaching: A didactical analysis of statistical tests). Madrid: Comares.

Vallecillos, A. (1999). Some empirical evidence on learning difficulties about testing hypotheses. *Bulletin of the International Statistical Institute: Proceedings of the Fifty-Second Session of the International Statistical Institute* (Tome LVIII, Book 2, pp. 201–204). Helsinki, Finland: International Statistical Institute.

Wild, C., & Pfannkuch, M. (1999). Statistical thinking in empirical enquiry. *International Statistical Review,* 67(3), 223–265.

Wilensky, U. (1995). Learning probability through building computational models. In D. Carraher y L. Meira (Eds.), *Proceedings of the 19th PME Conference* (Vol. 3, pp. 152–159). Recife, Brazil: Organizing Committee.

Wilensky, U. (1997). What is normal anyway? Therapy for epistemological anxiety. *Educational Studies in Mathematics,* 33, 171–202.

Chapter 12

DEVELOPING REASONING ABOUT SAMPLES

Jane M. Watson
University of Tasmania, Australia

INTRODUCTION

Although reasoning about samples and sampling is fundamental to the legitimate practice of statistics, it often receives little attention in the school curriculum. This may be related to the lack of numerical calculations—predominant in the mathematics curriculum—and the descriptive nature of the material associated with the topic. This chapter will extend previous research on students' reasoning about samples by considering longitudinal interviews with 38 students 3 or 4 years after they first discussed their understanding of what a sample was, how samples should be collected, and the representing power of a sample based on its size. Of the six categories of response observed at the time of the initial interviews, all were confirmed after 3 or 4 years, and one additional preliminary level was observed.

THE PROBLEM

Although appropriate sampling is the foundation of all inferential statistics, the topic rarely achieves a high profile in curriculum documents at the school level. Whether this is because the topic is more descriptive and less numerical than most in the mathematics curriculum or because it is acknowledged to be difficult for students to appreciate fully (National Council of Teachers of Mathematics [NCTM], 2000, p. 50) is unknown. Data collection is mentioned as part of Data Analysis and Probability in the NCTM's *Principles and Standards* but rarely with the emphasis—for example, on the importance of randomness (p. 326)—that might be expected. Perhaps the most salient reminder of the importance of sampling is found in the Australian Education Council's (AEC) *National Statement on Mathematics for Australian Schools* (1991) in the context of a general statement on statistical inference:

D. Ben-Zvi and J. Garfield (eds.),
The Challenge of Developing Statistical Literacy, Reasoning and Thinking, 277–294.

> The dual notions of sampling and of making inferences about populations, based on samples, are fundamental to prediction and decision making in many aspects of life. Students will need a great many experiences to enable them to understand principles underlying sampling and statistical inference and the important distinctions between a population and a sample, a parameter and an estimate. Although this subheading [Statistical Inference] first appears separately in band C [secondary school], the groundwork should be laid in the early years of schooling in the context of data handling and chance activities. (AEC, 1991, p. 164)

Related to this, upper primary students should "understand what samples are, select appropriate samples from specified groups and draw informal inferences from data collected" (p. 172), and high school students should "understand what samples are and recognize the importance of random samples and sample size, and draw inferences and construct and evaluate arguments based on sample data" (p. 179). Again it is noteworthy that calculations involving mathematical formulas are not involved in these statements, hence the reasoning involved may not be based on a preliminary mathematics skill base. Developing reasoning related to sampling may be associated with developing literacy and social reasoning skills rather than developing numeracy skills. This is potentially an unusual situation for the mathematics curriculum.

THE LITERATURE AND BACKGROUND

Except for research with college students on issues of sample size and representativeness that grew from the early work of Tversky and Kahneman (e.g., 1971, 1974), little research has taken place until recently on school students' understanding of sampling. In this context, however, reviewers (e.g., Shaughnessy, 1992; Shaughnessy, Garfield, & Greer, 1996) have suggested that school students are susceptible to the representativeness heuristic; that is, they have difficulty with the idea of variability in populations, have too much confidence in small samples, and do not appreciate the importance of sample size in random samples. In the early 1990s the research of Wagner and Gal (1991) with elementary students found that responses in comparing two groups depended on whether students assumed homogeneity or appreciated natural variation, whereas Rubin, Bruce, and Tenney (1991) found, with senior high school students, a tension between the reality of variability within samples and the need for sample representativeness. Although not specifically addressing sampling issues, Mokros and Russell (1995) in their study of students' understanding of average, discovered an increasing awareness of representativeness associated with the perceived need to measure an average to represent a set. In work with upper elementary students, Jacobs (1997, 1999) and Schwartz, Goldman, Vye, Barron, and The Cognition and Technology Group at Vanderbilt (1998) found that students preferred biased sampling methods, such as voluntary participation, due to perception of *fairness,* allowing everyone an opportunity and not forcing anyone to participate. Metz (1999) interviewed elementary students who had been involved in designing their own science experiments, finding many supporting the power of sampling for appropriate reasons and few succumbing to the "law of small numbers," that is, putting too much faith in small samples. There were, however, also many who argued against sampling due to the need to test all members of the population or due to the variability in the population.

The appreciation of the need to consider sampling and other aspects of statistical reasoning in social contexts, for example based in media reports, led Watson (1997) and Gal (2000) to suggest structures for considering student progress to reach the goal of statistical literacy for participation in decision making in society (Wallman, 1993). Watson's three-tiered hierarchy will be discussed below; and Gal's four-dimensional framework includes the importance of background knowledge, the skills to read and comprehend statistical information in context, a set of critical questions to apply in contexts, and the dispositions and beliefs that allow for questioning and acknowledgment that alternative explanations are possible.

The previous research most closely related to the current study is that of Watson, Collis, and Moritz (1995) based on surveys of 171 and interviews of 30 girls in a South Australian school and of Watson and Moritz based on large-scale longitudinal surveys of over 3,000 students (2000b) and interviews with 62 students (2000a) throughout the state of Tasmania. These studies were based on the three survey items in Figure 1 and the interview protocol in Figure 2. The analysis of Watson et al. (1995) also included a fourth part of the interview protocol with a sampling task comparing expected average values from samples with given population averages (Tversky & Kahneman, 1971). This task will not be considered as part of the current study.

Three theoretical frameworks were used as part of the earlier research. The first was related to the statistical content on sampling as reflected in the literature for students at various levels (e.g., Corwin & Friel, 1990; Landwehr, Swift, & Watkins, 1987; Moore, 1991; Orr, 1995). The second was a cognitive development taxonomy based on the structure of observed learning outcomes (SOLO) of Biggs and Collis (1982; 1991). The main interest in terms of analyzing responses that addressed sampling issues was the increased structural complexity shown. *Unistructural* (U) responses employed single elements of the tasks and did not recognize contradictions if they arose. *Multistructural* (M) responses used more than one element in a sequential fashion, often recognizing contradictions but unable to resolve them. *Relational* (R) responses integrated elements of the tasks to produce complete solutions free of contradictions. The third framework was Watson's (1997) three-tiered hierarchy of statistical literacy applied to sampling. Tier 1 related to understanding terminology associated with sampling. Tier 2 covered the application and understanding of sampling terminology as it occurs in context, particularly social contexts found in the media. Tier 3 was associated with the critical skills required to question claims about samples made without proper statistical foundation.

Q1. If you were given a "sample," what would you have?

Q2.

> ABOUT 6 in 10 United States high school students say they could get a handgun if they wanted one, a third of them within an hour, a survey shows. The poll of 2,508 junior and senior high school students in Chicago also found 15% had actually carried a handgun within the past 30 days, with 4% taking one to school.

(a) Would you make any criticisms of the claims in this article?

(b) If you were a high school teacher, would this report make you refuse a job offer somewhere else in the United States, say Colorado or Arizona? Why or why not?

Q3.

Decriminalize drug use: poll

SOME 96% of callers to youth radio station Triple J have said marijuana use should be decriminalized in Australia.

The phone-in listener poll, which closed yesterday, showed 9,924—out of the 10,000-plus callers—favored decriminalization, the station said.

Only 389 believed possession of the drug should remain a criminal offense.

Many callers stressed they did not smoke marijuana but still believed in decriminalizing its use, a Triple J statement said.

(a) What was the size of the sample in this article?

(b) Is the sample reported here a reliable way of finding out public support for the decriminalization of marijuana? Why or why not?

Figure 1. Sampling items from written surveys.

The statistical framework was a basis for all three earlier studies (Watson et al., 1995; Watson & Moritz, 2000a, 2000b). The Biggs and Collis (1982, 1991) taxonomy was used by Watson et al. and Watson and Moritz (2000b) as a major classification device. The 1995 study identified two U-M-R cycles for responses to the tasks set where the second cycle represented a consolidation of the idea of *sample* into a single construct and the increasingly complex application of it in the contexts presented (see Figures 1 and 2). In the large survey study (Watson & Moritz, 2000b), the first U-M-R cycle and a consolidation phase based on questioning bias in Items 3 and 4 in Figure 1 were identified. In the interview study (Watson & Moritz, 2000a) the taxonomy was used in conjunction with clustering techniques (Miles & Huberman, 1994) to identify six categories of performance with respect to the tasks in Figures 1 and 2. These categories were related to the three-tiered statistical literacy hierarchy (Watson, 1997) as shown in Figure 3. The hierarchy was also used with the survey outcomes in relation to the SOLO taxonomy to suggest the possibility of parallel development of U-M-R cycles within the three tiers once a basic unistructural definition provides a starting point for development.

1. (a) Have you heard of the word *sample* before?
Where? What does it mean?

(b) A newsperson on TV says:
"In a research study on the weight of Grade 5 children, some researchers interviewed a *sample* of Grade 5 children in the state."
What does the word *sample* mean in this sentence?

2. (a) Why do you think the researchers used a *sample* of Grade 5 children, instead of studying all the Grade 5 children in the state?

(b) Do you think they used a sample of about 10 children? Why or why not?
How many children should they choose for their sample? Why?

(c) How should they choose the children for their sample? Why?

3. The researchers went to 2 schools:
One school in the center of the city and 1 school in the country.
Each school had about half girls and half boys.

The researchers took a random sample from each school:
50 children from the city school
20 children from the country school

One of these samples was unusual: It had more than 80% boys.

Is it more likely to have come from
the large sample of 50 from the city school, or
the small sample of 20 from the country school, or
are both samples equally likely to have been the unusual sample?
Please explain your answer.

Figure 2. Three parts of the interview protocol for sampling.

Tier 1—Understanding Terminology

Small Samplers without Selection (Category 1)

- may provide examples of samples, such as food products
- may describe a sample as a small bit, or more rarely as a try/test
- agree to a sample size of less than 15
- suggest no method of selection, or an idiosyncratic method

Small Samplers with Primitive Random Selection (Category 2)

- provide examples of samples, such as food products
- describe a sample as either a small bit, or a try/test
- agree to a sample size of less than 15
- suggest selection by "random" means without description, or a simple expression to choose any, perhaps from different schools

Tier 2—Understanding Terminology in Context

Small Samplers with Pre-Selection of Results (Category 3)

- provide examples of samples, such as food products
- describe a sample as both a small bit, and a try/test
- agree to a sample size of less than 15
- suggest selection of people by weight, either a spread of fat and skinny, or people of normal weight

Equivocal Samplers (Category 4)

- provide examples and descriptions of samples
- may indicate indifference about sample size, sometimes based on irrelevant aspects
- may combine small size with appropriate selection methods or partial sensitivity to bias, or large sample size with inappropriate selection methods

Large Samplers with Random/Distributed Selection (Category 5)

- provide examples of samples, such as food products
- describe a sample as both a small bit, and a try/test
- may refer to term *average*
- suggest a sample size of at least 20 or a percentage of the population
- suggest selection based on a random process or distribution by geography

Tier 3—Critical Questioning of Claims Made without Justification

Large Samplers Sensitive to Bias (Category 6)

- provide examples of samples, sometimes involving surveying
- describe a sample as both a small bit, and a try/test
- may refer to the terms *average* or *representative*
- suggest a sample size of at least 20 or a percentage of the population
- suggest selection based on a random process or distribution by geography
- express concern for selection of samples to avoid bias
- identify biased samples in newspaper articles reporting on results of surveys

Figure 3. Characteristics of six categories of developing concepts of sampling with respect to the three tiers of statistical literacy (Watson, 1997).

The current study aimed to extend the research of these three studies by considering longitudinal interviews with 38 students who were interviewed 3 or 4 years after their original interview.

SUBJECTS AND METHODS USED

The subjects in the current study were 22 Tasmanian students interviewed 4 years after their original interview (19 from Watson & Moritz [2000a] and 3 from earlier pilot interviews) and 16 South Australian students interviewed 3 years later. During the intervening years students in both states had been exposed to mathematics influenced by the *National Statement* (AEC, 1991), but there was no intervention in relation to this research study in that time. The data set is limited to students who were still enrolled in the South Australian school or who could be traced to another school or university within Tasmania, and who had been interviewed on the sampling protocol in both interviews (7 students had incomplete data). A summary of the students' grades in the data set is given in Table 1. Grade 13 refers to first year at university.

Table 1. Distribution of 38 longitudinal interviews by state and grade

Tasmania		South Australia	
Grades	**Number**	**Grades**	**Number**
3 → 7	6	3 → 6	2
6 → 10	12	5 → 8	6
9 → 13	4	7 → 10	3
		9 → 12	5

When it is of interest to compare groups of students at different stages in school, the following three groups will be considered: Elementary, 8 students initially in Grade 3 and later in Grades 6 or 7; Middle School, 21 students initially in Grades 5, 6, or 7 and later in Grades 8 or 10; High School, 9 students initially in Grade 9 and later in Grades 12 or 13. These groups are based on the fact that elementary school ends in Grade 6 in Tasmania and Grade 7 in South Australia.

All students were interviewed using the protocol in Figure 2 as part of a longer interview including other concepts in the chance and data curriculum. Thirty-one students in Grade 12 and below were interviewed in their schools under conditions similar to the original interview. Three Grade 6/10 students and the four Grade 9/13 students were interviewed on the university campus and paid a small remuneration for coming to the campus. All interviews were videotaped and subsequently transcribed.

The method of analysis followed the model set by Watson and Moritz (2000a) in clustering responses into categories described in that earlier study. For 19 Tasmanian students, initial categories were assigned from the previous research. For the original South Australian data, three pilot interviews in Tasmania, and all longitudinal data, students were assigned to categories by the two researchers familiar with the data, based on a reading of all transcripts. After independently classifying the responses, there were four discrepancies between the researchers (representing 89% agreement), and these were decided after discussion. Not all students who participated in the longitudinal interviews were asked the two media questions (Q2 and Q3 in Figure 1). Where there was consequently some doubt about critical questioning and recognition of bias, this will be acknowledged.

RESULTS

The results are presented in two parts: a summary of the outcomes for the 38 students and examples of responses that illustrate the change taking place over the 3- or 4-year period.

Summary of Outcomes

Of the 38 longitudinal interviews and 19 initial interviews that were classified for the first time in relation to the six categories in Figure 3, two interviews were found difficult to classify. One of these was from a Grade 3 student who in the initial interview did not display any understanding of what a sample was. In the survey, for Q1 (Figure 1) she wrote, "a cube" and in the interview she gave the following responses:

> *S1*: [*Part 1a*] It means that you have an object or something. You could have a dice, or a board or a chalk. [*Part 1b*] Well, that they had a lot of children having these tests and they probably did well and they are probably are talking about the ones that got them all right. [*Part 2a*] Some people. [*Why?*] Well mostly, they couldn't go around, they couldn't put it on television, some people might miss out and also if they went around looking for people and telling everybody they wouldn't come in because they probably had something to do on that day. [*Part 2b*] Maybe 12. [*Why?*] Well most people wanted to have a turn and they probably really wanted to have this interview or something and well I'd say they would have about 12 and they would get a fair chance. [*Part 2c*] Well I'd get a sheet and say what is 100 and well, something, and you are only allowed to guess and the person nearest guess or if there were two I would probably say well you two people had the closest number so I would let you go.

This student was classified as Prestructural (Category 0) with respect to the concept of sampling, using imaginative stories as part of her reasoning.

The other response that was unusual was from a longitudinal interview of a Grade 6/10 student. This was the only instance of a student insisting on a population view of the interview questions, although knowing the basic idea of what a sample is.

> *S2*: [*Part 1a*] Yes, like in the grocery store, you can sample something, you can try it before you buy. Like a part of, whatever. [*Part 1b*] A few children. [*Part 2a*] I don't know, I don't know what they were thinking. I think they should interview everyone because then they would get it more correct. [*How many*

should they choose?] I think they should choose everyone because otherwise you won't get it right. Because you might choose 10 lightweighted people and that would be wrong then, wouldn't it? Because there might be a lot of fat people in the school.

Although a stronger statement than that made by one nonlongitudinal student in the initial data set (Watson & Moritz, 2000a, p. 62–3, S15), it was also placed in the Equivocal category.

Table 2 contains a summary of students' classifications at the two times, recorded by whether they were elementary school students (E), middle school (M), or high school (H). The representation displays the improved performance of 78% of the students who could improve, the ceiling effect for four high school students, and the decreased performance for four students (13%). All of the elementary students improved. The greatest possible improvement (from category 1 to 6) occurred for two of the middle school students. Of those below the highest category initially, 12% performed at the same level later. Of those whose performance deteriorated or stayed the same, in each case half were middle school and half were high school students.

Table 2. Initial and longitudinal performance for elementary (E), middle school (M), and high school (H) students

	Initial Category							
Final Category	**0**	**1**	**2**	**3**	**4**	**5**	**6**	**Total**
1		M						1
2		MM						2
3	E	EEE		M	MM			7
4			M	H		M		3
5		E	EM	EMM	MM	MM	HH	12
6		MM		EH	MHH	MM	HHHH	13
Total	1	9	3	7	7	5	6	38

Examples of Performance on the Two Interviews

The examples provided in this section will show improved performance, diminished performance, and unchanged outcomes.

Improvement

The student, S1, whose response was judged to be prestructural on the initial interview, was later classified in Category 3 (Small Samplers with Preselection) in Grade 7.

S1: [*Part 1a*] You can get a sample as, in tasting sample they give you something and you feel it or whatever you do with it. [*Part 1b*] They took maybe 4 or 5 children. [*Part 2a*] Some people might be the same weight. [*Part 2b*] Depending on how much there is, say there were 7 fat children and 7 skinny children. Probably ask about 3 skinny and maybe 4 fat. [*Part 2c*] And the weight I guess they just thought well this person and this person they look very different and this person is sort of in between those people and so ...

Several other elementary and middle school students gave similar longitudinal responses.

Student, S2, who in Grade 10 insisted that all students be "sampled," had earlier in Grade 6 given a Category 2 response (Small Samplers with Primitive Random Selection).

S2: [*Part 1a*] Science ... To take something from somewhere and test it. [*Part 2b*] About 10 because it would be shorter. [*Part 2c*] Any children, just pick any! One from one school, or a couple from a school.

Student S3 was a middle school student whose responses changed from Category 1 to 6 over the 3-year period:

S3: (*Grade 5*) [*Part 1a*] Sample could be as in food you could try and ... it could be just something you try before the big thing comes out. [*Part 1b*] ... a few not all ... [*Part 2a*] Because it probably would have taken too long. [*Part 2b*] They could have done that many children. [*Part 2c*] It doesn't really matter. They could choose any 10 because they would all be different.

S3: (*Grade 8*) [*Part 2a*] Well studying all the Grade 5 children would take quite a while to do, so using this sample you can get the basic idea. ... If you just take them randomly, it just gives you a basic idea on the rest of them. [*Part 2b*] I think 10 children is probably a bit too few, because they might all be skinny or they all might [be] slightly more overweight, so if you use say a whole class or a couple of classes, maybe even a hundred students, the higher number the more chance you're getting of equally weight or um. Like if they were slightly less in weight or slightly higher in weight, so you've got more of a chance people in both areas. [*Part 2c*] [*How 'random'?*] Sort of not set, so like you wouldn't choose everybody who looks like really skinny or everybody who looks really overweight. You could just go through and right and say you, you and you, without really paying any attention to why you are really saying that, just taking the people because they are there. Not picking them for like how big or how much weight was on them. [*Part 3*] I'd say that one the sample of 20, because you've got less people, and so if you just took a sample of 20 people, you might have more boys than girls. ... You'd have um, yeah, the percentage would be higher from a lower number than it would in a higher number.

Although not asked the media questions, this student was sensitive to bias in describing sampling and able to suggest the smaller class in Part 3.

Three students' responses changed from Category 3 to Category 5 over time; one of those was the following student, quoted in Watson and Moritz (2000a, p. 58, S7), with an elaborate story of preselecting small and tall people. Four years later, in Grade 7, she responded as follows:

S4: [*Part 1a*] It is part of, a little bit of, to try. [*Part 2a*] Because what they get from the sample they can probably figure that it is probably going to be the same for most of every one else. [*Part 2b*] Umm, probably more than 10 because in [the state] there are a lot of Grade 5 children and 10 could be different from the rest. Probably about maybe 200. [*Part 2c*] Just have them pick a number out of a hat.

Except for the comment that "10 could be different from the rest," the student did not use other opportunities in the survey or interview (e.g., Part 3) to display sensitivity to relative sample size and bias.

One of the Equivocal Samplers in the initial interviews (S14 of Watson & Moritz, 2000a, p. 62) improved to the highest category 4 years later in Grade 10.

S5: [*Survey Q2b*] This evidence is only from Illinois, with no mention of Colorado or Arizona. It would make me investigate where my job offer came from. [*Survey Q3b*] Triple J is a youth station, therefore only surveying only one age group from the population … not a reliable sample ... if people have no interest in the topic … they will not respond. … A compulsory referendum of the whole population would be required to be reliable. Even then, many … would be too young to vote. [Part *1b*] A sample, say there's 10,000 Grade 5 kids in [the state], or something, they could have only interviewed 100 of them or like a fairly small number of them, I mean enough to be able to obtain reasonable results from. But for the sample they may have only taken a class then like a class from one area of the state, which would have been less accurate, or they could have taken like some Grade 5 kids from all over the state which would have been a better sample. So a sample in this sentence is just a small amount of the Grade 5 kids in [the state]. [*Part 3*] If they take a random sample then there is more chance of a smaller sample that you will have an inequality in the amount in the ratios say, because it is just a smaller sample.

Diminished Performance

The only falls in category of response were from Category 5 to Category 4 or 3. For a student initially in Grade 6, this resulted in an Equivocal response in Grade 10.

S6: (*Grade 6*) [*Part 1a*] A sample is a small portion of something big. [*Part 1b*] I would say it means, for a school, um, a sample of the school would maybe, I would think mean from all, a little, say 2, from each class. [*Part 2b*] It might not be enough people … um to like actually … know what the people in that class, they might have to take a few more to know what that grade … um, is, like, the children from that grade like, what they behave like, and what they like and all. [*How many?*] 30.

S6: (*Grade 10*) [*Part 1a*] … Shampoo or something … You try it … [*Part 1b*] … Just a handful … [*Part 2a*] Um, well order, picking, picking order should perhaps be random, maybe because I don't know, just because it gives it more true sort of, end results, if you pick a rich person or a poor person or something like that you know. It's just like, vary it a bit, so go random … [*Part 2b*] … I suppose there'd probably be a fair few different weights maybe, um, so I suppose, yeah, ten is fair … well maybe pick more people just because there could be a lot of varied weight, you know … like ten or so children is you know, just the same as … fifty, sort of, … ten children would kind of be the same, different types, fifty maybe or something.

This student had great difficulty coming to a decision about sample size but in discussing selection suggested students from "every school" chosen at random, "names out of a hat or something."

One of the Equivocal Samplers from the initial interviews (S16 of Watson & Moritz, 2000a, p. 63) continued to be equivocal about a small sample size and appeared to suggest preselection of results (Category 3):

S7: [*Part 2b*] It doesn't matter, I don't think. [*Part 2c*] Just get all different size, forms. [*Why?*] To be, to make it fair. If you just picked 5 fat ones, you would say everyone was fat.

Unchanged Performance

Eight students gave responses that were classified in the same category at both interviews. One middle school student was considered a Small Sampler without Selection (Category 1) both times.

S8: (*Grade 5*) [*Survey Q1*] A packet of something. [*Part 1a*] A sample of grain. [*Part 1b*] A bunch of them. [*Part 2a*] Because it might have been too many or they might have just wanted to pick out the smart ones. [*Part 2b*] A different number. [*How many?*] About 5. [*Why?*] Because I think that for them to use 10 is too many, maybe 5 at a time. [*Part 2c*] Maybe they are interested in something they do.

S8: (*Grade 8*) [*Part 1a*] ... Food you can taste ... [*Part 1b*] ... About five children in [the state]. [*Part 2b*] Probably about ten because it's a um, probably because it's not too many like if you had 23 or something then you'd be getting a bit too big and um, if you write data and stuff on it, ten children wouldn't be that many. [*Part 2c*] Um, suppose it doesn't really matter. They can just pick them out.

One of the students who gave Category 5 responses each time was a Grade 7 student initially. Although suggesting a stratified sample larger than 10, she did not recognize bias in Item Q2 (Figure 1).

S9: (*Grade 7*) [*Survey Q1*] A little bit of something like a test. A little bit of shampoo or maybe a teabag. [*Survey Q2a*] I think they should tell us about it so that if we know someone in the same position we can stop them doing it. [*Survey Q2b*] Yes because one day I might get into a fight with one of the students and he or she might shoot me or even kill me. ... It's too dangerous too be teaching at a school with those sorts of people. It would be a very scary life and I don't think I'd like it. [*Part 2b*] Umm I think more than that 'cos there's lots of Grade 5s in [the state]. They could have got a bit from each school. [*Why?*] Because some might be out in the country and they might be a different weight. They might be city ... don't know, just to get it a bit different. [*Part 3*] ... Probably the [city] school 'cos there's more people from it.

S9: (*Grade 10*) [*Part 2b*] They would have used more than that. For all of [the state], that wouldn't even be from each school or anything, because you need like different people from different areas and different schools and things like that, not from schools. [*How many?*] About a couple of hundred or something like that, or maybe even a bit more depending on [how many] children there are. [*Part 2c*] Just randomly somehow. I don't know, just [choose] their names! Just choose them somehow. [*Why random?*] So they are not all the same, they live somewhere different, they come from [inaudible] backgrounds, different schools, like different religions maybe even, things like that. [*Part 3*] It's hard to tell. It could of like come from the [city] one because there's more people, but ... then again it could have come from the country school because if it's selected by random, you can't really tell ... like if each school had half girls and half boys, it would probably be like equal or ...

Although not surveyed the second time, this student did not take up opportunities to suggest possible bias and gave interview responses very similar to earlier.

One of the four students in Category 6 each time was the following student, who although not asked the survey questions the second time was consistent in the understanding displayed and sensitivity to bias.

S10: (*Grade 9*) [*Survey Q1*] An average part of a larger group that represents that larger group as a whole to be analyzed/studied. [*Survey Q2a*] It claims a sample of the U.S.A. students when there was only a sample of Chicago. [*Survey 3b*] Because there is not a fair representation of the population would listen to Triple J or bother to call. [*Part 2b*] I don't think they would have used 10 children because it's not a fair amount to judge the whole of [the state] ... but a lot more than 10 or 100. ... I'd go for about one quarter. [*Part 2c*] They should randomly choose. They shouldn't have any preference from one Grade 5 to another. [*Part 3*] It would have come from the 20 in the country school because there was more ... fairer amount from the [city] school because there were 50 ... the more you have in your sample the better the research will be ...

S10: (*Grade 13*) [*Part 2b*] 10 children ... would be ... too small to make any conclusions. ... They should have chosen about ... 5% of the number of children ... around that figure. [*Part 2c*] They should somehow choose them randomly, draw their names out of a hat ... and not just pick them from the one [city] school or something. And they should also make sure that they get a similar number of girls and boys to measure and that they get a similar number of ages. There may be one or two years variation but it's really not that important since they are all the same grade ... if you take them by what they look, skinny or heavy, then you are pretty sure that the skinny ones will weigh less, the weighty ones would weigh more. ... I think that you would be influencing your results beforehand. [*Part 3*] I'd expect that the 80% boys that are randomly chosen would be from the country school ... not because there are more boys in the country but the number of children, the more that you have in the sample, the more the distribution would be similar to the population, so the smaller the more likely that it is not. So therefore the country one since it has 20 instead of 50 like the city one, would be more likely.

DISCUSSION

The improved outcomes for students' responses to questions on sampling after 3 or 4 years are encouraging in terms of increased representation in higher categories, particularly a doubling in Categories 5 and 6. Whether life experiences or the mathematics curriculum is responsible is impossible to determine. Some of the older students' responses to Part 3 of the interview protocol suggest that instruction on sample size may have been a factor. One quote, however, from a student in Grade 10 at the time of the second interview indicates that if the mathematics curriculum is having an influence, it is not always absorbed.

> *S11*: [*Part 1a*] It means like a small dose or something. Like everywhere, like a small sample of perfume or like to try something in a supermarket or something. [*Are there ways you might use it in maths or somewhere else talking about school subjects?*] You would use it like in science to like test a sample of something but I don't think you would use it in maths. You would but, it is not like the word average, not like something you would use all the time.

This view may reflect that of Derry, Levin, Osana, and Jones (1998) about the lack of calculations associated with statistical reasoning. This may mean that some students (or teachers) fail to realize that ideas about sampling are important.

The students who participated in this longitudinal study are likely to have participated as well in other longitudinal studies of topics such as average (Watson & Moritz, 2000c), beginning inference (Watson, 2001), and pictographs (Watson & Moritz, 2001). Although the criteria for classification were not identical in the studies, they were hierarchical in nature. Of 43 students interviewed longitudinally on average, for example, there were no students who performed at a lower level in their second interview; 12 performed at the same level, but 4 of these could not improve. Hence of those who could display improved understanding, 79% did so on the topic of average. This is nearly the same percentage as in the current study. The difference in the percent for diminished performance (13% here compared to 0% for average) may reflect the greater emphasis of the school curriculum on the topic of average in the middle and high school years, the years of the gap between interviews for these students. Also learning calculations associated with averages may have reinforced general understanding of average, something unlikely to have happened with the topic of sampling.

Several limitations are associated with the design and implementation of this study. The interview format was time-consuming and hence a limited number of students could be involved. This and the consequent dropout rate meant that data from only 38 students could be analyzed in the longitudinal study. There was also no control over the distribution of the remaining students across grades. Although students represented two different Australian states and public and private education, most of those interviewed were females, and it would be desirable to have an even more representative group of students. The ideal, however, is rarely achieved in educational research; and given the range of understanding displayed, it is felt that a rich picture of developing understanding has been gained that will be of benefit to educational planners. If numbers are not the only criteria for research respectability, then the descriptions of understanding sampled should compensate for some of the other limitations.

Further questions arising from this research might involve the monitoring of learning experiences that occur during the time gap between interviews or the planning and implementation of a specific program aimed at improving understanding. Such a program might be based on the outcomes observed in this study, particularly with respect to ideas for moving students' understanding into Tier 3 of the statistical literacy hierarchy (Watson, 1997). Such further research, however, is potentially very expensive unless sponsored within an educational system committed to providing the support necessary for well-documented implementation. Outside researchers will find it very difficult.

One of the issues in studying students' development of statistical understanding over time is whether cross-cohort studies are sufficient, or if longitudinal interviews are necessary. The major advantage of longitudinal interviews is the constancy of the individual and hence the confidence in the change observed as actual for that person. On the other hand, longitudinal studies usually cannot control for many factors that can influence outcomes—for example, school curriculum, which may be different for different students; and dropout rates, which for older students may skew outcomes to higher categories. Cross-cohort studies carried out simultaneously do not suffer the last two disadvantages, and if enough students are sampled in a representative manner, then confidence in the outcomes in terms of a developmental model is quite high. In the current study, it is not possible to make direct comparisons of the distribution of outcomes at the same grade levels in the different years. The 21 students in Grades 8 and 10 in their second interviews, did not perform as well generally as the 20 in the Grade 9 cohort originally (Watson & Moritz, 2000a). The different states and education systems represented in the later data set may contribute to its greater variation and somewhat lower level of performance. The expense and difficulty of conducting adequate longitudinal studies, and general trends for improvement observed from them, however, suggest that cross-cohort studies may be an acceptable substitute for most purposes.

IMPLICATIONS

One of the interesting educational issues to arise from a study of these interviews on samples and sampling is the dilemma for students in catering for variation that they know exists in the population. As noted by Wild and Pfannkuch (1999), this "noticing and acknowledging variation" is an aspect of the general consideration of variation that is fundamental to statistical thinking. It is a critical starting point. Some students, usually but not always younger, select a sample to *ensure* variation. Obviously the idea of using random selection to allow for a chance process to ensure appropriate variation is a fairly sophisticated idea. Combined with various forms of stratification as noted by some students, random selection caters for the needs that might occur in studying the weight of Grade 5 children in a state. Although most students are either in Category 3 or at least Category 5, a few have great difficulty distinguishing between allowing for variation and forcing it to occur, and suggest methods for both. One Grade 6 student, for example, suggested "choose an average person" as well as "one from each school [for] a wider variety of students." It is important for discussion to take place in the classroom to distinguish these situations explicitly for students. Perhaps class debates could be used to address the issue. As noted by Metz (1999), even a large proportion of students who have been involved in devising and carrying out their own research are not convinced of the power of sampling. The range of views she reported would be an excellent starting point for discussion.

As well as confirming the six categories of response reported in Watson and Moritz (2000a), this study identified a student who had not yet entered Category 1 or Tier 1 of the statistical literacy hierarchy. In terms of movement among tiers over the period of the study, no one reverted to Tier 1 understanding. Hence once students became involved in relating the idea of sample to a context, they did not lose the ability to do so. Of the 19 students originally responding in Tier 2, 37% were able to respond in Tier 3, three or four years later, whereas 20% of those originally in Tier 1, responded later in Tier 3. That the movement to Tier 3 was not stronger, and that 2 of 6 originally in Tier 3 dropped to Tier 2, is disappointing but perhaps not unexpected. It may reflect the lack of emphasis in mathematics classrooms, and in subjects other than mathematics, on bias in media reporting and other settings.

The observations in this study in terms of longitudinal change among the three tiers of understanding (Watson, 1997) reflect those of Watson and Moritz (2000a) in terms of cohort differences. The importance of emphasizing variation and representativeness in the transition from Tier 1 to Tier 2, and the recognition of bias in the transition from Tier 2 to Tier 3, is supplemented by the realization that for some younger children, examples of samples with appropriate associated meaning will be needed to introduce the idea of sample to students for Tier 1 understanding. Recognizing how structurally complex the construction of meaning of "sample" is (Watson & Moritz, 2000b, Table 2) implies that talking about "shampoo" is not sufficient. The idea of representation must be supplemented and distinguished from "*just* like the real thing." Student S2, for example, even at Grade 10, appeared to have a view of sampling from the grocery store that implied a perfect representation of the population and which then necessitated choosing all students from the Grade 5 population in order to "get it right."

Returning to the statement from the AEC (1991) on the importance of sampling and making inferences about populations, it is certain that the first is the foundation for the second. In perusing some of the responses reported in this study, it must be said that a more concerted effort is required throughout the middle and high school years in order to consolidate this foundation for many students.

ACKNOWLEDGMENTS

This research was funded by an Australian Research Council grant (No. A79800950) and a small grant from the University of Tasmania. Jonathan Moritz conducted the longitudinal interviews and corroborated the classifications of the author.

REFERENCES

Australian Education Council. (1991). *A national statement on mathematics for Australian schools.* Carlton, Vic.: Author.

Biggs, J. B., & Collis, K. F. (1982). *Evaluating the quality of learning: The SOLO taxonomy.* New York: Academic Press.

Biggs, J. B., & Collis, K. F. (1991). Multimodal learning and the quality of intelligent behavior. In H. A. H. Rowe (Ed.), *Intelligence: Reconceptualization and measurement* (pp. 57–76). Hillsdale, NJ: Erlbaum.

Corwin, R. B., & Friel, S. N. (1990). *Statistics: Prediction and sampling. A unit of study for grades 5–6 from used numbers: Real data in the classroom.* Palo Alto, CA: Dale Seymour.

Derry, S. J., Levin, J. R., Osana, H. P., & Jones, M. S. (1998). Developing middle-school students' statistical reasoning abilities through simulation gaming. In S. P. Lajoie (Ed.), *Reflections on statistics: Learning, teaching and assessment in grades K–12* (pp. 175–195). Mahwah, NJ: Erlbaum.

Gal, I. (2000). Statistical literacy: Conceptual and instructional issues. In D. Coben, J. O'Donoghue, & G. E. Fitzsimons (Eds.), *Perspectives on adults learning mathematics: Research and practice* (pp. 135–150). Dordrecht, The Netherlands: Kluwer Academic Publications.

Jacobs, V. R. (1997, March). *Children's understanding of sampling in surveys.* Paper presented at the Annual Meeting of the American Educational Research Association, Chicago.

Jacobs, V. R. (1999). How do students think about statistical sampling before instruction? *Mathematics in the Middle School, 5,* 240–263.

Landwehr, J. M., Swift, J., & Watkins, A. E. (1987). *Exploring surveys and information from samples.* Palo Alto, CA: Seymour.

Metz, K. E. (1999). Why sampling works or why it can't: Ideas of young children engaged in research of their own design. In F. Hitt & M. Santos (Eds.), *Proceedings of the 21st annual meeting of the North American Chapter of the International Group for the Psychology of Mathematics Education* (Vol. 2, pp. 492–498). Cuernavaca, Mexico: PME.

Miles, M. B., & Huberman, A. M. (1994). *Qualitative data analysis: An expanded sourcebook* (2nd ed.). Thousand Oaks, CA: Sage.

Mokros, J., & Russell, S. J. (1995). Children's concepts of average and representativeness. *Journal for Research in Mathematics Education, 26*(1), 20–39.

Moore, D. S. (1991). *Statistics: Concepts and controversies* (3rd ed.). New York: Freeman.

National Council of Teachers of Mathematics (2000). *Principles and standards for school mathematics.* Reston, VA: Author.

Orr, D. B. (1995). *Fundamentals of applied statistics and surveys.* New York: Chapman and Hall.

Rubin, A., Bruce, B., & Tenney, Y. (1991). Learning about sampling: Trouble at the core of statistics. In D. Vere-Jones (Ed.), *Proceedings of the Third International Conference on Teaching Statistics.* Vol. 1 (pp. 314–319). Voorburg: International Statistical Institute.

Schwartz, D. L., Goldman, S. R., Vye, N. J., Barron, B. J., & The Cognition and Technology Group at Vanderbilt. (1998). Aligning everyday and mathematical reasoning: The case of sampling assumptions. In S. P. Lajoie (Ed.), *Reflections on statistics: Learning, teaching and assessment in grades K–12* (pp. 233–273). Mahwah, NJ: Erlbaum.

Shaughnessy, J. M. (1992). Research in probability and statistics: Reflections and directions. In D. A. Grouws (Ed.), *Handbook of research on mathematics teaching and learning* (pp. 465–494). New York: NCTM & Macmillan.

Shaughnessy, J. M., Garfield, J., & Greer, B. (1996). Data handling. In A. J. Bishop, K. Clements, C. Keitel, J. Kilpatrick, & C. Laborde (Eds.), *International handbook of mathematics education, Part 1* (pp. 205–237). Dordrecht, The Netherlands: Kluwer Academic Publishers.

Tversky, A., & Kahneman, D. (1971). Belief in the law of small numbers. *Psychological Bulletin, 76*(2), 105–110.

Tversky, A., & Kahneman, D. (1974). Judgment under uncertainty: Heuristics and biases. *Science, 185,* 1124–1131.

Wagner, D. A., & Gal, I. (1991). *Project STARC: Acquisition of statistical reasoning in children* (Annual Report: Year 1, NSF Grant No. MDR90-50006). Philadelphia: Literacy Research Center, University of Pennsylvania.

Wallman, K. K. (1993). Enhancing statistical literacy: Enriching our society. *Journal of the American Statistical Association, 88*(421), 1–8.

Watson, J. M. (1997). Assessing statistical literacy using the media. In I. Gal & J. B. Garfield (Eds.), *The Assessment Challenge in Statistics Education* (pp. 107–121). Amsterdam: ISO Press and the International Statistical Institute.
Watson, J. M. (2001). Longitudinal development of inferential reasoning by school students. *Educational Studies in Mathematics, 47,* 337–372.
Watson, J. M., Collis, K. F., & Moritz, J. B. (1995, November). *The development of concepts associated with sampling in grades 3, 5, 7 and 9.* Paper presented at the Annual Conference of the Australian Association for Research in Education, Hobart.
Watson, J. M., & Moritz, J. B. (2000a). Developing concepts of sampling. *Journal for Research in Mathematics Education, 31,* 44–70.
Watson, J. M., & Moritz, J. B. (2000b). Development of understanding of sampling for statistical literacy. *Journal of Mathematical Behavior, 19,* 109–136.
Watson, J. M., & Moritz, J. B. (2000c). The longitudinal development of understanding of average. *Mathematical Thinking and Learning,* 2(1&2), 11–50.
Watson, J. M., & Moritz, J. B. (2001). Development of reasoning associated with pictographs: Representing, interpreting, and predicting. *Educational Studies in Mathematics, 48,* 47–81.
Wild, C. J., & Pfannkuch, M. (1999). Statistical thinking in empirical enquiry. [And Discussions]. *International Statistical Review, 67,* 223–265.

Chapter 13

REASONING ABOUT SAMPLING DISTRIBUTIONS

Beth Chance[1], Robert delMas[2], and Joan Garfield[2].
California Polytechnic State University, USA[1], and University of Minnesota, USA[2]

INTRODUCTION

This chapter presents a series of research studies focused on the difficulties students experience when learning about sampling distributions. In particular, the chapter traces the seven-year history of an ongoing collaborative research project investigating the impact of students' interaction with computer software tools to improve their reasoning about sampling distributions. For this classroom-based research project, three researchers from two American universities collaborated to develop software, learning activities, and assessment tools to be used in introductory college-level statistics courses. The studies were conducted in five stages, and utilized quantitative assessment data as well as videotaped clinical interviews. As the studies progressed, the research team developed a more complete understanding of the complexities involved in building a deep understanding of sampling distributions, and formulated models to explain the development of students' reasoning.

THE PROBLEM

Many published research reports, as well as popular media accounts, utilize ideas of statistical confidence and significance. Consequently, a large proportion of the introductory statistics courses at the tertiary level is concerned with statistical inference. While many students may be able to carry out the necessary calculations, they are often unable to understand the underlying process or properly interpret the results of these calculations. This stems from the notoriously difficult, abstract topic of sampling distributions that requires students to combine earlier course topics such as sample, population, distribution, variability, and sampling. Students are then

D. Ben-Zvi and J. Garfield (eds.),
The Challenge of Developing Statistical Literacy, Reasoning and Thinking, 295–323.

asked to build on these ideas to make new statements about confidence and significance. However, student understanding of these earlier topics is often shallow and isolated, and many students complete their introductory statistics course without the ability to integrate and apply these ideas. Our experience as teachers of statistics suggests that the statistical inference calculations that students perform later in a course tend to become rote manipulation, with little if any conceptual understanding of the underlying process. This prevents students from being able to properly interpret research studies.

To address this problem, research and education literature has suggested the use of simulations for improving students' understanding of sampling distributions (e.g., Behrens, 1997; Davenport, 1992; Glencross, 1988; Schwarz & Sutherland, 1997; Simon, 1994). Many of these articles discuss the potential advantage of simulations to illustrate this abstract idea by providing multiple examples of the concept and allowing students to experiment with all of the variables that form the concept. In particular, technology allows students to be directly involved with the "building up" of the sampling distribution, focusing on the process involved, instead of presenting only the end result. Recently, numerous instructional computer programs have been developed that focus on use of simulations and dynamic visualizations to help students develop their understanding of sampling distributions and other statistical concepts: ConStatS (Cohen, 1997), HyperStat (Lane, 2001), Visual Statistics (Doane, Tracy, & Mathieson, 2001), StatPlay (Thomason & Cummings, 1999), StatConcepts (Newton & Harvill, 1997), ExplorStat (Lang, Coyne, & Wackerly, 1993), and ActivStats (Velleman, 2003).

Despite this development of software programs, little has been published that evaluates the effectiveness of simulation activities to improve students' reasoning about statistics. Some papers have cited anecdotal evidence that students are more engaged and interested in learning about statistics with such simulations (e.g., Simon, 1994), but even fewer studies have gathered and presented empirical data, especially in the context of the college statistics classroom (see Mills, 2002). Of the empirical studies that have been conducted, most demonstrated only very modest, if any, gains in student learning (Schwartz, Goldman, Vye, & Barron, 1997; Well, Pollatsek, & Boyce, 1990).

For example, Earley (2001) found that an instructor-led demonstration using the Sampling SIM (delMas, 2001) program was not sufficient to "convince" students of various features of the Central Limit Theorem. They could recognize facts, but were not able to consistently apply their knowledge. However, Earley noted evidence that the students referred to the images from the program later in the course and used them as a building block when the course proceeded to hypothesis testing. Saldanha and Thompson (2001) documented the difficulties high school students exhibited during two teaching experiments about sampling distributions. These included use of computer simulations to investigate what it means for the outcome of a stochastic experiment to be unusual. They found students had difficulty grasping the multilevel, stochastic nature of sampling distributions and often did not sufficiently participate in the teaching activity. Studies by Hodgson (1996) revealed that simulations may actually contribute to the formation of misconceptions (e.g., the belief that inference required multiple samples). As a consequence, Hodgson and

Burke (2000) suggest ways of ensuring that students attend to the more salient features of simulation activities and highlight the importance of pre-organizers, ongoing assessment, debriefing, and follow-up exercises.

There is at least one exception to the findings of only modest gains by empirical studies. Sedlmeier (1999), using an adaptive algorithms perspective, designed software based on a "flexible urn" model to train students on sampling distribution problems. The adaptive algorithms perspective argues that the urn model is similar to frequency information that people deal with regularly and to which the human mind has adapted through evolution. Translating sampling distribution problems into the urn model is thought to make the task more understandable and facilitate reasoning. Sedlmeier found significant immediate and long-term effects from the flexible urn training. One question that is not addressed in Sedlmeier's studies is whether students develop an abstract understanding of sampling distributions through these activities or remain dependent on translation to the urn model.

BEGINNING A SERIES OF CLASSROOM-BASED RESEARCH STUDIES

To investigate the potential impact of simulation software on students' understanding of sampling distributions, the Sampling Distribution program, a precursor of the Sampling SIM program (delMas, 2001), was developed. Initial development of this software was guided by literature in educational technology and on conceptually enhanced simulations (e.g., Nickerson, 1995; Snir, Smith, & Grosslight, 1995). An activity was created to guide the students' interaction with the simulation software based on ideas from literature in learning and cognition (e.g., Holland, Holyoak, Nisbett, & Thagard, 1987; Perkins, Schwartz, West, & Wiske, 1995). Assessment tasks were designed to determine the extent of students' conceptual understanding of sampling distributions.

The three classroom researchers began using the software, activity, and assessments in different settings: a small, private university (University of the Pacific), a College of Education, and a Developmental Education College (the latter two both at the University of Minnesota). These were fairly standard algebra-based introductory courses, presumed to be students' first exposure to the material. The courses used common introductory textbooks (Moore & McCabe, 2002; Moore, 2000; and Siegel & Morgan, 1996), and included numerous classroom activities and uses of technology. A primary goal of the classroom research was to document student learning of this challenging topic, while providing feedback for further development and improvement of the software and the learning activity. Four questions guided the investigation: *how* the simulations could be utilized more effectively, *how* to best integrate the technology into instruction, *why* particular techniques appeared to be more effective, and *how* student understanding of sampling distributions was affected by use of the program. Five sequential research studies were conducted, each building on the previous work. These are described in the following sections.

First Study: Assessing the Impact of Instruction with Simulation Software

An initial version of the instructional activity asked students to use the Sampling Distribution program to change settings such as the population shape and sample size and to summarize the results for the different empirical sampling distributions they observed. Graphics-based test items were used to determine whether students could demonstrate a visual understanding of the implications of the Central Limit Theorem for a sample mean. Each item presented a population distribution and required students to choose which of five empirical distributions of sample means best represented a potential sampling distribution for a specified sample size.

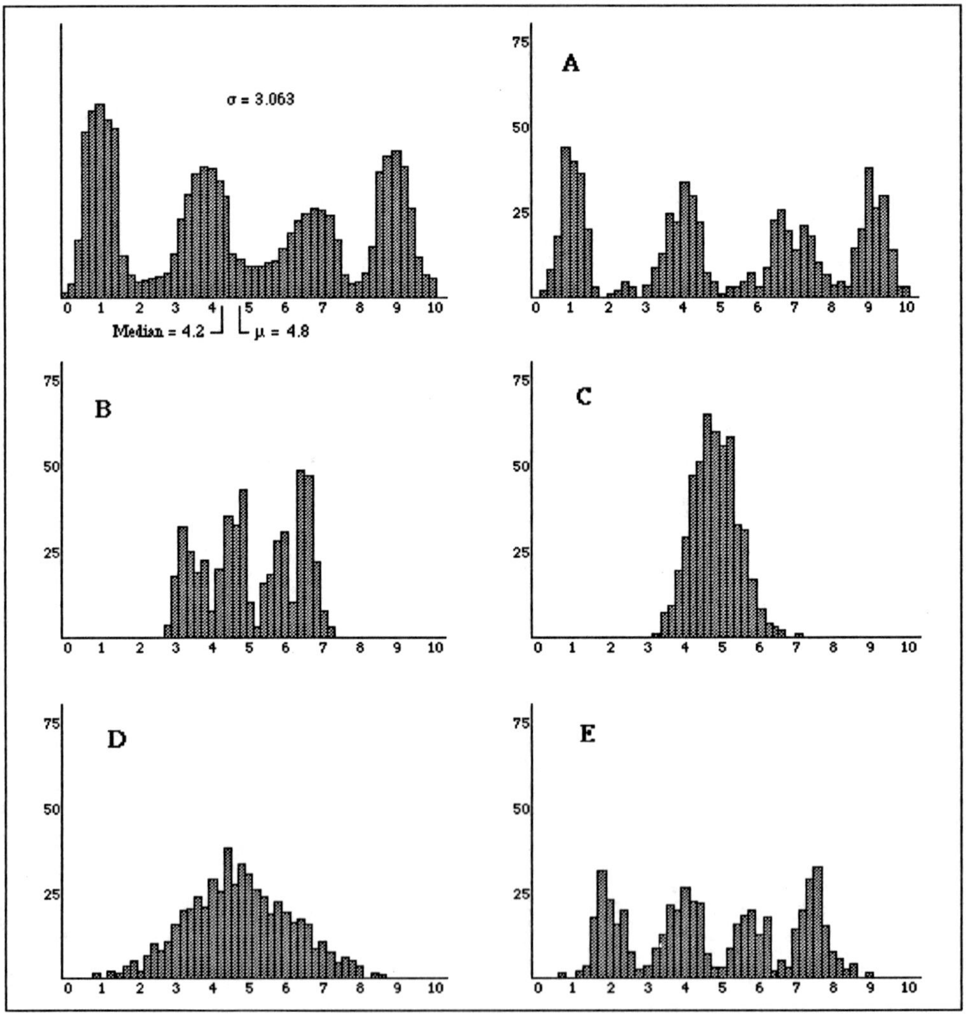

Figure 1. A graph-based item. (Correct answers: D, C)

For example, the item in Figure 1 asked students which distribution of sample means they thought best represented an empirical distribution for 500 samples of size $n = 4$ and also for size $n = 25$. Students were asked to justify their choice of graphs and explain their reasoning in writing. These responses were then categorized so that future instruments asked students to select which statement best described their own reasoning. Students were given these test instruments before using the program and immediately after using the program. Comparing the pre- and posttest scores isolated the change in students' understanding from interacting with the program and activity.

While there were some positive changes, several students still did not appear to be developing correct reasoning about sampling distributions. See delMas, Garfield, and Chance (1998) for more details of the program and instructional activities. Reflection on these results led to further development of the program and instructional activities.

Second Study: Applying a Conceptual Change Approach

Despite the software's capability to provide an excellent visualization of the abstract process of creating sampling distributions, students were still having difficulty understanding and applying the Central Limit Theorem. Research on conceptual change theory in science education offered a different approach to this problem (e.g., Posner, Strike, Hewson, & Gertzog, 1982; Lord, Ross, & Lepper, 1979; Jennings, Amabile, & Ross, 1982; Ross & Anderson, 1982). An attempt was made to build on this theory in redesigning the activity to engage students in recognizing their misconceptions and to help them overcome the faulty intuitions that persisted in guiding their responses on assessment items. In the new activity, students were first asked to give their response to the graphical test items, as in Figure 1, for five different populations, each at two different sample sizes, and then to use the Sampling SIM program to produce an empirical sampling distribution under the same conditions. They were then asked to compare the simulation results to their earlier responses and comment on whether their answer agreed or disagreed (and if so how) with what the program revealed about the behavior of the sample means. This predict/test/evaluate model forced students to more directly confront the misconceptions in their understanding, which resulted in statistically significant improvements in their performance on the posttest (delMas, Garfield, & Chance, 1999a).

Third Study: Conceptual Analysis of Prior Knowledge and Misconceptions

While many students did demonstrate better understanding after this revised activity, many still exhibited misconceptions as indicated by responses on an immediate posttest as well as final exam items (delMas, Garfield, & Chance, 1999b). Since the topic of sampling distributions requires students to integrate many concepts from earlier in the course, gaps in the students' prerequisite knowledge were considered as a plausible cause. For example, when discussing the variability of the sampling distribution and how variability decreases as the sample size increases, it appeared that some students were not able to fully understand or identify variability, nor to properly read a histogram when presented in a different context. Therefore, this study involved a series of conceptual analyses related to student understanding of sampling distributions. These analyses were based on the experiences and observations of the classroom researchers, contributions of colleagues, and analyses of students' performance on assessment items.

The first analysis produced a thorough list of what students should know before learning sampling distributions (Garfield, delMas, & Chance, 2002; see Table 1). This list guided the development of a set of pretest questions that were given to students and discussed in class before instruction (see http://www.gen.umn.edu/faculty_staff/delmas/stat_tools/, click the MATERIALS button, and scroll to the Sampling Distributions Activity section). Using the items to diagnose areas that students did not understand provided some review and remediation to students before proceeding to the new topic of sampling distributions. This analysis led to more detailed descriptions of what is meant by "understanding sampling distributions," including a detailed list of the necessary components of understanding (Table 2), a list of what students should be able to do with their knowledge of sampling distributions (Table 3), and a list of common misconceptions that students exhibit about sampling distributions (Table 4). These lists guided revisions of the activity, pretests, and posttests. This analysis was helpful in identifying types of correct and incorrect understanding to look for in students' reasoning, and enabled more detailed examination of individual student conceptions via clinical interviews.

Table 1. Prerequisite knowledge to learning about sampling distributions

- *The idea of variability.* What is a variable? What does it mean to say observations vary? Students need an understanding of the spread of a distribution in contrast to common misconceptions of smoothness or variety.
- *The idea of a distribution.* Students should be able to read and interpret graphical displays of quantitative data and describe the overall pattern of variation. This includes being able to describe distributions of data; characterizing their shape, center, and spread; and being able to compare different distributions on these characteristics. Students should be able to see between the data and describe the overall shape of the distribution, and be familiar with common shapes of distributions, such as normal, skewed, uniform, and bimodal.
- *The normal distribution.* This includes properties of the normal distribution and how a normal distribution may look different due to changes in variability and center. Students should also be familiar with the idea of area under a density curve and how the area represents the likelihood of outcomes.
- *The idea of sampling.* This includes random samples and how they are representative of the population. Students should be comfortable distinguishing between a sample statistic and a population parameter. Students should have begun considering or be able to consider how sample statistics vary from sample to sample but follow a predictable pattern.

Table 2. What students should understand about sampling distributions

- A sampling distribution of sample means (based on quantitative data) is a distribution of all possible sample means (statistics) for a given sample size randomly sampled from a population with mean μ and standard deviation σ. It is a probability distribution for the sample mean.
- The sampling distribution for means has the same mean as the population.
- As the sample size (n) gets larger, the variability of the sample means gets smaller (a statement, a visual recognition, and predicting what will happen or how the next picture will differ).
- Standard error of the mean is a measure of variability of sample statistic values.
- The building block of a sampling distribution is a sample statistic.
- Some values of statistics are more or less likely than others to be drawn from a particular population.
- The normal approximation applies in some situations but not others.
- If the normal approximation applies, then the empirical rule can be applied to make statements about how often the sample statistic will fall within, say, 2 standard deviations of the mean.
- Different sample sizes lead to different probabilities for the same statistic value (know how sample size affects the probability of different outcomes for a statistic).
- Sampling distributions tend to have the shape of a normal distribution rather than the shape of the population distribution, even for small samples.
- As sample sizes get very large, all sampling distributions for means look alike (i.e., have the same shape) regardless of the population from which they are drawn.
- Averages are more normal and less variable than individual observations.
- Be able to distinguish between a distribution of observations in one sample and a distribution of $\overline{x}$ statistics (sample means) from many samples (sample size n greater than 1) that have been randomly selected.

Table 3. What students should be able to do with their knowledge of sampling distributions of the sample mean

- Describe what a sampling distribution would look like for different populations and sample sizes (based on shape, center and spread, and where most of the values would be found).
- Interpret and apply areas under the (theoretical sampling distribution) curve as probability statements about sample means.
- Describe which values of the sample mean are likely, and which are less likely. This may include ability to apply the empirical rule to the distribution of sample means.
- Describe the size of the standard error of the mean and how or when it changes.
- Describe the likelihood of different values of the sample mean. In particular, make statements about how far a sample statistic is likely to vary from the population proportion. For example, explain how often the sample mean should fall within two standard deviations of the population mean, and whether a random set of outcomes is unusual based on given population characteristics.
- Describe the mean of the sample means for different-shaped populations.

Table 4. Some common student misconceptions

- Believe sampling distribution should look like the population (for sample size $n > 1$).
- Think sampling distribution should look more like the population as the sample size increases (generalizes expectations for a single sample of observed values to a sampling distribution).
- Predict that sampling distributions for small and large sample sizes have the same variability.
- Believe sampling distributions for large samples have more variability.
- Do not understand that a sampling distribution is a distribution of sample statistics.
- Confuse one sample (real data) with all possible samples (in distribution) or potential samples.
- Pay attention to the wrong things, for example, heights of histogram bars.
- Think the mean of a positive skewed distribution will be greater than the mean of the sampling distribution for samples taken from this population.

Fourth Study: Student Interviews and a Developmental Model

To gather more detailed information about how students' conceptions of related concepts (e.g., distribution, variability) as well how they actually develop reasoning about sampling distributions, several guided interviews were conducted. The interviews were also designed to capture students' interaction with the Sampling SIM program in an individualized setting (Garfield, 2002). The students were enrolled in a graduate-level introductory statistics course. Interviews, which lasted from 45 to 60 minutes, asked students to respond to several open-ended questions about sampling variability while interacting with the Sampling SIM software. The interviews were videotaped, transcribed, and viewed many times to determine students' initial understanding of how sampling distributions behave and how feedback from the computer simulation program helped them develop an integrated reasoning of concepts. While conducting and later reviewing these interviews, the authors noted some differences between students as they progressed throughout the interview and activity. These findings initially suggested a developmental model might describe the stages students appeared to progress through in going from faulty to correct reasoning. Based on the work of Jones and colleagues (Jones et al., 2000; Jones et al., 2001; Mooney, 2002), who had proposed developmental models of statistical thinking and reasoning in children, a framework was developed that describes stages of development in students' statistical reasoning about sampling distributions. An initial conception of the framework is as follows (Garfield, delMas, & Chance, 1999):

Level 1—Idiosyncratic Reasoning The student knows words and symbols related to sampling distributions, uses them without fully understanding them, often incorrectly, and may use them simultaneously with unrelated information.

Level 2—Verbal Reasoning The student has a verbal understanding of sampling distributions and the implications of the Central Limit Theorem, but cannot apply this to the actual behavior of sample means in repeated samples. For example, the student can select a correct definition, but does not understand how key concepts such as variability and shape are integrated.

Level 3—Transitional Reasoning The student is able to correctly identify one or two characteristics of the sampling process without fully integrating these characteristics. These "characteristics" refer to three aspects of the Central Limit Theorem: understanding that the mean of the sampling distribution is equal to the population mean, that the shape of the sampling distribution becomes more normal as the sample size increases, and that the variability in the sample means decreases as the sample size increases. A student who understands only one or two characteristics might state only that large samples lead to more normal-looking sampling distributions, or that larger samples lead to narrower sampling distributions.

Level 4—Procedural Reasoning The student is able to correctly identify the three characteristics of the sampling process but does not fully integrate them or understand the predictable long-term process. For example, the student can correctly predict which sampling distribution corresponds to the given parameters, but cannot explain the process, and does not have full confidence when predicting a distribution of sample means from a given population for a given sample size.

Level 5—Integrated Process Reasoning The student has a complete understanding of the process of sampling and sampling distributions, in which rules and stochastic behavior are coordinated. For example, students can explain the process in their own words, describing why the distribution of sample means becomes more normal and has less variability as the sample size increases. They also make predictions correctly and confidently.

Fifth Study: Defining Dimensions of Reasoning

Having described these levels of student reasoning, it was important to validate the levels through additional interviews across the three environments. A nine-item diagnostic assessment was developed to identify students who were potentially at different levels of statistical reasoning (see the Appendix for items 5–9). The test contained graph- and text-based questions, and asked students to rank their level of confidence in their answers. The first item tested students' understanding of the relationship between sample size and the variability of a sample estimate. The second and third items required students to apply their knowledge of the standard error of the mean. Students were expected to have some understanding of the empirical rule as well. The fourth and fifth problems were graph-based items that assessed students' ability to make correct predictions about the behavior of sampling distributions, as well as their ability to identify reasonable descriptions and comparisons of distributions. The sixth through eighth items required students to apply their understanding of the Central Limit Theorem. The final item assessed students' definitional knowledge of the Central Limit Theorem.

The assessment was administered to 105 undergraduates at the University of Minnesota currently enrolled in an introductory statistics course that utilized the Sampling SIM software. At the start of the second half of the course, these students used Sampling SIM to complete an activity on sampling distributions, and then took the diagnostic test at the start of the following class session. Nine statistics majors at Cal Poly who were enrolled in a senior-level capstone course but had never interacted with Sampling SIM also completed the diagnostic test. Altogether, the 114 students showed substantial variation in their responses to the questions. With respect to the two graph-based problems (items 4 and 5; see Appendix), only 10% made correct choices for both graphs in both problems, and another 22% made correct choices for both graphs in only one of the problems. Many students (47%) made choices that were not correct, but indicated an understanding that sampling variability decreases as sample size increases. Of the remaining students, 19% made choices for at least one problem that suggested a belief that sampling variability increases with increases in sample size, and two of the students chose the same graph for both sample sizes.

Not all of the student responses to the questions about the distribution shape and relative variability in problems 4 and 5 were consistent with their graph choices. Concerning shape, only 49% of the students made choices that were completely consistent. There were two questions about shape for each problem (questions c and g) with a total of four questions between the two problems. The average percentage of consistent shape choices per student varied from 0% to 100% with an average of 80% (*SD* = 24.8). Regarding variability comparisons, an even smaller percentage of students (33%) made completely consistent choices. Students were asked to make three comparisons of variability for each problem (questions d, h, and i) for a total of six comparisons between the two problems. The percentage of consistent variability comparisons varied from 0% to 100% with an average of 72% (*SD* = 29.6).

On average, students correctly answered 61% of the remaining non-graph items of the diagnostic test (*SD* = 16.2). Most of the students (88%) answered the first problem correctly regarding the relationship between sample size and estimation accuracy. The students did not fare as well with the second and third problems that involved an understanding of the standard error of the mean and application of the empirical rule. Only 4% answered both questions correctly, and another 7% answered at least one of the items correctly. On the three items that required application of the Central Limit Theorem (items 6, 7, and 8 in the Appendix), most of the students correctly answered either all three items (22%) or two of the three (39%). They also demonstrated a definitional understanding of the Central Limit Theorem in that 55% correctly answered all four questions in problem 9, while another 26% answered three of the four questions correctly.

Of the 114 students who took the diagnostic test, 37 signed a consent form indicating they would be willing to participate in an interview. All students who gave consent received an invitation to participate in an interview, and nine accepted. Statistics on the nine students' diagnostic test performance are presented in Table 5.

These nine students represent a fair amount of variation in performance across the items on the diagnostic test.

Table 5. Diagnostic test performance of undergraduates who participated in an interview

Student	**Non-Graph Items**					**Graph-Based Items**				
	Sample Size Item 1	SEM Application Items 2 & 3	Central Limit Theorem: Application Items 6–8	Central Limit Theorem: Definition Item 9	All Items	Choice Pattern: Item 4	Choice Pattern: Item 5	Average Confidence	Agreement with Graphs: Shape Comparisons	Agreement with Graphs: Variability Comparisons
Kelly	Correct	100%	100%	100%	100%	G	C	80.0	100%	100%
Jack	Correct	0%	67%	75%	60%	C	C	76.3	100%	83%
Mitzy	Correct	0%	33%	100%	60%	G	C	70.0	75%	100%
David	Correct	0%	0%	50%	30%	G	C	66.3	75%	100%
Karen	Correct	0%	67%	75%	60%	L-S	G	53.8	100%	67%
Marie	Correct	100%	33%	75%	70%	L-S	L-S	70.0	100%	67%
Martha	Correct	50%	33%	50%	50%	L-S	*	77.5	50%	50%
Susan	Correct	0%	33%	75%	30%	S-L	L-S	86.3	100%	100%
Elaine	Incorrect	0%	33%	75%	40%	L-S	S-L	27.5	50%	100%

Legend.

C Correct choice of graph for both sample sizes.

G Good or reasonable choice of graphs; graph for smaller sample size is like population shape with less variance than the population, graph for larger sample size is bell-shaped with less variance than the small sample size graph.

L-S Neither C nor G, but graph for larger sample size has less variance than graph for smaller sample size.

S-L Graph for larger sample size has more variance than graph for smaller sample size.

* The student did not answer this item.

With the hope of adding more variability to the interview pool, a group of master's-level students enrolled in a graduate-level introductory statistics course at the University of Minnesota were also invited to participate in an interview. Items from one of their course exams that were similar to items on the diagnostic test were identified. Students' performance on these items, along with the students' grades, were used to select four students who would potentially represent a variety of levels in statistical reasoning. All four of the graduate students participated in an interview.

An initial interview protocol (e.g., Clement, 2000) was developed, and the problems and script were piloted with three of the undergraduate students. In the initial protocol, students were asked to solve only a sampling distribution question (see part 3 of the final version of the interview protocol). After reviewing the pilot interviews, it became clear that responses to the individual question did not provide enough information to clearly understand the reasons for students' responses. More information was needed on students' understanding of basic statistical definitions, concepts, and processes to better understand their statements and choices. Consequently the number of tasks was expanded from one to four, and the interview script and technique were revised to better probe students' reasoning. Interviews took 30–45 minutes, after which students were debriefed.

Identical interview scripts were used at each location. The final interview protocol consisted of four parts:

Part 1—Students were asked to describe the population distribution presented in Figure 2. They were then given empty axes and asked to draw a representation of two samples from the population, one for a sample of size $n = 4$ and one for a sample of size $n = 16$.

Part 2—Students were given 5 graphs representing possible empirical sampling distributions for 500 samples (see Figure 3). They were asked to judge which graph(s) best corresponded to empirical sampling distributions for samples of size 4 and for samples of size 16.

Part 3—Participants were asked to make true/false judgments for a series of statements about samples and sampling distributions, as shown in Figure 4.

Part 4—They were shown the same population used in Parts 1 and 2 and asked to select which graph in Figure 5 best represented a potential sample of size 50 from that population.

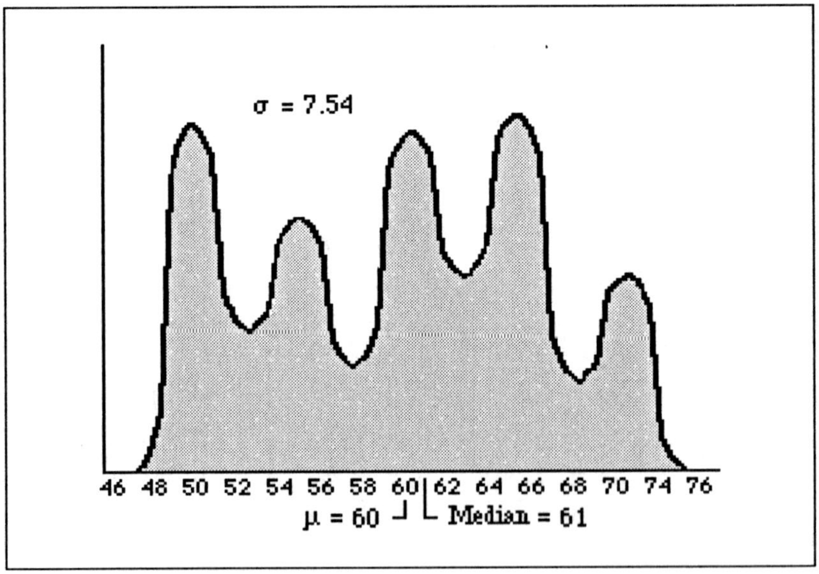

Figure 2. Population distribution used in the student interviews.

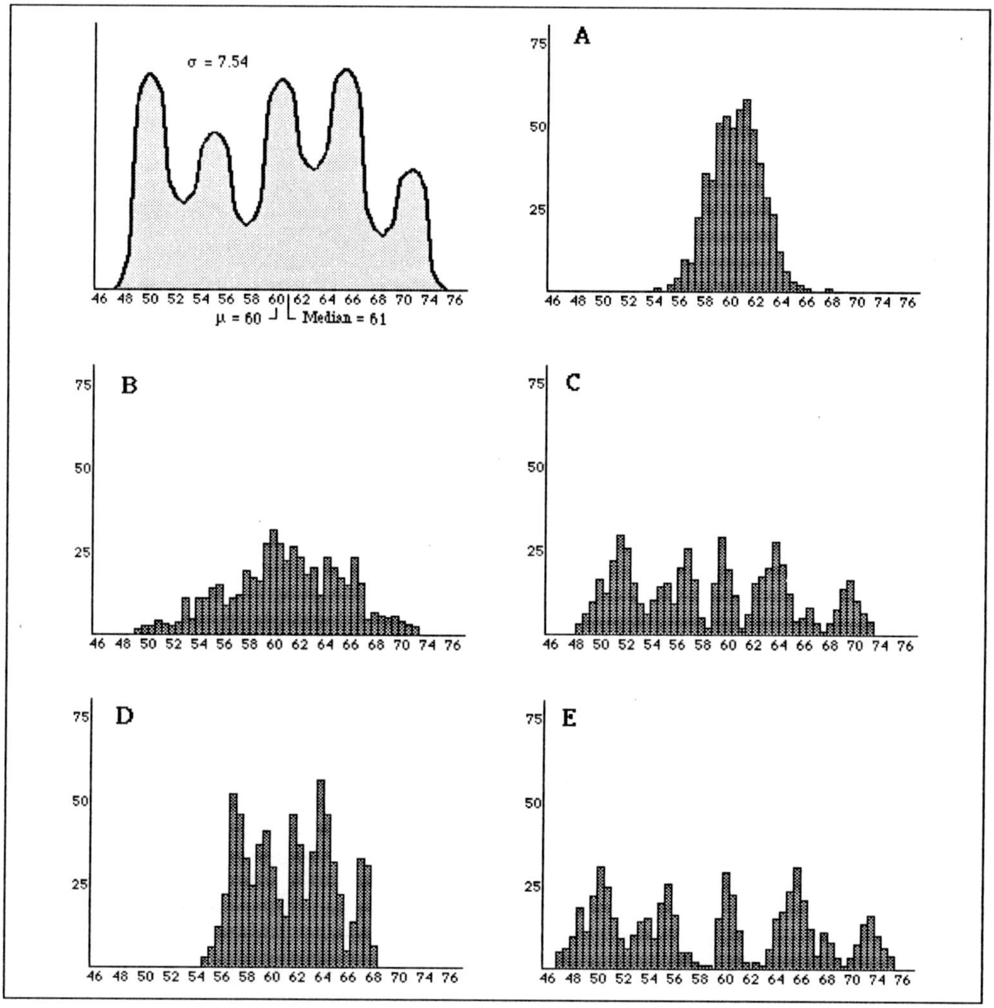

Figure 3. Population distribution and potential empirical sampling distributions.

1.	As the sample size increases, the samples look more like the normal distribution, each sample will have the same mean as the population, and each sample will have a smaller standard deviation than the population.	TRUE FALSE
2.	As the sample size increases, the sampling distribution of means looks more like the population, has the same mean as the population, and has a standard deviation that is similar to the population.	TRUE FALSE
3.	As the sample size increases, the sampling distribution of means looks more like the normal distribution, has a mean that is the same as the population, and a standard deviation that is smaller than the population standard deviation.	TRUE FALSE

Figure 4. Item used in part 3 of the student interviews. (Correct answers: false, false, true)

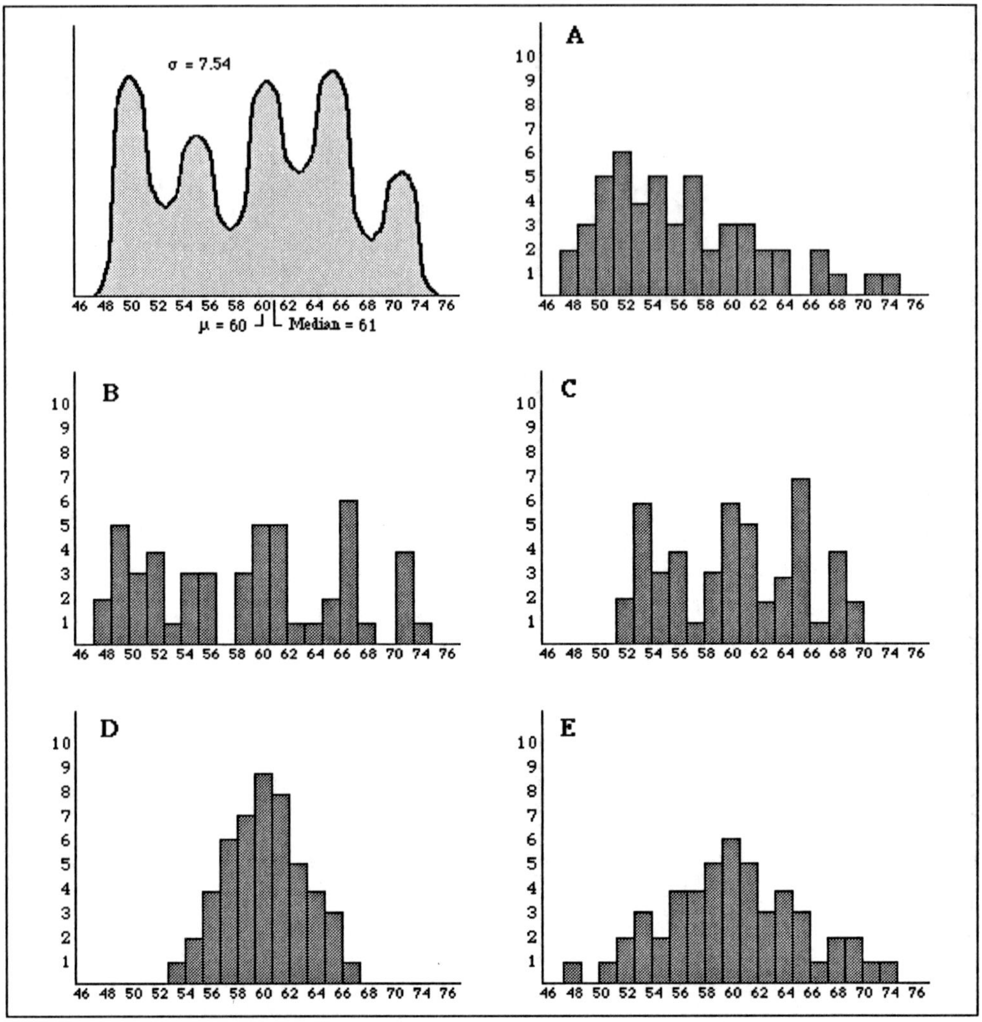

Figure 5. Population distribution and potential empirical samples.

At the end of each part of the interview, students were asked to explain why they had made their particular statements or choices and to indicate their confidence in their choices.

Initially the videotapes were reviewed for evidence that students were at the different levels of reasoning described earlier, as suggested by their written test results. However, it was difficult to place students into a particular level of reasoning when observing the videos. The data from the posttest and the clinical interviews did not support the idea that a student is at one particular level or stage of statistical reasoning as initially hypothesized. Instead, these interviews suggested that students' reasoning is more complex and develops along several interrelated dimensions. Building on other theoretical perspectives (e.g., Perkins, Crismond, Simmons, & Unger, 1995; Case, 1985; Biggs & Collis, 1982), the following "components" or "dimensions" of statistical reasoning behavior are proposed:

1. *Fluency*—how well the student understands and uses appropriate terminology, concepts, and procedures
2. *Rules*—the degree to which a student identifies and uses a formal rule to make predictions and explanations
3. *Consistency*—the presence or absence of contradictory statements
4. *Integration*—the extent to which ideas, concepts, and procedures are connected
5. *Equilibrium*—the degree to which a student is aware of any inconsistencies or contradictions in his or her statements and predictions
6. *Confidence*—the degree of certainty in choices or statements

The videotapes were reexamined to identify students who exhibited the extremes of each dimension. Some examples follow.

Example 1—From part 1 of the interview, students were asked to describe the population displayed. Two students, Jack and Allie, gave noticeably different responses.

Jack: Um, some of it's, you know, up, some of it's down. There's differences, so it's not like, you know, perfect. Skewed in a way, I guess, you could say ... I can tell you like the median and stuff.

Allie: Okay, it's a multimode, uh graph, distribution, and it's pretty symmetric in one sense. I see that the average is 60 and the median is pretty close, so that's it's ... (inaudible) ... pretty much normal, like a symmetric distribution. Standard deviation is 7.5, which is actually pretty big if you look at the scores here, which means there's lots of variation in the population.

Allie is able to use relevant statistical terms much more freely than Jack. She immediately jumps to characteristics of shape, center, and spread on her own, providing an evaluation of the amount of variability in the population and focusing on the multimodal nature as a distinctive feature of the distribution.

Example 2—In part 2 of the interview, Karen and Allie correctly chose graph B for $n = 4$ and graph A for $n = 16$ (Figure 3). They were then asked if they could explain why the sampling distributions behaved this way.

Karen: Because I just remember learning in class that it goes to a ... when you draw, it goes to a normal distribution, which is the bell-shaped curve. So, I just look at graphs that are tending toward that ... That's just how ... I don't know like the why, the definition. That's just what I remember learning.

Allie: If you keep plotting averages; if you keep taking averages, the outliers are going to actually be less, um, have less big effect on your data. So you're actually always dropping out those outliers. So it's getting more and more … they call it also, I think, regression toward the mean. I don't know if that term actually is used in this kind of situation, but since you're already … you're always taking averages, the outliers get less efficient. No, no, that's not a word I'm looking for. Um, will have less effect, I'll just use that word, so it will get more and more narrower.

While both students made correct predictions, it would be difficult to classify them at different levels of understanding. Karen appears to have partial mastery of the formal terminology related to sampling distributions (normal distribution, the bell-shaped curve), but is not able to explain in her own terms the process behind the sampling distribution of the mean. Allie has more trouble finding the standard terms and incorrectly applies the phrase "regression to the mean," but seems to be viewing the sampling distribution as a long-term process and to have more understanding of the influences of the sample size on that process. Thus, students with different mastery levels of the formal terminology can also show different abilities to integrate the concepts and understand their statements.

Example 3—In part 1 of the interview, students were asked to draw a second empirical sampling distribution when the sample size was changed from $n = 4$ to $n = 16$.

Betty: Still keep the mean at 60 like I did with $n = 4$, but it's going to have a … oh, I'm not an artist I'm a nurse. It's going to have a higher bell shape and a narrower distribution. The standard deviation will be bigger for $n = 4$ than 16. So the standard deviation is narrower, it has a normal distribution shape, a normal density curve, and the mean isn't going to move anywhere. The center of the density curve is always going to be in the mean of the population.

Betty appears to be able to focus on both the shape and the spread of the sampling distribution, using both dimensions in making her choices. When asked about a sample of size 50 (part 4), she again appeals to the idea that "If it's randomly selected it should hopefully be going more toward a normal distribution," and considers graph D as too normal for a sample size of 50 and so chooses graph E (both incorrect choices; see Figure 3). She did not appear to consider variation in selecting the graphs, despite her earlier comments on variability; and she was not able to clearly differentiate between the sample and the sampling distribution, thus exemplifying the student who does not have complete integration of different components of the concept.

Example 4—From part 2 of the interview (see Figure 3), Martha attempted to correct an earlier response.

Martha: I'm going to go for C for $n = 4$ and then 16 for ... $n = 16$ for A. (laughs) Yeah. And partially because ... with $n = 4$, I'm thinking you're going to have a larger range ... yeah, a larger range for $n = 4$ than you would for $n = 16$. Because before I was guessing and I thought that the standard deviation for a larger sample would be closer to the original than the standard deviation for $n = 4$.

Further discussion with this student reveals that she has not quite settled on whether the standard deviation of the sampling distribution will decrease as sample size n increases, or will approach the population standard deviation. She recognizes her inconsistency and continues to try to reconcile these two "rules" in subsequent parts of the interview.

DISCUSSION

Sampling distributions is a difficult topic for students to learn. A complete understanding of sampling distributions requires students to integrate and apply several concepts from different parts of a statistics course and to be able to reason about the hypothetical behavior of many samples—a distinct, intangible thought process for most students. The Central Limit Theorem provides a theoretical model of the behavior of sampling distributions, but students often have difficulty mapping this model to applied contexts. As a result, students fail to develop a deep understanding of the concept of sampling distributions and therefore often develop only a mechanical knowledge of statistical inference. Students may learn how to compute confidence intervals and carry out tests of significance, but they are not able to understand and explain related concepts, such as interpreting a p-value.

Most instructors have welcomed the introduction of simulation software and web applets that allow students to visualize the abstract process of drawing repeated random samples from different populations to construct sampling distributions. However, our series of studies revealed that several ways of using such software were not sufficient to affect meaningful change in students' misconceptions of sampling distributions. Despite the ability of a software program to offer interactive, dynamic visualizations, students tend to look for rules and patterns and rarely understand the underlying relationships that cause the patterns they see. For example, students noticed that the sampling distribution became narrower and more normal as the sample size increased, but did not necessarily understand why this was happening. Therefore, when asked to make predictions about plausible distributions of samples for a given sample size, students would resort to rules, often misremembered or applied inconsistently, rather than think through the process that might have generated these distributions. As a result, we often noticed students' confusion when asked to distinguish between the distribution of one sample of data and the distribution of several sample means.

By experimenting with different ways of having students interact with a specially designed simulation program, we have explored ways to more effectively engage students in thinking about the processes and to construct their own understanding of the basic implications of the Central Limit Theorem. Our research has identified several misconceptions students have about sampling and sampling distributions, and has documented the effects of learning activities that are designed to directly confront these misconceptions. For example, having students make predictions about distributions of sample means drawn from different populations under different conditions (such as sample size), and then asking them to use the technology to determine the accuracy of their predictions, appears to improve the impact of the technology on their reasoning. By forcing students to confront the limitations of their knowledge, we have found that students are more apt to correct their misconceptions and to construct more lasting connections with their existing knowledge framework. These learning gains appear to be significantly higher than from using the technology solely for demonstrations by the instructor or asking students to record and generalize specific observations made from the software.

Part of the problem in developing a complete understanding of sampling distributions appears to be due to students' less than complete understanding of related concepts, such as distribution and standard deviation. We have found our own research progressing backward, studying the instruction of topics earlier in the course and the subsequent effects on students' ability to develop an understanding of sampling distributions. For example, initially we explored student understanding of the effect of sample size on the shape and variability of distributions of sample means. We learned, however, that many students did not fully understand the meanings of *distribution* and *variability.* Thus, we were not able to help them integrate and build on these ideas in the context of sampling distributions until they better understood the earlier terminology and concepts. We are now studying ways to help students better understand the basic ideas of distribution, variability, and models to see if this will facilitate student learning about sampling distributions.

In the process of studying students' reasoning about sampling distribution, we identified several possible dimensions of student reasoning. These dimensions provide an initial vocabulary for describing student behavior and for comparing students. Accurate placement of students along the different dimensions will facilitate prescription of specific interventions or activities to help students more fully develop their reasoning. Interviews with students already suggest some interesting relationships. For example, students with the least amount of fluency (inaccurate definitions, misuse of terms) had the most difficulty reasoning about sampling distributions. Some students were very consistent in their reasoning, while others made contradictory and inconsistent statements. Among those who were inconsistent, some were aware of the inconsistencies in their remarks and others were not. It may be that students in a state of disequilibrium are more motivated to learn about the sampling distribution process and more receptive to instruction. There is also evidence that students can develop and follow rules to correctly predict the behavior of sample means and still be unable to describe the process that produces a sampling distribution. This suggests that they have not fully integrated information about sampling distributions, despite their ability to correctly predict behavior.

A current goal of our research program is to explore the following set of questions related to these dimensions of student reasoning:

- How are the dimensions related to each other?
- How does instruction affect each dimension? If instruction affects one dimension, are other dimensions also affected (in positive or negative ways)?
- How universal are the dimensions? Do they play a role in other types of statistical reasoning?
- Which dimensions are more important to the development of statistical reasoning?

We believe these dimensions extend beyond the topic of sampling distributions and can provide a window into how students' more general statistical reasoning develops.

To improve our ability to place students along each dimension, to explore the relationships of the dimensions, and to document how these dimensions are affected by particular learning activities, there is a clear need for new assessment tools. The tools we developed in the course of studying students' conceptions of sampling distributions are quite different from the types of assessment typically used to evaluate students learning of this topic. Figures 6 and 7 display two items we have used to assess students' ability to apply their understanding of sampling distributions to problems with a context. While these multiple-choice items are quick to score, they require students to *apply* their knowledge and allow deeper diagnostic analysis of student responses than many traditional items. Additional assessment tools are needed to better reveal the complexity of students' reasoning about sampling distributions.

Scores on a particular college entrance exam are NOT normally distributed. A distribution of scores for this college entrance exam is presented in the figure below. The distribution of test scores is very skewed toward lower values with a mean of 20 and a standard deviation of 3.5.

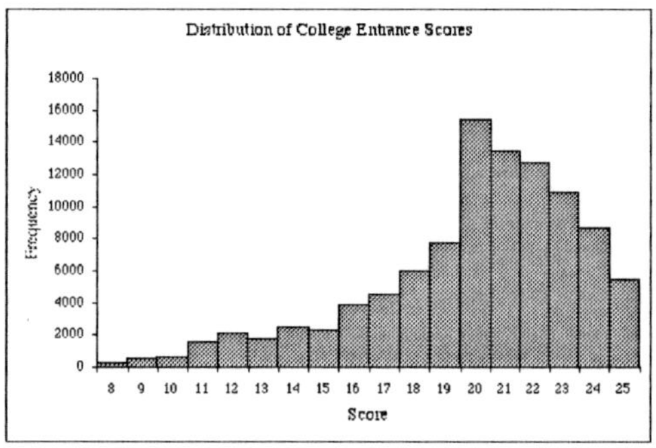

A research team plans to take a simple random sample of 50 students from different high schools across the United States. The sampling distribution of average test scores for samples of size 50 will have a shape that is: **(CIRCLE ONE)**

a. very skewed toward lower values.
b. skewed toward lower values, but not as much as the population.
c. shaped very much like a normal distribution.
d. It's impossible to predict the shape of the sampling distribution.

Explain your choice in detail: ______________________________

Figure 6. College entrance exam item. (Correct answer: C)

American males must register at a local post office when they turn 18. In addition to other information, the height of each male is obtained. The national average height for 18-year-old males is 69 inches (5 ft. 9 in.). Every day for one year, about 5 men registered at a small post office and about 50 men registered at a large post office. At the end of each day, a clerk at each post office computed and recorded the average height of the men who registered there that day.

Which of the following predictions would you make regarding the number of days on which the average height for the day was more than 71 inches (5 ft. 11 in.)?

a. The number of days on which average heights were over 71 inches would be greater for the small post office than for the large post office.
b. The number of days on which average heights were over 71 inches would be greater for the large post office than for the small post office.
c. There is no basis for predicting which post office would have the greater number of days.

Explain your choice and feel free to include sketches in your explanation.

Figure 7. Post office item (based on an item from Well et al., 1990). (Correct answer: A, since there will be more sampling variability with a smaller sample size.)

IMPLICATIONS

The few pages given in most textbooks, a definition of the Central Limit Theorem, and static demonstrations of sampling distributions are not sufficient to help students develop an integrated understanding of the processes involved, nor to correct the persistent misconceptions many students bring to or develop during a first statistics course. Our research suggests that it is vital for teachers to spend substantial time in their course on concepts related to sampling distributions. This includes not only the ideas of sampling, distributions of statistics, and applications of the Central Limit Theorem but also foundational concepts such as distribution and variability. Focus on these early foundational concepts needs to be integrated throughout the course so students will be able to apply them and understand their use in the context of sampling distributions.

While technological tools have the potential to give students a visual and more concrete understanding of sampling distributions, mere exposure to the software is unlikely to significantly change students' deep understanding. More careful implementation of the technology needs to be conducted to ensure students reach the highest learning potential. The following recommendations stem from our own research and research on conceptually enhanced simulations tools:

- *Use the technology to first explore samples and compare how sample behavior mimics population behavior.* Instructional time needs to be spent to allow students to become more familiar with the idea of sampling, to visually see how individual samples are not identical to each other or identical to the population, but that they do follow a general model. Furthermore, students will then have sufficient knowledge of the software so that it is a more effective tool instead of another distraction when they move to the more complicated statistical concept of sampling distribution.
- *Provide students with the experience of physically drawing samples.* Activities such as having students take samples of colored candies from a bowl, or using a random-number table to select observations from a population list, give them a meaningful context to which they can to relate the computer simulations. Otherwise, the computer provides a different level of abstraction and students fail to connect the processes.
- *Allow time for both structured and unstructured explorations with the technology.* Students need to be guided to certain observations, but they also need freedom to explore the concepts and to construct and test their own knowledge. Some students will require a higher number of discrediting experiences before they will correct their knowledge structure. Student exploration and establishment of disequilibrium can also make them more receptive to follow-up instruction.
- *Discuss the students' observations after completing the activity.* Students need opportunities to describe their observations and understandings. This can take place either in the classroom or through student writing assignments. These discussions allow the instructor to highlight the most crucial details that students need to pay attention to, so that students do not feel overwhelmed with too much information or disconnected pieces of knowledge, or focus on unimportant details such as the heights of the simulated distributions rather than their shape and spread.
- *Repeatedly assess student understanding of sampling distributions.* It is important to carefully assess what students are learning and understanding about sampling distributions at multiple times following the class activities. Students need many opportunities to test their knowledge, and feedback should be frequent and increasingly rigorous. Students' knowledge will be tenuous and needs to be reinforced. Additional assessment tools also need to be developed and used.
- *Build on students' understanding of sampling distributions later in the course.* It is important to build on these concepts throughout subsequent units on inference. Instructors should actively refer to students' tactile and technological experiences with sampling distributions as they introduce ideas of confidence and significance.

We hope that continued exploration of students' reasoning as they interact with simulation software will lead to better ways to help students develop a complete understanding of the process of sampling distribution. Once this has been achieved, students will be better able to develop an understanding of concepts related to statistical inference, such as statistical confidence and statistical significance, and should be more successful in their statistics courses.

REFERENCES

Behrens, J. T. (1997). Toward a theory and practice of using interactive graphics in statistics education. In J. B. Garfield & G. Burril (Eds.), *Research on the role of technology in teaching and learning statistics: Proceedings of the 1996 International Association for Statistics Education (IASE) roundtable* (pp. 111–122). Voorburg, The Netherlands: International Statistical Institute.

Biggs, J. B., & Collis, K. F. (1982). *Evaluating the quality of learning: The SOLO taxonomy.* New York: Academic Press.

Case, R. (1985). *Intellectual development: From birth to adulthood.* New York: Academic Press.

Clement, J. (2000). Analysis of clinical interviews. In A. Kelly & R. Lesh (Eds.), *Handbook of research design in mathematics and science education* (pp. 547–589). Mahwah, NJ: Erlbaum.

Cohen, S. (1997). *ConStatS: Software for Conceptualizing Statistics.* Tufts University: Software Curricular Studio. Retrieved April 23, 2003, from http://www.tufts.edu/tccs/services/ css/ConStatS.html

Davenport, E. C. (1992). Creating data to explain statistical concepts: Seeing is believing. In *Proceedings of the Section on Statistical Education of the American Statistical Association* (pp. 389–394).

delMas, R. (2001). *Sampling SIM* (version 5). Retrieved April 23, 2003, from http://www.gen.umn.edu/ faculty_staff/delMas/stat_tools

delMas, R., Garfield, J., & Chance, B. (1998). Assessing the effects of a computer microworld on statistical reasoning. In L. Pereira-Mendoza, L. S. Kea, T. W. Kee, & W. Wong (Eds.), *Proceedings of the Fifth International Conference on Teaching Statistics* (pp. 1083–1089), Nanyang Technological University. Singapore: International Statistical Institute.

delMas, R., Garfield, J., & Chance, B. (1999a). Assessing the effects of a computer microworld on statistical reasoning. *Journal of Statistics Education, 7*(3). Retrieved April 23, 2003, from http://www.amstat.org/publications/jse/secure/v7n3/delmas.cfm

delMas, R., Garfield, J., & Chance, B. (1999b). Exploring the role of computer simulations in developing understanding of sampling distributions. Paper presented at the Annual Meeting of the American Educational Research Association, Montreal, Canada.

Doane, D. P., Tracy, R. L., & Mathieson, K. (2001). *Visual Statistics, 2.0.* New York: McGraw-Hill. Retrieved April 23, 2003, from http://www.mhhe.com/business/opsci/doane/show_flash_intro.html

Earley, M. A. (2001). Improving statistics education through simulations: The case of the sampling distribution. Paper presented at the *Annual Meeting of the Mid-Western Educational Research Association*, Chicago, IL.

Garfield, J. (2002). The challenge of developing statistical reasoning. *Journal of Statistics Education, 10*(3). Retrieved April 23, 2003, from http://www.amstat.org/publications/jse/

Garfield, J., delMas, R., & Chance, B. (1999). Developing statistical reasoning about sampling distributions. Presented at the First International Research Forum on Statistical Reasoning, Thinking, and Literacy (SRTL), Kibbutz Be'eri, Israel.

Garfield, J., delMas, R., & Chance, B. (2002). *Tools for teaching and assessing statistical inference* [website]. Retrieved April 23, 2003, from: http://www.gen.umn.edu/ faculty_staff/delmas/stat_tools/index.htm

Glencross, M. J. (1988). A practical approach to the Central Limit Theorem. In R. Davidson & J. Swift (Eds.), *Proceedings of the second international conference on teaching statistics* (pp. 91–95). Victoria, B.C.: Organizing Committee for the Second International Conference on Teaching Statistics.

Hodgson, T. R. (1996). The effects of hands-on activities on students' understanding of selected statistical concepts. In E. Jakubowski, D. Watkins, & H. Biske (Eds.), *Proceedings of the Eighteenth Annual Meeting of the North American Chapter of the International Group for the Psychology of Mathematics Education* (pp. 241–246). Columbus, OH: ERIC Clearinghouse for Science, Mathematics, and Environmental Education.

Hodgson, T. R., & Burke, M. (2000). On simulation and the teaching of statistics. *Teaching Statistics, 22*(3).

Holland, J. H., Holyoak, K. J., Nisbett, R. E., & Thagard, P. R. (1987). *Induction: Processes of inference, learning, and discovery.* Cambridge, Mass.: MIT Press.

Jennings, D., Amabile, T., & Ross, L. (1982). Informal covariation assessment: Data-based versus theory-based judgments. In D. Kahneman, P. Slovic, & A. Tversky (Eds.), *Judgment under uncertainty: Heuristics and biases* (pp. 211–230). Cambridge, UK: Cambridge University Press.

Jones, G. A., Langrall, C. W., Mooney, E. S., Wares, A. S., Jones, M. R., Perry, B., et al. (2001). Using students' statistical thinking to inform instruction. *Journal of Mathematical Behavior, 20,* 109–144.

Jones, G. A., Thornton, C. A., Langrall, C. W., Mooney, E., Perry, B., & Putt, I. (2000). A framework for characterizing students' statistical thinking. *Mathematical Thinking and Learning, 2,* 269–308.

Lane, D. M. (2001). *HyperStat.* Retrieved April 24, 2003, from http://davidmlane.com/hyperstat/

Lang, J., Coyne, G., & Wackerly, D. (1993), *ExplorStat—Active Demonstration of Statistical Concepts,* University of Florida. Retrieved April 24, 2003, from http://www.stat.ufl.edu /users/dwack/

Lord, C., Ross, L., & Lepper, M. (1979). Biased assimilation and attitude polarization: The effects of prior theories on subsequently considered evidence. *Journal of Personality and Social Psychology, 37,* 2098–2109.

Mills, J. D. (2002). Using computer simulation methods to teach statistics: A review of the literature. *Journal of Statistics Education, 10*(1). Retrieved April 24, 2003, from: http://www.amstat.org/publications/jse/v10n1/mills.html

Mooney, E. S. (2002). A framework for characterizing middle school students' statistical thinking. *Mathematical Thinking and Learning, 4*(1), 23–63.

Moore, D. (2000). *Basic Practice of Statistics.* New York: Freeman.

Moore, D., & McCabe, G. (2002). *Introduction to the Practice of Statistics* (4th ed.). New York: Freeman.

Newton, H. J., & Harvill, J. L. (1997). *StatConcepts: A visual tour of statistical ideas* (1st ed.). Pacific Grove, CA: Brooks/Cole.

Nickerson, R. S. (1995). Can technology help teach for understanding? In D. N. Perkins, J. L. Schwartz, M. M. West, & M. S. Wiske (Eds.), *Software goes to school: Teaching for understanding with new technologies* (pp. 7–22). New York: Oxford University Press.

Perkins, D. N., Crismond, D., Simmons, R., & Unger, C. (1995). Inside understanding. In D. N. Perkins, J. L. Schwartz, M. M. West, & M. S. Wiske (Eds.), *Software goes to school: Teaching for understanding with new technologies* (pp. 70–87). New York: Oxford University Press.

Perkins, D. N., Schwartz, J. L., West, M. M., & Wiske, M. S. (Eds.). (1995). *Software goes to school: Teaching for understanding with new technologies.* New York: Oxford University Press.

Posner, G. J., Strike, K. A., Hewson, P. W., & Gertzog, W. A. (1982). Accommodation of a scientific conception: Toward a theory of conceptual change. *Science Education, 66*(2), 211–227.

Ross, L., & Anderson, C. (1982). Shortcomings in the attribution process: On the origins and maintenance of erroneous social assessments. In D. Kahneman, P. Slovic, & A. Tversky (Eds.), *Judgment under uncertainty: Heuristics and biases* (pp. 129–152). Cambridge, UK: Cambridge University Press.

Saldanha, L. A., & Thompson, P. W. (2001). Students' reasoning about sampling distributions and statistical inference. In R. Speiser & C. Maher (Eds.), *Proceedings of The Twenty-Third Annual Meeting of the North American Chapter of the International Group for the Psychology of Mathematics Education* (pp. 449-454), Snowbird, Utah. Columbus, Ohio: ERIC Clearinghouse.

Schwarz, C. J., & Sutherland, J. (1997). An on-line workshop using a simple capture-recapture experiment to illustrate the concepts of a sampling distribution. *Journal of Statistics Education, 5*(1). Retrieved April 24, 2003, from http://www.amstat.org/publications/jse/v5n1/schwarz.html

Schwartz, D. L., Goldman, S. R., Vye, N. J., Barron, B. J., & the Cognition Technology Group at Vanderbilt. (1997). Aligning everyday and mathematical reasoning: The case of sampling assumptions. In S. Lajoie (Ed.), *Reflections on statistics: Agendas for learning, teaching and assessment in K-12* (pp. 233–273). Hillsdale, NJ: Erlbaum.

Sedlmeier, P. (1999). *Improving statistical reasoning: Theoretical models and practical implications.* Mahwah, NJ: Erlbaum.

Siegel, A., & Morgan, C. (1996). *Statistics and data analysis: An introduction* (2nd ed.). New York: Wiley.

Simon, J. L. (1994). What some puzzling problems teach about the theory of simulation and the use of resampling. *American Statistician, 48*(4), 290–293.

Snir, J., Smith, C., & Grosslight, L. (1995). Conceptually enhanced simulations: A computer tool for science teaching. In D. N. Perkins, J. L. Schwartz, M. M. West, & M. S. Wiske (Eds.), *Software goes to school: Teaching for understanding with new technologies* (pp. 106–129). New York: Oxford University Press.

Thomason, N., & Cummings, G. (1999). *StatPlay.* School of Psychological Science, La Trobe University, Bandora, Australia. Retrieved April 24, 2003, from: http://www.latrobe.edu.au/ psy/cumming/statplay.html

Velleman, P. (2003). *ActivStats*, Ithaca, NY: Data Description, Inc. Retrieved April 24, 2003, from http://www.aw.com/catalog/academic/product/1,4096,0201782456,00.html

Well, A. D., Pollatsek, A., & Boyce, S. J. (1990). Understanding the effects of sample size on the variability of the mean. *Organizational Behavior and Human Decision Processes, 47,* 289–312.

APPENDIX

Questions 4 through 9 from Sampling Distributions Posttest—Spring 2001

4) The distribution for a population of test scores is displayed below on the left. Each of the other five graphs labeled A to E represent possible distributions of sample means for random samples drawn from the population.

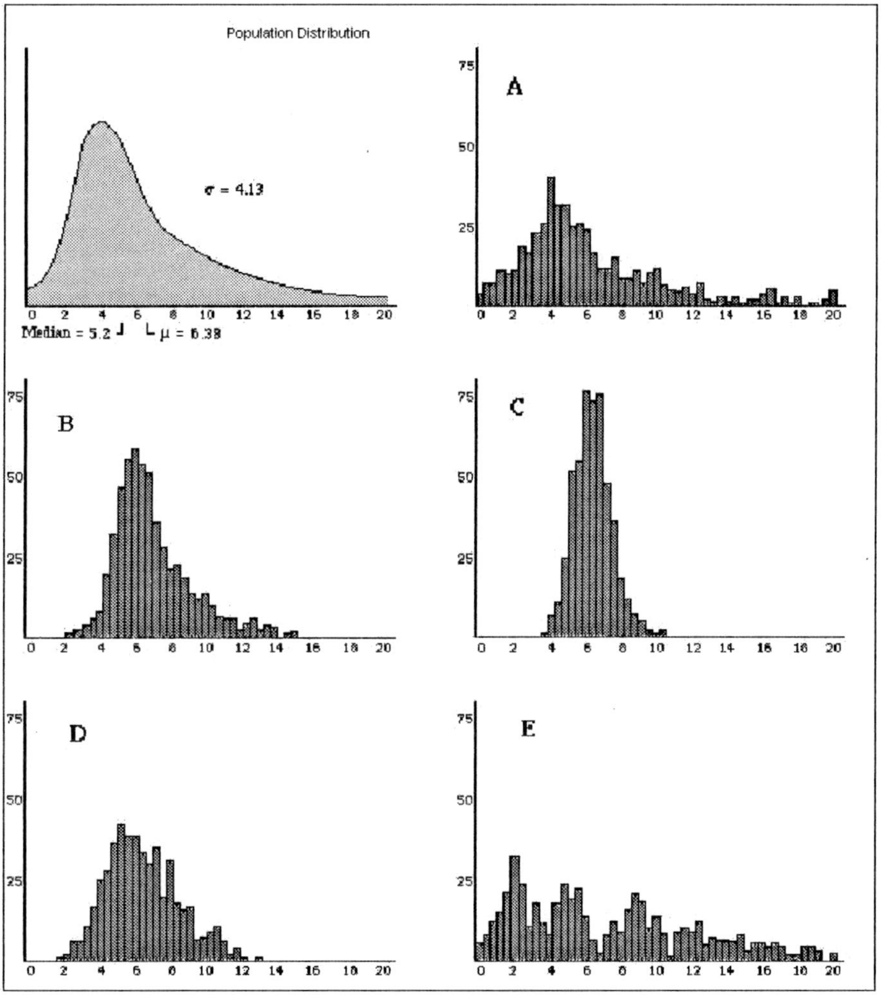

Figure 1. Population Distribution.

4a) Which graph represents a distribution of sample means for 500 samples of size 4? (Circle one.) A B C D E

4b) How confident are you that you chose the correct graph? (Circle one of the values below.)
20% 25% 30% 35% 40% 45% 50% 55% 60% 65% 70% 75% 80% 85% 90% 95% 100%

- Answer each of the following questions regarding the sampling distribution you chose for question 4a.

4c) What do you expect for the shape of the sampling distribution? (Check only one.)
Shaped more like a NORMAL DISTRIBUTION.
Shaped more like the POPULATION.

- Circle the word between the two vertical lines that comes closest to completing the following sentence.

4d) I expect the sampling distribution to have	less the same more	VARIABILITY than / as the POPULATION

4e) Which graph represents a distribution of sample means for 500 samples of size 16? (Circle one.) A B C D E

4f) How confident are you that you chose the correct graph? (Circle one of the values below.)
20% 25% 30% 35% 40% 45% 50% 55% 60% 65% 70% 75% 80% 85% 90% 95% 100%

- Answer each of the following questions regarding the sampling distribution you chose for question 4e.

4g) What do you expect for the shape of the sampling distribution? (Check only one.)
Shaped more like a NORMAL DISTRIBUTION.
Shaped more like the POPULATION.

- Circle the word between the two vertical lines that comes closest to completing the following sentences.

4h) I expect the sampling distribution to have	less the same more	VARIABILITY than / as the POPULATION

4i) I expect the sampling distribution I chose for question 4e to have	less the same more	VARIABILITY than / as the sampling distribution I chose for question 4a.

5. The distribution for a second population of test scores is displayed below on the left. Each of the other five graphs labeled A to E represent possible distributions of sample means for random samples drawn from the population.

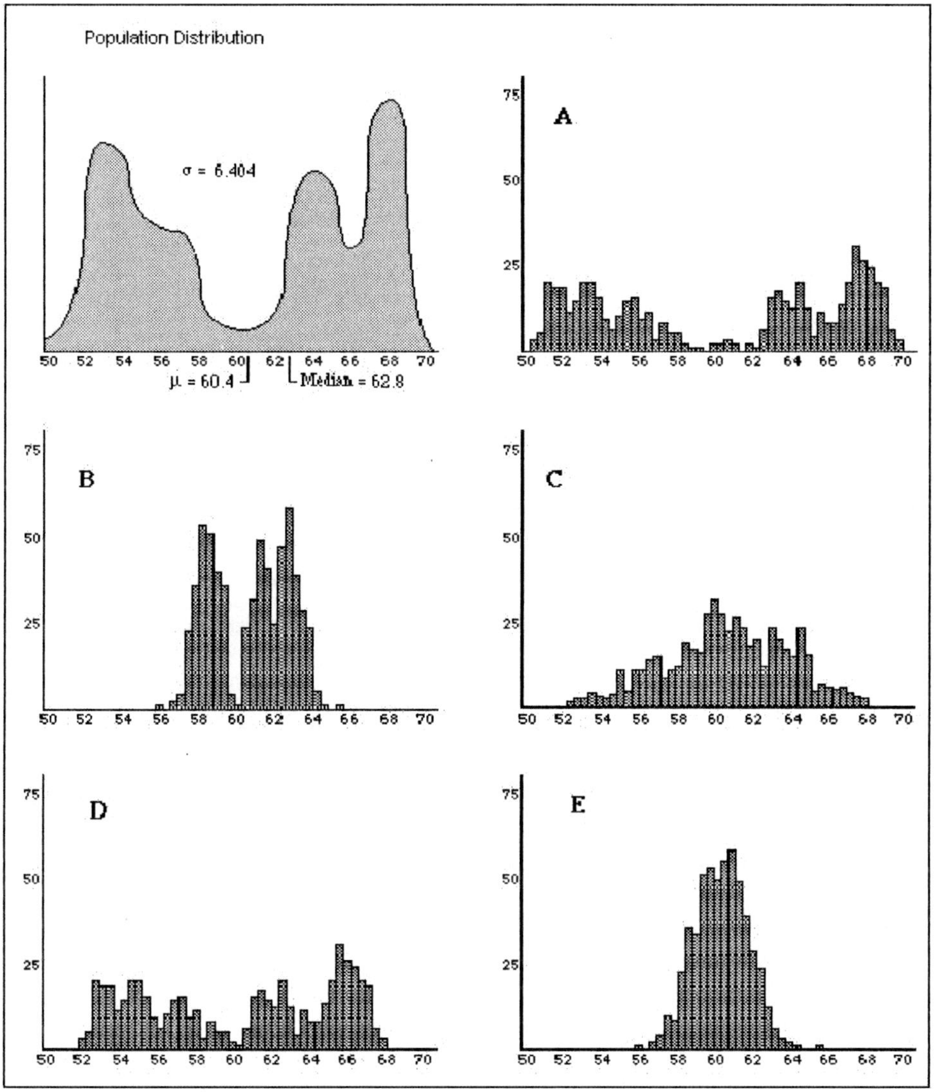

Figure 2. Population Distribution.

5a) Which graph represents a distribution of sample means for 500 samples of size 4? (Circle one.) A B C D E

5b) How confident are you that you chose the correct graph? (Circle one of the values below.)
20% 25% 30% 35% 40% 45% 50% 55% 60% 65% 70% 75% 80% 85% 90% 95% 100%

- Answer each of the following questions regarding the sampling distribution you chose for question 5a.

5c) What do you expect for the shape of the sampling distribution? (Check only one.)
Shaped more like a NORMAL DISTRIBUTION.
Shaped more like the POPULATION.

- Circle the word between the two vertical lines that comes closest to completing the following sentence.

5d) I expect the sampling distribution to have	less the same more	VARIABILITY than / as the POPULATION

5e) Which graph represents a distribution of sample means for 500 samples of size 16? (Circle one.) A B C D E

5f) How confident are you that you chose the correct graph? (Circle one of the values below.)
20% 25% 30% 35% 40% 45% 50% 55% 60% 65% 70% 75% 80% 85% 90% 95% 100%

- Answer each of the following questions regarding the sampling distribution you chose for question 5d.

5g) What do you expect for the shape of the sampling distribution? (Check only one.)
Shaped more like a NORMAL DISTRIBUTION.
Shaped more like the POPULATION.

- Circle the word between the two vertical lines that comes closest to completing the following sentences.

5h) I expect the sampling distribution to have	less the same more	VARIABILITY than / as the POPULATION

5i) I expect the sampling distribution I chose for question 5e to have	less the same more	VARIABILITY than / as the sampling distribution I chose for question 5a.

6. The weights of packages of a certain type of cookie follow a normal distribution with mean of 16.2 oz. and standard deviation of 0.5 oz.
   Simple random samples of 16 packages each will be taken from this population. The sampling distribution of sample average weight (⊠ these will still be $\overline{x}$'s) will have:
   a. a standard deviation greater than 0.5
   b. a standard deviation equal to 0.5
   c. a standard deviation less than 0.5
   d. It's impossible to predict the value of the standard deviation.

7. The length of a certain species of frog follows a normal distribution. The mean length in the population of frogs is 7.4 centimeters with a population standard deviation of .66 centimeters.
   Simple random samples of 9 frogs each will be taken from this population. The sampling distribution of sample average lengths (the average, ⊠) will have a mean that is:
   a. less than 7.4
   b. equal to 7.4
   c. more than 7.4
   d. It's impossible to predict the value of the mean.

8. Scores on a particular college entrance exam are NOT normally distributed. The distribution of test scores is very skewed toward lower values with a mean of 20 and a standard deviation of 3.5.
   A research team plans to take simple random samples of 50 students from different high schools across the United States. The sampling distribution of average test scores (the average, ⊠) will have a shape that is:
   a. very skewed toward lower values.
   b. skewed toward lower values, but not as much as the population.
   c. shaped very much like a normal distribution.
   d. It's impossible to predict the shape of the sampling distribution.

9. Consider any possible population of values and all of the samples of a specific size (n) that can be taken from that population. Below are four statements about the sampling distribution of sample means. For each statement, indicate whether it is TRUE or FALSE.
   a. If the population mean equals μ, the average of the sample means in a sampling distribution will also equal μ. TRUE FALSE
   b. As we increase the sample size of each sample, the distribution of sample means becomes more like the population. TRUE FALSE
   c. As we increase the sample size of each sample, the distribution of sample means becomes more like a normal distribution. TRUE FALSE
   d. If the population standard deviation equals σ, the standard deviation of the sample means in a sampling distribution is equal to $\sigma/\sqrt{n}$ TRUE FALSE

Correct Answers:

4. (a) D, (c) normal, (d) less, (e) C, (g) normal, (h) less, (i) less
5. (a) C, (c) normal, (d) less, (e) E, (g) normal, (h) less, (i) less
6. C
7. B
8. C
9. true, false, true, true

PART III

INSTRUCTIONAL, CURRICULAR AND RESEARCH ISSUES

Chapter 14

PRIMARY TEACHERS' STATISTICAL REASONING ABOUT DATA

William T. Mickelson and Ruth M. Heaton
University of Nebraska—Lincoln, USA

OVERVIEW

This study offers a descriptive qualitative analysis of one third-grade teacher's statistical reasoning about data and distribution in the applied context of classroom-based statistical investigation. During this study, the teacher used the process of statistical investigation as a means for teaching about topics across the elementary curriculum, including dinosaurs, animal habitats, and an author study. In this context, the teacher's statistical reasoning plays a central role in the planning and orchestration of the class investigation. The potential for surprise questions, unanticipated responses, and unintended outcomes is high, requiring the teacher to "think on her feet" statistically and react immediately to accomplish content objectives as well as to convey correct statistical principles and reasoning. This study explores the complexity of teaching and learning statistics, and offers insight into the role and interplay of statistical knowledge and context.

THE PROBLEM

With the call for more statistics in the elementary curriculum (NCTM, 2000), there is a need to consider ways to make statistics not only accessible and understandable to K–6 teachers but also useful to their teaching practice. Recently, the idea of teachers designing and implementing statistical investigations as a means to teach topics across the elementary curricula, as well as the statistical skills and reasoning involved in collecting, organizing, summarizing, and interpreting data has been examined (Heaton & Mickelson, 2002; Lehrer & Schauble, 2002). Statistical investigation within the established elementary curriculum has the potential to lend meaningful and purposeful contexts for data collection, summarization, and interpretation of data that are similar to the notion of authentic pedagogy and

D. Ben-Zvi and J. Garfield (eds.),
The Challenge of Developing Statistical Literacy, Reasoning and Thinking, 327–352.

assessment advocated by Newmann & Wehlage (1997). This context of purposeful investigation into nonstatistical curriculum topics is often absent in many predeveloped or "canned" statistical activities, or when statistical content is viewed only as isolated topics and the end of learning within a mathematics curriculum (Lehrer & Schauble, 2002). In mathematics education, the goal of teaching mathematical topics in meaningful ways (NCTM, 2000) clearly places great intellectual demands on the teacher (Fennema & Nelson, 1997; Heaton, 2000; Schifter, 1996), and mathematics educators are trying to better understand and respond to these demands of practice. Analogously, a better understanding is needed of teachers' conceptions of statistics (Shaughnessy, 1992), the pedagogical content knowledge required for teaching (Shulman, 1986), and how statistical knowledge is used by teachers in teaching (Ball, Lubienski, & Mewborn, 2001) if we want to understand and support their efforts.

BACKGROUND

The process of statistical investigation (Friel & Bright, 1997; Graham, 1987) is a central topic in national statistics education guidelines in the United States. The NCTM Standards (2000) state that students should:

> formulate questions that can be answered using data … learn how to collect data, organize their own or others' data, and display the data in graphs and charts that will be useful in answering their own questions. This Standard also includes learning some methods for analyzing data and some ways of making inferences and conclusions from data. (p. 48)

Schaeffer, Watkins, and Landwehr (1998), and Friel and Bright (1998), all argue against teaching statistics in isolation from other areas of the curriculum. They support extending and integrating statistics with other subjects and argue that such an approach is an ideal way to improve students' knowledge of both the content area and the process of statistical investigation.

Despite the growing number of studies of students' reasoning about statistical information, only recently has attention been paid to teachers' understanding of statistical ideas. The methodology used to study teachers' knowledge has focused primarily on giving teachers isolated statistics problems (Watson, 2001; Watson, 2000) or studying teachers' understanding of statistics based on their work as learners of statistical content in a professional development setting (Confrey & Makar, 2001) and assessing their competency. Implied in this approach is the aim to help teachers acquire a competency of statistics comparable to proficient learners. Our study approaches an investigation of teacher knowledge differently. In this study, we examine a teacher's knowledge of data and distribution as it appears in the records of a teacher's practice, complementing previous work on teacher knowledge in statistics education by situating how and for what purposes a teacher reasons about statistical ideas while teaching. The empirical findings of this study support and explicate the report of what elementary teachers need to learn related to the topic of statistics found in *The Mathematical Education of Teachers* (CBMS, 2001) and contribute to meeting a portion of the agenda for research in mathematics education found in the RAND Report (Ball, 2002). Our work addresses the need to "distinguish teachers' knowledge from the knowledge held and used by others who draw from that same discipline in their work" (Ball, 2002, p. 16), specifically around the topic of reasoning with data and distribution. What this means is that the reasoning about data and distribution done by a teacher applying statistical reasoning in teaching takes a form and has complexities different from students merely learning to reason about data and distribution.

SUBJECT AND METHOD

This study employs a descriptive qualitative analysis to examine one third-grade teacher's statistical reasoning about data and distribution (Creswell, 1998) in the context of three classroom investigations with students. A profile of the teacher is followed by descriptions of the context and investigations, and methods of data collection and analysis.

The Teacher and Context

The third-grade teacher featured in this study, Donna (pseudonym), has been teaching elementary school for 16 years. She was highly regarded by her principal and peers, being described as "extremely thoughtful and planful in all that she does, structured and organized yet flexible" (Interview, 2/2/01). Because of her interest and attitude toward the notion of statistical investigation as a means for teaching content across the elementary curriculum, we purposefully selected Donna for this study. Previously, Donna was a participant in the American Statistical Association's (ASA) Quantitative Literacy Project (Scheaffer, 1988), where she experienced and learned statistics through hands-on activities designed to teach statistical concepts. Furthermore, she has supported children's participation in ASA project competitions with help from members of the local ASA chapter.

We initially interacted with Donna as facilitators of a professional development workshop on merging statistical investigation with topics of the K–6 curriculum. During the workshop, Donna routinely demonstrated highly competent statistical reasoning skills about data and distribution, successfully completing data collection, graphing, and interpretation activities such as, "Is Your Shirt Size Related to Your Shoe Size," and "Gummy Bears in Space: Factorial Designs and Interactions" (Schaeffer, Gnanadesikan, Watkins, & Witmer, 1996). Collaboration with Donna carried over into the subsequent academic year when she agreed to participate in this study. Between August and December 2000, Donna created and implemented seven units that merged the process of statistical investigation with topics in the third-grade curriculum. These investigations, detailed in Table 1, vary in degree of complexity with regard to framing the problem, the analysis, and the teacher's familiarity with conducting the investigation or one similar.

During most of these investigations, Donna continued to exemplify correct statistical reasoning. She routinely guided her class through the identification of salient variables, collection and organization of relevant data, summarization of data into correct graphical representations, interpretation of findings in context, and the correct instruction of statistical concepts like distribution and its importance in making predictions.

Donna's Steven Kellogg Author Study, Animal Habitats, and Dinosaur Investigations were selected for presentation in this study, for several reasons. First, each had a larger purpose beyond teaching the process of statistical investigation, with clear learning goals about content. Second, constructing and implementing the investigations to suit the teacher's content area learning goals challenged the teacher's statistical reasoning ability. Finally, these three activities highlighted different strengths and limitations in the teacher's statistical knowledge use in teaching. For each of these examples, description of the activity is given and the statistical issues articulated and analyzed. Therefore, what this study represents are the best efforts of a highly experienced and competent teacher, with previous training and interest in statistics, to teach in ways that merge statistical investigation with the study of topics in the elementary curriculum.

Table1. Statistical investigations of participating teacher

Investigation Topic	**Dates**	**Duration**	**Purpose**
Getting to Know You	Aug. 21 – Sept. 1	2 weeks	For students to get acquainted with each other at the beginning of the semester. Variables: favorite foods, books, and pets. Students in class make up sample. Bar graphs used to summarize data. Statistical lesson on distribution and prediction.
Animal Habitats	Oct. 2–20	3 weeks	To study natural habitats of animals and compare to human-made habitats of zoo animals. Described in text.
Steven Kellogg Author Study	Oct. 9–13	1 week	To study the writing characteristics of a children's author. Described in text.
Election	Nov. 1–7	1 week	To do a poll and compare school results with district, city, and national outcomes. Variables: for/against candidates and local issues on ballot. Sample selected from school at large. Bar graphs and percentages used to summarize data. Statistical lesson: graphing and prediction.
Math	Nov. 1–7	1 week	Lessons to read and interpret different types of graphs.
Cold Remedies	Nov. 13–14	1 week	Health and social studies related topic to survey families about cold remedies. Variables: family demographics, preferred cold remedies (homemade versus store-bought drugs), prevention. Sample included students' extended families and neighbors. Multiple comparative bar graphs to summarize and interpret data. Statistical lesson: variability, comparing graphs by differences in distribution.
Dinosaurs	Nov. 6–30	3 weeks	To apply research skills to learn about dinosaurs, create graphs and draw conclusions. Described in text.

Merging Statistical Investigation with Curriculum Topics

In each statistical investigation, to the extent possible, Donna situated student learning in a much broader context so that the investigation was more than data collection and graphing. In the broader context, she connected the investigation to numerous other curriculum standards and content learning outcomes that she was required to cover and teach over the course of a year. During an interview, Donna commented on her goals and the overall nature of the Dinosaur Investigation. She stated,

> "The thing for me was to correlate the curriculum." This means she took into account the entire scope of district learning standards and objectives, looking for ways to simultaneously address as many as possible. In summary, Donna stated, "So you can tell that with this we cover just about every part of the curriculum in one way or another, and all these main objectives that I need to meet. And I did them all basically thinking about this project and then branching out from there" (Interview 2/15/01).

Data Collection

Friel and Bright (1998) noted in the study of their own effort at teacher education, "much may be ascertained about teachers' understanding of [statistics] content by watching them teach" (p. 106). We embraced this perspective and remained observers during the data collection phase, not attempting to intervene or influence what happened. Any statistics-related question initiated by the teacher during the data collection phase was promptly answered. Concerns we might have had, however, were retained until the conclusion of this phase when we debriefed the teacher. This approach is consistent with Jaworski's (1994) idea of investigative teaching.

All lessons involving the process of statistical investigation were videotaped. The video camera was placed in the subject's classroom for the entire semester. One of the researchers tried to be present during each lesson; however, there were times when this was not possible. In these few instances, the teacher ran a video camera stationed on a tripod throughout the class period. Data sources collected in the classroom context include videotapes and transcripts of classroom interactions, classroom artifacts including student work and teacher produced materials, researcher field notes, and the teacher's written plans and reflections. Additional data include audiotaped interviews and transcripts, and products from the summer workshop. Data from the classroom context documents the teacher's current practice, while the summer workshop data gives evidence of the statistical skills and abilities the teacher had prior to conducting statistical investigations with her third-grade students.

Data Analysis

Data analysis focuses primarily on records of practice from the classroom statistical investigations. Initially, classroom artifacts and end products from each of the statistical investigations were examined to look for evidence of the teacher's statistical reasoning about data and distribution. The teacher's choices for organizing and representing data are embedded in the artifacts. From these we infer the teacher's statistical knowledge and reasoning. We supplement our analysis with the teacher's own language as she teaches, found in transcripts of classroom interactions, and as she reflects on teaching, found in transcripts of interviews. Additionally, selected transcripts were analyzed for evidence of the teacher's use of statistical knowledge. In several instances the teacher's statistical knowledge, as evidenced during the summer workshop and other classroom investigations like Getting to Know You (Table 1), were compared to findings of teacher knowledge within the classroom data. Instances of disparity in findings of teacher knowledge from these two contexts prompted detailed analysis of artifacts as well as associated videotapes and transcripts, and gives evidence about the importance of context in teacher knowledge acquisition, its application, and research. Observation notes and the teacher's written plans and reflections offered opportunities to triangulate findings.

RESULTS

For the three investigations studied, we describe the investigation. We also articulate, analyze, and infer Donna's statistical reasoning about data and distribution.

The Steven Kellogg Author Study Investigation

Throughout the elementary years, children are commonly introduced to a variety of authors and illustrators through Author Studies. According to Donna, the purpose of an Author Study is

> To have them begin to see writing styles … I want them to be able to say, this is his voice, this is what he does … I want them to know about his style … We've been learning about character traits and that's real hard for them. (Interview, 10/5/00)

Steven Kellogg is one author and illustrator whose works are part of the third-grade curriculum set by the district and the object of this investigation.

After reading five books written or illustrated by Steven Kellogg, Donna posed two questions to her students. She asked, "What are the things that are most true about Steven Kellogg?" and, "What are we observing from reading his books?"

These questions launched the statistical investigation and generated the following brainstormed list of ideas generated by students:

- Likes to draw
- Fantasy/imagination
- Likes animals
- No pets as a child
- Drawings—accuracy and detail
- Puts Pinkerton in many books
- Thought bubbles
- Signs
- Uses his pets in his stories
- Dedicates his books to his family

This list became the basis for further data collection. Based on this list, Donna reported,

> I made graphing forms for each student, typed a list of all Kellogg's books in the room, and assigned each book a letter. Students fill in spaces on their form for author characteristics by using the letter for that book. That way when we compile our results as a class we won't record a book twice ... The part I see as difficult and yet exciting for me is to figure out how to help them compile their information. If we had unlimited time we could hammer more of this out together. (Interview 10/10/00)

The first task was to make the transition from individual student data to a whole class set of data. Having individual students first collect their own data accomplished several objectives. Students could make choices about specific books read, many Kellogg books could be read by the class as a whole, and each student had responsibility for making judgments about author characteristics. About compiling the class data, Donna said:

> When they were all done reading, I just did this as a class exercise. They'd raise their hand. I'd say, okay, does anybody have a book that used thought bubbles? And then as quickly as we could, we put this together. (Interview 2/00)

Figure 1 is a reproduction of the final product. It simultaneously represents both the organizational structure and summary of this data. This graph became the means by which Donna was able to converse with students about the differing qualities or traits observed in Kellogg's work and how often those qualities or traits were noticed in his writing or illustrations. In Figure 1, the horizontal axis is a list of traits similar to those brainstormed by students. The vertical axis is the number of books that exhibit the trait. The identity of each book is preserved through the letter codes Donna assigned. The codes are used in the development of the body of the bar graph. Notice that there is no intended ordering of the letters; rather, it is a function of which students raised their hands and the order the teacher called on them.

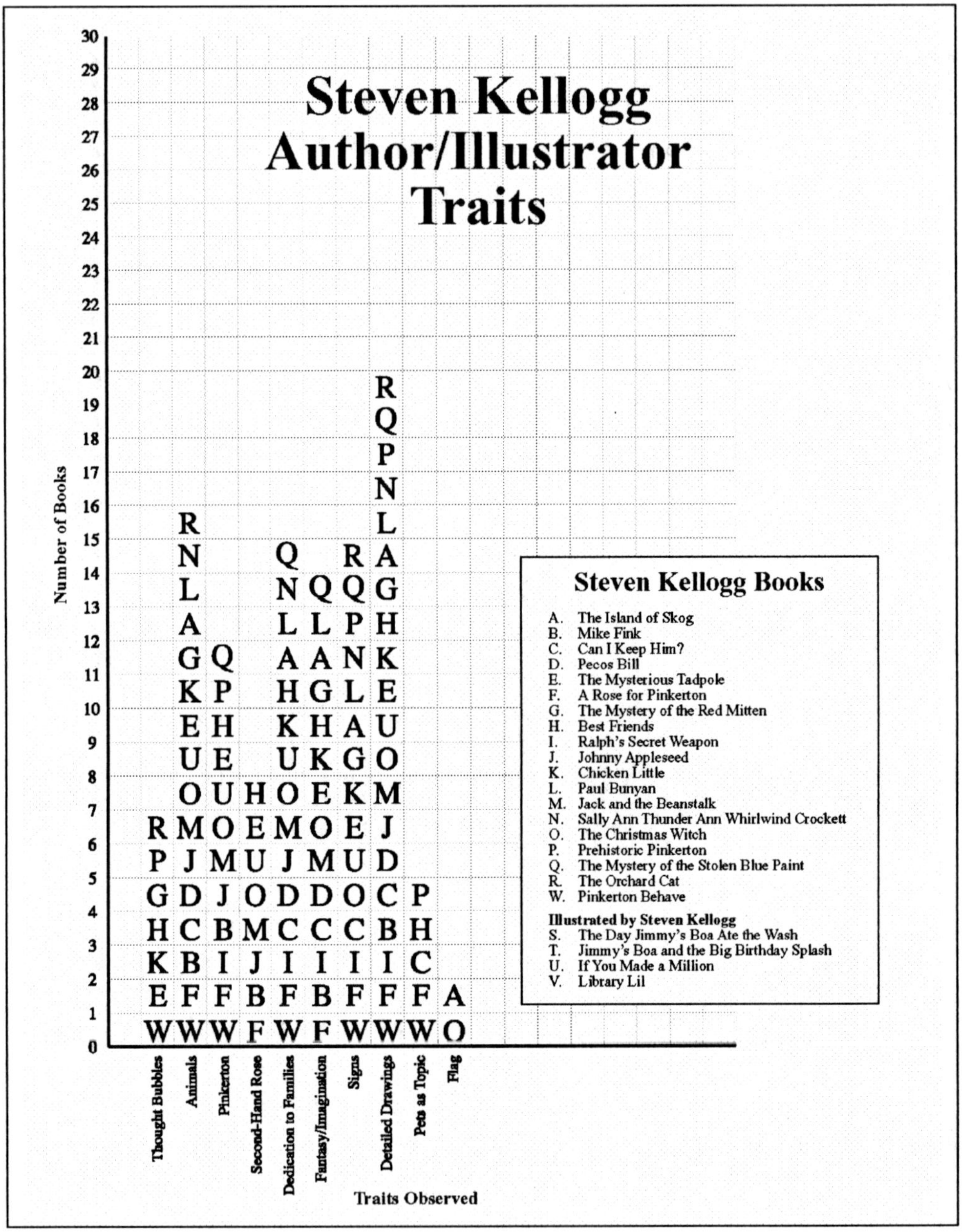

Figure 1. Graphical artifact of author study investigation.

Statistical Issues and Reasoning during the Author Study Investigation

Inferring Donna's understanding of data and distribution from what is present and missing is one mechanism to analyze the graph of Figure 1. The data summarized in this graph do allow Donna and the class to answer the basic investigation question, "What are we observing from reading his (Kellogg's) books?" Donna used the graph in Figure 1 during class in three ways. First, the class discussed the list of observed traits to understand the varying techniques and content Kellogg includes in his books. Second, Donna's statistical questioning helped students determine which of the observed traits was more or less prevalent in the books that were read. Finally, the class used the graph to refer to and describe individual books. They accomplished this by focusing on a letter and determining which columns (or traits) contain that letter and which do not. As such, Donna sees the investigation as being successfully completed. The type of analysis and discussion the class had with the summarized data in Figure 1 is consistent with the typical types of questions posed and answered when analyzing graphs in lessons in their mathematics textbook. From the evidence presented thus far, Donna's statistical reasoning includes correct identification of the case (or experimental unit) of this study, care in preventing double counting since the case is not a person, facilitation of a comprehensive sample (census) covering Kellogg's work, and a graphical summarization of data indicating a conception of distribution as tallies or counts of a trait.

A closer look at the graph in Figure 1 reveals that the salient features of a bar graph, namely the mutually exclusive or exhaustive condition that defines categories, are missing. The intent of a bar graph is to display the distribution of data for a categorical variable (Utts, 1999; Bright & Friel, 1998). Donna uses the 10 observed traits stemming from the brainstorming session as the categories in the graph. In doing this, each of Kellogg's books is counted in multiple categories because the traits are not mutually exclusive or exhaustive. As such, Figure 1 is not a true bar graph; what this implicitly and instructionally does is give students a limited and incorrect perspective on bar graphs and what information they convey. By the nature of the graph, Donna was precluded from being able to teach and discuss the statistical concept of distribution of a categorical variable, resulting in questions focused only on identifying traits with the most and least counts. In addition, the graph does not help answer the investigation question, "What are the things that are most true about Steven Kellogg?" since the graph does not give information on the alternatives to the listed traits. For example, Thought Bubbles may not be the most prevalent (i.e., most true) mechanism to convey character's thoughts since the alternatives are not given. Also, Figure 1 does not address the overarching purpose of an Author Study in Donna's conception, namely, conveying the "writing style" of the author.

One could infer from this data that Donna's reasoning about data and distribution is at a low or naïve level since Figure 1 conveys a conception of data that is merely counts, the bar graph is not technically correct, and it does not address the overarching purpose of the investigation. This interpretation, however, is contrary to Donna's performance in the contexts of the summer workshop and the other classroom investigations like Getting to Know You (Table 1). In these contexts, Donna created technically correct bar graphs and taught the concept of distribution. The discrepancy between Donna's classroom performance during the Author Study and prior performance in other contexts raises the questions, what are the contextual differences, and how did the context influence her reasoning? The main difference is that the Author Study investigation does not have a well-defined or naturally occurring implicit set of variables connected to the study. For example, if Donna wanted to focus on writing styles, she would have to define *style* and operationalize how to measure it. Donna's pedagogical decision to give ownership to students through their brainstorming of observed traits resulted in the data for this investigation being tallies and counts, which was consistent with and addressed the investigation, "What are we observing from reading Kellogg's books?" The subsequent process of data collection, summarization, and discussion being so similar to her prior experience teaching from a textbook gave Donna no reason to question the investigation in any deeper sense. Lack of experience with the process of statistical investigation and the overall success of the investigation in terms of meeting her content expectations may have contributed to her oversight of not checking to see if the result addressed the original purpose of the investigation (NCTM, 2000). At the same time, there was not a well-defined, compelling, content-driven need to derive additional meaning from the graph of Figure 1 that would prompt Donna to make this connection between purpose and product of the investigation.

What was lost was an excellent opportunity to illustrate how to define and create multiple variables, construct multiple graphs on different aspects of the books, and to synthesize information across multiple graphs to characterize Kellogg as a writer and illustrator. Despite these difficulties, the Author Study was a dramatic improvement over most author studies that focus on personal information about the author without ever attending directly to the author's writing style, and many of the nonstatistical learning goals Donna had for students were achieved through this approach. Donna's statistical reasoning about data and distribution is at a relatively low level, particularly in terms of the lack of discussion on distribution; however, her reasoning is contextually influenced and not consistent with her statistical reasoning in other similar contexts.

The Animal Habitat Investigation

Animal Habitats is a unit Donna routinely teaches. Typically, students do library research and presentations on the habitat of an animal of their choice. During the year of this case, Donna planned to merge her usual unit on Animal Habitats with a statistical investigation to evaluate human-made animal habitats found at the local children's zoo. Donna obtained a list of animals residing at the zoo, and students each selected one to study. First, students researched their animal's natural habitat in the library. This phase was guided by the four components of an animal's habitat initially studied in science class, specifically focusing on climate, food and water, shelter, and space. Other topics researched by students included the animal's main form of protection, ways it reproduces, places it lives, and other facts of interest.

Library inquiry was followed by a statistical investigation evaluating the local zoo. The statistical investigation was based on the following questions: "Is your animal being treated at the zoo like you think it should be based on your research on how they live in the wild?" and, "Is the zoo doing a good job providing a proper habitat for your animal?" To quantitatively address these questions, Donna and the class used their library research to develop a rating scale. Relying on previous experience with rating scales from their self-evaluations of writing assignments in language arts, Donna and the students made a table with numbers 1–5 running across the top and the four components of habitat (space, climate, shelter, food and water) listed down the far left side. For each component, students described both the best and worst possible conditions. The best conditions were likened to what is found in an animal's natural habitat, the worst were far from resembling the natural habitat. As students generated ideas, Donna wrote descriptors under the "1" for worst and "5" for best. Donna realized that "the best" descriptors represented students' understanding of the best conditions in the wild. This prompted reframing the discussion to be about "the best" that could be obtained in zoo exhibits (see Table 2).

Table 2. Beginnings of a rating scale

HABITAT NEEDS	**1**	**2**	**3**	**4**	**5**
Food and Water	Pellets Polluted Water Not enough				Natural food Clean, pure water Plenty
Space	Too many animals Small cage Cement floor				Plenty of room Soil, plants, floor
Shelter	Wrong type No shelter				Correct type Appropriate shelter
Climate	Too hot, natural (i.e., open to outdoors) Too cold Too changeable				Same as (i.e., same as their original natural habitat)

In planning the zoo field trip, Donna decided students should not concern themselves with applying the rating scale during the zoo visit. Rather, she wanted them to observe the habitat, take field notes, and carefully consider what they had observed. Donna gave each student a small notebook with directions for taking detailed descriptive field notes. The students' field notes became the primary data source for the statistical investigation and were converted into the numerical rating scores.

When the students transferred their observation notes to numerical ratings, Donna reviewed how they could use the numbers of the rating scale to represent the quality of their zoo animal's habitats. Helping students realize that numbers in a rating scale have an assigned meaning was Donna's way of teaching number sense (Lajoie, 1998). One student likened the scale to a report card and framed the task as assigning the zoo a grade for each observed component of habitat. To further help students transfer observations to ratings of quality, Donna used faces with different types of expressions to illustrate different levels of quality. A smiling face was associated with 5, a frowning face associated with 1, and an expressionless face associated with 3. Each student read over his or her observation notes and rated each component of habitat in their particular animal's zoo exhibit.

In preparation for the statistical analysis, Donna and the class decided on additional variables to help evaluate how well the zoo was caring for the animals. These variables were size of the animal (small, medium, or large) and animal's habitat in wild natural settings (grassland, rain forest, desert, forest, and inland waterway). Donna helped students identify variables and delineate attributes to draw conclusions about whether

> "they (the zoo) do better in one habitat than another? Better with large animals or small animals?" (Interview, 10/5/00).

To facilitate data collection and organization, Donna distributed four 3×5 colored cards of the same color to each student. The purpose of the colored cards was to represent the different types of habitat in the wild. Students wrote their animal's name on each of four cards and placed a symbol in the lower right-hand corner to represent the size of their animal. Donna constructed a huge table on paper that crossed the habitat needs of the animals by levels of the rating scale. She hung it at the front of the room, and each student placed their four colored cards in the table to indicate a rating within each habitat category. Figure 2 represents the final table developed by the class. From Figure 2, the class attempted to observe patterns in the data and draw inferences about the zoo's performance in caring for the animals residing there. Toward the end of the unit, Donna invited the zoo director to visit and discuss student findings and questions.

Statistical Issues and Reasoning during the Animal Habitat Investigation

In the context of the Animal Habitat investigation, unlike in the Author Study, Donna understands the need to create variables to evaluate the zoo's performance in creating animal habitats. In developing the rating scale, Donna transferred knowledge from a similar situation she experienced during language arts, demonstrated knowledge about the content and concept of a rating scale, and was pedagogically and statistically sophisticated in developing this knowledge with children. Donna used multiple mechanisms, like description, smiling faces, and analogy, to convey the meaning of the rating scale used to self-evaluate writing by students to help them develop number sense in relationship to a new scale. She concerned herself with the potential distribution of scores—as demonstrated by her revision of the scale when it became clear that the scale, as originally conceived, would not show much variability in the data. Finally, she was concerned about objectivity demonstrated by having students translate written field notes into rating scores instead of immediately applying the rating rubric at the site. Donna exhibits exemplary reasoning about data, teaching students how to create new variables to suit investigator purposes and how people attribute meaning to numbers.

A potentially curious juxtaposition occurs where the focus of the investigation changes from creating variables and collecting data, to organizing, summarizing, and interpreting the data. Lehrer & Schauble (2002) stress the importance of organizing and structuring the data in a table format. In such a structured data table, the rows typically correspond to the individual animals (cases) and columns correspond to the different variables—like food and water rating, shelter rating, climate rating, space rating, habitat of origin category, and size of animal category. Table 3 illustrates how the Animal Habitat data could have been organized following these principles. Figure 2 represents how Donna structured and summarized the data for interpretation with the class. The teacher used a table to organize data from the Animal Habitat investigation. The organizing structure to the table, however, was the rating scale. With color coding (i.e., shading) for the habitat of origin and symbols for the size of the animal, the data structure in Figure 2 attempts to capture the totality of data across all variables. Donna made no other reorganization or graphical summarization of the data during the remainder of the investigation.

The decision at the beginning of the statistical analysis to structure the data table using the numbers of the rating scale as the organizational feature of the table raises potential questions about Donna's conception of the rating scale, variables, data, and the process of structuring, organizing, and graphing data. The temptation, as in the Author Study, is to quickly infer that Donna has a deficit in statistical understanding or a naïve, low level of statistical reasoning about data and distribution. This oversimplifies the context of the Animal Habitat investigation and the complexity of understanding Donna's statistical reasoning in the context of teaching. In the Dinosaur Investigation (see next section), Donna implements a structured data table consistent with the recommendation of Lehrer and Schauble (2002). During the summer workshop, Donna successfully completed two data collection and graphing activities that employed multiple variables. From this evidence, we know she is capable of correctly structuring a data table and graphing complex data.

For evidence of how Donna is reasoning about data and distribution, we turn to her comments during an interview:

> And the one thing we can do by looking at this (Figure 2) is say, well, how do you think the zoo did? Well, they thought they did pretty well because they have a whole lot more cards in the fours and fives. I wanted them to come up with did they do better in one habitat than another. Better with large animals or small animals? And that did not come out exactly the way that I wanted. (Interview, 10/21/00)

	1	2	3	4	5
SPACE	Gila Monster (m) Otter (m)	NG Singing Dog (m) Blood Python (m) Dwarf Crocodile (m) Tamarin (s)	Tree Kangaroo (m) Meerkat (s) Bald Eagle (m) Standing Gecko (s) Amur Leopard (l)	Zebra Mice (s) Spec. Bear (l) Boa (m) Bactrian Camel (l)	Baboon (l) Poison Arrow Frog (s) Iguana (m)
CLIMATE	Otter (M)	Spec. Bear (l) Tamarin (s)	Gila Monster (m)	NG Singing Dog (m) Poison Arrow Frog (s) Iguana (m) Bactrian Camel (l)	Amur Leopard (l) Standing Gecko (s) Zebra Mice (s) Blood Python (m) Tree Kangaroo (m) Dwarf Crocodile (m) Baboon (l) Bald Eagle (m) Meerkat (s) Boa (m)
SHELTER	Otter (M)	Zebra Mice (s) Bald Eagle (m)	Meerkat (s) Tamarin (s) Tree Kangaroo (m) Dwarf Crocodile (m) Spec. Bear (l) Golden Pheasant (m)	Gila Monster (m) Amur Leopard (l) Standing Gecko (s) Bactrian Camel (l)	NG Singing Dog (m) Blood Python (m) Baboon (l) Boa (m) Poison Arrow Frog (s) Iguana (m)
FOOD & WATER		Zebra Mice (s)	Tree Kangaroo (m) Bald Eagle (m)	NG Singing Dog (m) Dwarf Crocodile (m) Standing Gecko (s)	Otter (m) Gila Monster (m) Baboon (l) Iguana (m) Bactrian Camel (l) Spec. Bear (l) Amur Leopard (l) Poison Arrow Frog (s)

Legend

Forest | Grasslands | Desert | Inland Waterway | Rainforest

Figure 2. Animal habitat evaluation data.

Table 3. Structured data table for the investigation of animal habitat at the local children's zoo

Animal Name	Size	Habitat Origin	Space Rating	Climate Rating	Shelter Rating	Food & Water Rating
Gila Monster	Medium	Desert	1	3	4	5
Otter	Medium	Inland waterway	1	1	1	5
NG Singing Dog	Medium	Grasslands	2	4	5	4
Blood Python	Medium	Rain forest	2	5	5	5
Dwarf Crocodile	Medium	Rain forest	2	5	3	4
Tamarin	Small	Rain forest	2	2	3	3
Tree Kangaroo	Medium	Rain forest	3	5	3	3
Meerkat	Small	Desert	3	5	3	3
Bald Eagle	Medium	Forest	3	5	2	3
Standing Gecko	Small	Rain forest	3	5	4	4
Amur Leopard	Large	Grasslands	3	5	4	5
Zebra Mice	Small	Desert	4	5	2	2
Speckled Bear	Large	Grasslands	4	2	3	5
Boa	Medium	Rain forest	4	5	5	5
Bactrian Camel	Large	Desert	4	4	4	5
Baboon	Large	Grasslands	5	5	5	5
Poison Arrow Frog	Small	Rain forest	5	4	5	5
Iguana	Medium	Rain forest	5	4	5	5

Looking closely at Figure 2, if we ignore shading and the animal size indicator, then focus on one habitat quality, like space, the data summarization is that of a bar graph. The numbers of the rating scale indicate membership in an evaluative category. Furthermore, by crossing the numbers of the rating scale with habitat components (space, climate, shelter, food and water), as in a contingency table, Figure 2 simultaneously conveys the distributions of four bar graphs. This is a very creative means of organizing and summarizing the data so that the main investigation question can be addressed. Donna's desire to have students "come up with did they [zoo] do better in one habitat than another" is met because the table allows performance comparison across all four components of habitat. At the same time, this approach is not effective for the natural habitat in the wild and size of the animal variables, which require a restructuring or regraphing of the data table to make these comparisons.

In the Animal Habitat investigation, considerable class time was devoted to library research, development, and student understanding of the rating scale, data collection, and application of the rating scale. The timing of the visit by the zoo director had a major impact on the data organization and analysis. In fact, Donna did not perceive herself or the students as done with the analysis at the time of the zoo director visit. Donna states:

> I had hoped to bring more closure to it (Animal Habitat investigation) but (1) we ran out of time due to end of the quarter testing and Red Ribbon anti-drug lesson, (2) it

seemed anticlimactic after our visit from the zoo director, (3) I couldn't decide how to revisit the rating scale (i.e., data table) and make it meaningful to the students. My instincts said it was time to stop. (Interview, 10/21/00)

In a journal entry (10/12/00), Donna comments on the data table of Figure 1 and the overall investigation:

We just weren't able to draw many conclusions from our rating scale. Maybe too much information? I'm sure the process could have been improved but I really feel we've gained from the muddling through this. The children and I have been exposed to new ways of doing research, asking questions, and thinking.

The "too much information" comment refers to the color coding (shading) for wild habitat variable and numbers indicating animal size that made the data table busier and hard to interpret. Complementing this perspective, Donna later commented on the other variables of the investigation included but not analyzed in the data table:

Maybe the numbers didn't show it, but it opened up discussion. As we discussed it, they (the students) realized that it was easier to do a good job with the camel who lives in cold and hot climates, just like our state, and therefore it could be outside and it was okay, better than an animal from the rain forest that had to live inside, especially a large animal from the rain forest, those kind of things. And so we did get some information out of the table. (Interview 10/21/00)

For Donna's class, this is a very different type of discussion and process than what usually happened before and after a zoo trip.

Within the Animal Habitat investigation, Donna's reasoning about data and distribution is mixed and contextually driven. Donna exhibits exemplary teaching of data and variable creation to suit the purpose of the investigation, while the product of the investigation (Figure 2) is more complex. Under the time pressure imposed by the visit of the zoo director, Donna uses basic counts within categories (level of rating scale) to summarize the data. Indicating a very flexible understanding of data organization and summarization, Donna augments the count data with a creative use of a pseudo-contingency table format to display four distributions for comparative purposes in one table. The comparison of the components of habitat to evaluate zoo performance was at a rudimentary level, "because they (one component) have a whole lot more cards in the fours and fives." Decisions at the beginning of the investigation, like those to evaluate zoo performance relative to animal size and wild habitat of origin, were initially appealing and interesting. Under time constraints, however, these variables were secondary, their analysis forgone for the sake of the main investigation question, yet they were the source of confusion as Donna tried to preserve this information in the data table. In addition to the Animal Habitat investigation being an exemplar of authentic teaching and learning, it also illustrates the degree to which a teacher's reasoning about data and distribution is inextricably connected to the pedagogical decisions the teacher makes.

The Dinosaur Investigation

The Dinosaur unit was one Donna also had conducted numerous times in the past with her third-grade classes. Similarly, the statistical investigation component was a natural extension of her previous mode of teaching the topic. Her written plans included the following as the goals of this investigation:

> To apply research skills to learn about the characteristics of dinosaurs. To create a graph of one of those characteristics in a cooperative group situation. Students will then use that graph and the graphs of other groups to classify and draw conclusions about additional characteristics of dinosaurs. (Teacher planning document)

In many ways, the Dinosaur Investigation resembles the type of data analysis activity a teacher might find in a textbook. In fact, there is a similar published data analysis activity on dinosaurs in which a table of North American dinosaur data is provided for students to construct various types of graphs (Young, 1990).

Donna's variation on the dinosaur theme had students collect data through research at the library. Students selected and were responsible for a specific dinosaur to research. The teacher maintained a list of dinosaurs with information known to be available in the school library, and the students used the list in the selection process. Donna prepared the students for library research by structuring their note taking around the dinosaur's height, weight, and diet; what era they lived in; and where they lived. These variables were purposely selected by the teacher to teach specific ideas and concepts about the dinosaurs. Specifically, she wanted students to learn that (a) the sizes of dinosaurs changed between eras, (b) the size of dinosaurs is related to diet, and (c) dinosaurs were widely dispersed around the world during all of the eras studied. As Donna stated before one class period, "they need to learn dinosaurs are not always these huge meat-eating creatures that sometimes they think they are" (researcher field notes).

The information found during the library research became the quantitative data for this investigation. Donna helped the class compile and organize the dinosaur information from all students into a structured data table. Students transferred information from notes to note cards, and attached the cards to a huge data table rolled out on the floor of the classroom. Table 4 reproduces the original 19-foot-long data table. The data table was used to construct multiple graphs to address the dinosaur content learning goals of the investigation. Figures 3 and 4 are reproductions of two graphs developed by the whole class.

In addition to creating graphs, students mapped locations of different dinosaurs on a world map, wrote dinosaur reports, and created dioramas as ways of representing what they had learned. As a culminating event, all Dinosaur Investigation products were displayed throughout the classroom. During a visit by fourth graders, the third-grade students were expected to explain their work.

Statistical Reasoning during the Dinosaur Investigation

Up front, Donna defined specific concepts about dinosaurs that she wanted students to learn as a result of the investigation. She took a more structured approach to the statistical aspects of the study than she had used in the previous investigations. Here, Donna identified the variables of the study while students collected and organized the data in a structured data table (Table 3). The data table was correctly constructed with dinosaurs as cases and the characteristics of the dinosaurs (e.g., heights, weights, eras, diet) as variables. Finally, specific learning outcomes about dinosaurs were connected to statistical analyses through the interpretation of multiple graphs.

The content learning goals Donna had for students necessitated an analysis that examined relationships between variables. In this situation and context, the analyses were graphical, with each graph representing the data from two or more variables. Donna used the graphs in Figures 2 and 3 to convey answers to the investigation questions on the relationships between the dinosaur size and era in which they lived, and the dinosaur size and diet, respectively. Figure 3 has the named dinosaurs listed on the x-axis with weight in tons on the y-axis. The dinosaur names listed on the x-axis are sorted by the categories of the Era variable. The Eras are color coded (hatching) so that data points between adjacent dinosaurs within an Era are connected via a hatched line. In Figure 4, the graph is identical to that of Figure 3 except for a switch of the categorical comparative variable from Era Lived to Diet. The order in which the dinosaur names are listed on the x-axis remains unchanged from the graph of Figure 3, no longer sorting dinosaurs relative to the categorical variable of diet.

As in the previous two investigations, initial examination of these two graphs raises potential questions about Donna's knowledge of data, distribution, and graphing. The analysis of her reasoning, evidenced by products of this investigation, needs to be tempered by a more in-depth consideration of the context of the case. This investigation marks the first and only investigation using continuous variables that Donna conducts with children. Donna made the pedagogical decision to give students ownership of individual dinosaurs, similar to the data collection methods she employed in the other investigations. When a colleague suggested the use of line graphs to represent the data, Donna saw a means for students to see their individual contribution to the overall graph and proceeded to implement the colleague's suggestion.

The graph in Figure 3 does address variability in the data, as can be easily seen in the jaggedness of the line, or the lack thereof. Since the individual dinosaurs were sorted by Era, the comparison of groups is possible through the line graph. Furthermore, line graphs are precursors to bar graphs in the development of statistical reasoning about data (Bright & Friel, 1998) and are not inappropriate for this situation and context. Unfortunately, Donna loses sight of the need to sort the data relative to the categories of the comparison variable in Figure 4. This causes confusion for the class and ends the investigation.

Table 4. Class data table for Dinosaur Investigation

Prehistoric Animal	**When?**	**Where?**	**Size? (Height, Length, Weight)**	**Diet?**	**Researcher**
Pteranodon	Triassic	Every continent except Antarctica	33 pounds, 6 feet, 30 foot wingspan	Meat	Drew
Plesiosaurs Ichthyolosaurus	Jurassic and Cretaceous	Europe and North America	Ichthyolosaurus—30 feet; weight 200 pounds	Meat	Noah B
Archaeopteryx	Jurassic	Europe	1.5 feet wingspan; 1 foot length (beak to tail); 11–18 ounces	Meat	Mitchell
Apatosaurus	Jurassic	North American and Europe	30–40 tons; 70 feet	Plant	Amanda
Compsognatnus	Jurassic	North America	3 pounds; 30 inches; 7 pounds	Meat	Trey
Brachiosaurus	Jurassic and Cretaceous	Colorado, Algeria, Tanzania, Europe, Africa, North America	75–80 feet; 40 feet in height; 66 tons	Plant	Allison
Stegosaurus	Late Jurassic	North America and Europe	5 feet long; 11 feet tall; 2 tons	Plant	Steffie
Allosaurus	Jurassic	North America, Africa, and Australia	Height: 35; Weight: 2–3 tons; Length: 39 ft long	Meat	Keaton
Dienonychus	Cretaceous	North America	10 feet; Weight: 175 lbs.	Meat	Logan
Ankylosaurus	Cretaceous	North and South America, Europe, and Asia	17 feet long, 6 feet wide, 4 foot high, and weighed about 5 tons	Plant	Dylan
Triceratops	Cretaceous	Western North America	30 feet long, 3 feet length; all 30–35 feet; weight: 6 tons	plant	Joshua
Tyrannosaurus (T-Rex)	Late Cretaceous	North America, Asia, China	Weight: 7 tons; Length: 18 1/2 feet; Height: 18 feet	Meat	Taylor
Pachgycephal osaurus	Cretaceous	North America	Weight: 950 lbs; Length: 15 feet; Height: 8 feet	Plant	Remi
Parasaurolophas	Cretaceous	South America, North America, Asia, Europe	33 feet; skull 10 inch thick; 3 tons	Plant	Shasta
Corythosaurus	Cretaceous	North America	33 feet long; 4–5 tons	Plant	Kody
Glyptodon	Pleistocene	South America, North America	Long: 10 feet; Weight: 2 tons; Tall: 5 feet	Plant	Marshall
Smilodon	Pleistocene	North America, South America	Height: 4 at shoulder; Length: 10 feet; Weight: 500–600 lbs	Meat	Stephanie
Eahippus Hyracotheriun	Eocene	Asia	2 feet 60 (cm); 15–20 lbs; 89 inches	Plant	Hannah
Mammoths	Pleistocene	Asia	11–14 feet; 5 tons	Plant	Heather

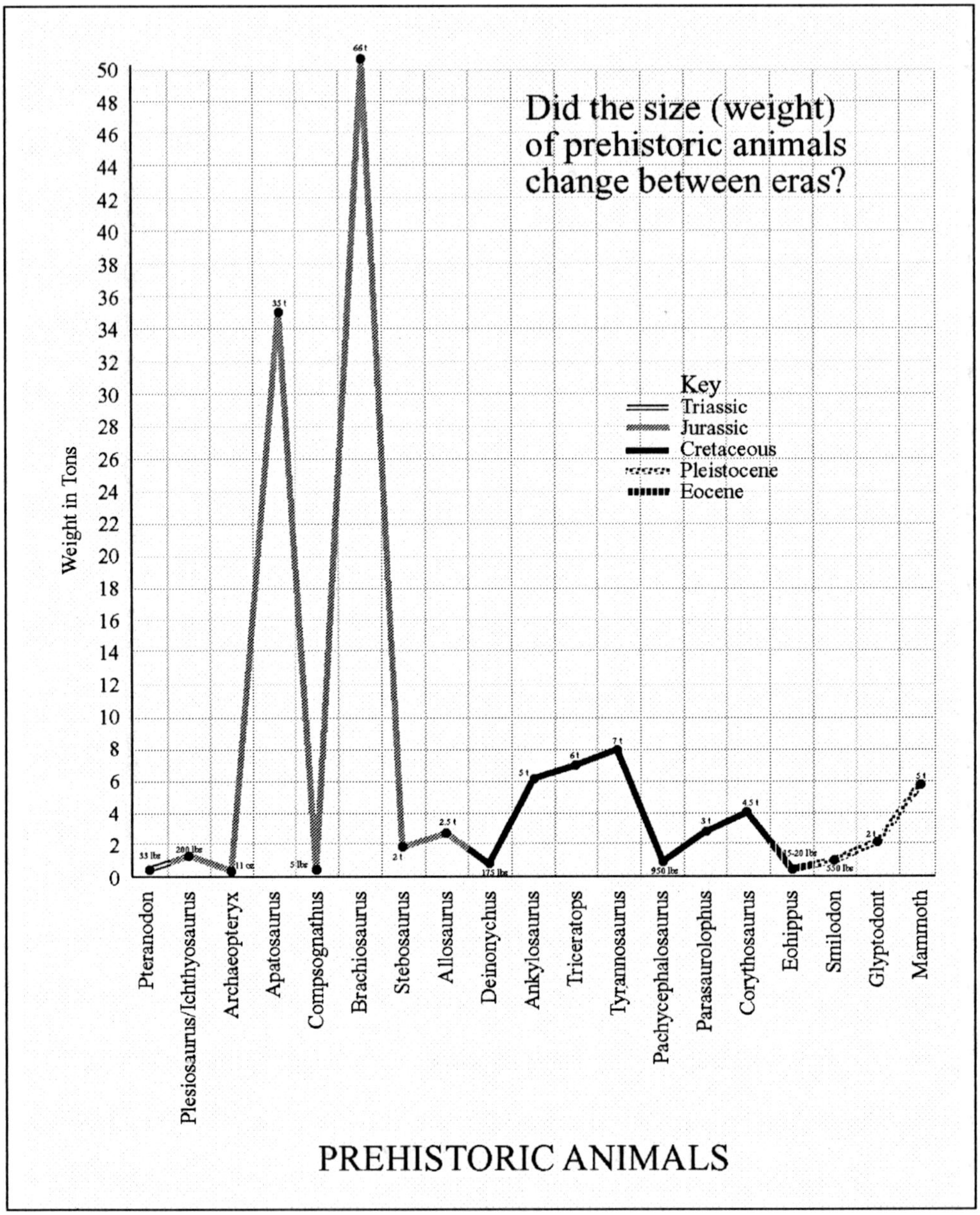

Figure 3. Artifact of Dinosaur Investigation comparing size of dinosaurs across eras.

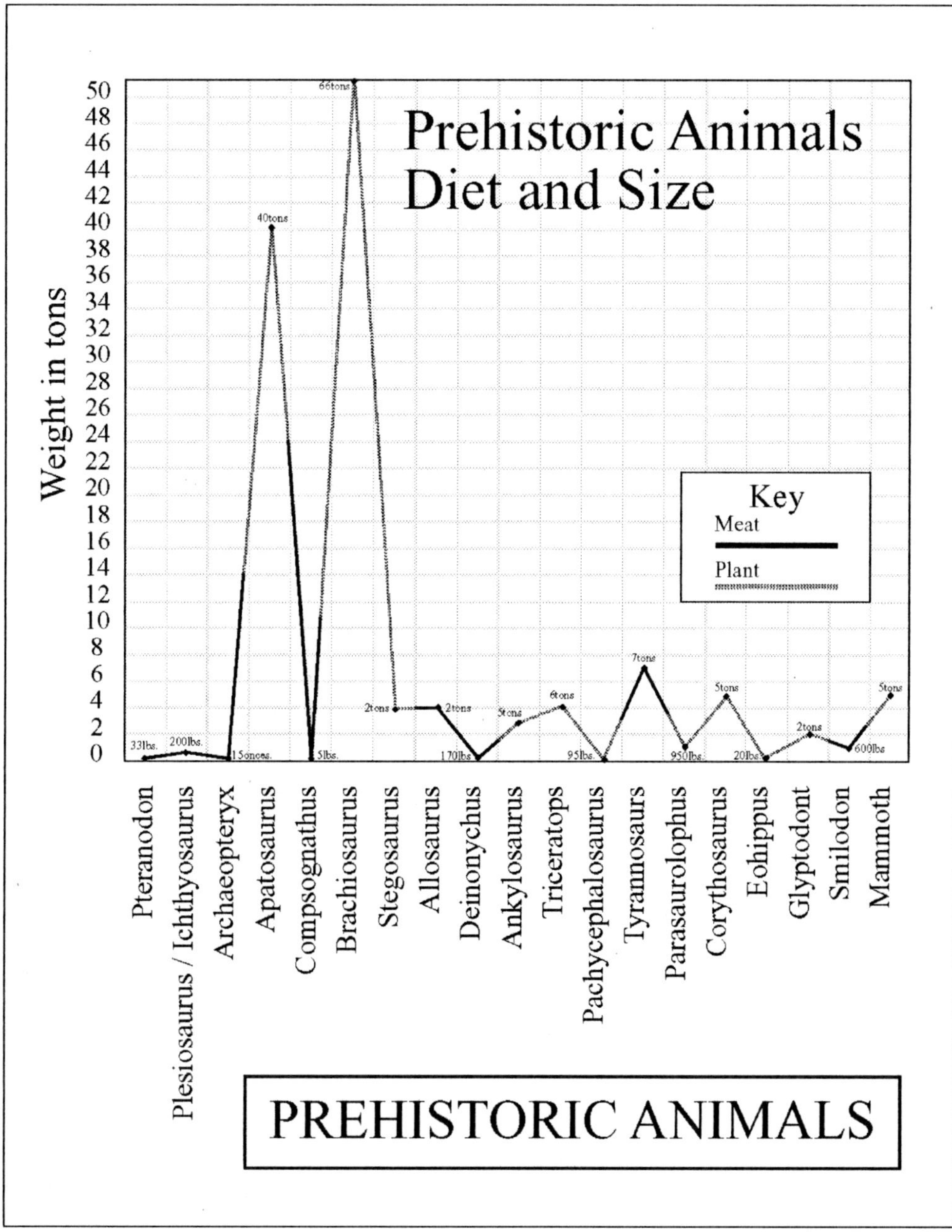

Figure 4. Artifact of Dinosaur Investigation comparing size of dinosaurs by diet.

How the Dinosaur Investigation concluded is explainable when considering some potentially contributing factors. First, basic manipulation of data, like sorting, is often overlooked in statistics education, especially when analyses are conducted by hand. Donna had no prior experience with sorting a data table, and we have no evidence that the task occurred to her. Second, as in the other investigations, Donna felt pressed for time and found herself using "cracks in the day" to try to make even more progress. She often used the students' snack time to discuss the graphs of the investigation. Third, Donna, being a highly organized and effective teacher, wanted to facilitate the investigation as expeditiously as possible. In her preparations, she constructed templates for graphing dinosaur data prior to class time, not recognizing the need for sorting data. The class completed both graphs in the same session. Without computers with graphing software, it was not possible to quickly reconfigure the graphing task. Donna found herself in a situation where she expected a successful resolution to the investigation; but instead, she was unable to help students draw a reasonable conclusion from Figure 4. She became unsure what to do next, given time constraints.

Given the outcome of the Dinosaur Investigation, it is arguable that Donna's reasoning about data and distribution is very low level, coupled with the added problem of not being able to distinguish between different types of graphs and know when a certain type of graph is more applicable. Donna did not try to reconfigure or redo the dinosaur graphs herself "on the fly" with students. She was unwilling to risk creating more graphs that could be equally confusing or unproductive, or to use a lot more class time for this project. She also did not pose the problem she saw with the graphs to the students. She could have made it a class task to take the data table and create various types of graphs for this data. Again, time was a contributing factor in the decision. Instead, she ended the investigation; but for the first and only time during the entire sequence of investigations, she directed specific questions to the authors about the graphical techniques she used, her implementation, and her statistical content knowledge. What is particularly interesting about the Dinosaur Investigation, however, was that in the self-contained context of the Gummy Bears in Space (Scheaffer et al., 1996) activity during the summer workshop, Donna successfully created complex graphs that combined categorical and continuous variables. When reminded of this and the similarity of the graphing tasks, Donna cringed and exclaimed, "this is why I need a statistician in the closet" (Interview, 12/1/00).

DISCUSSION

Donna's statistical reasoning about data and distribution was examined in the context of how she applied this knowledge in the action of conducting applied statistical investigations to help her third-grade students learn about other topics in the curriculum. What is interesting and perplexing is that Donna exhibits strong statistical reasoning skills in one contextual setting, but that same knowledge or skill does not necessarily transfer to all of her teaching work with children. For example, she creates important variables connected to the purpose of the investigation for the Animal Habitat investigation to evaluate the performance of the zoo, but does not do this well in the Steven Kellogg Author Study. Similarly, she develops a well-structured data table in the Dinosaur Investigation, but fails to do this in the Author

Study and Animal Habitat investigations. In another example, Donna constructs complex multivariable graphs during the summer workshop, but fails to adequately graph the data of the Dinosaur Investigation. Finally, she teaches the concept of distribution and its importance in making predictions during some classroom investigations like Getting to Know You, and Cold Remedies (see Table 1), but fails to include this type of discussion in the Author Study and offers only rudimentary coverage in the Animal Habitat investigation. Stating this in another way, Donna exhibits both exemplary and naïve, or basic, statistical reasoning about data and distribution, depending on the context of the investigation. The pattern emerging in her reasoning is that the more open-ended and unstructured the investigation, the more Donna relies on the basics of statistical reasoning about data, namely, tallies and counts. Conversely, the more the investigation resembles the activities and context of her prior teaching practice, the more comfortable she becomes in teaching sophisticated concepts like distribution.

Donna was selected to represent a best-case scenario to illustrate how competent, experienced teachers can incorporate the process of statistical investigation into their teaching practice. We had every expectation that her statistical knowledge and graphing ability would be perfectly suited to the task asked of her. When faced with the challenge of implementing open-ended statistical investigations into content ideas of her curricula, Donna seemingly took the following approach. First, she connected the curriculum topic with the investigation to make "space," covered standards, and ensured a positive learning experience. Second, she mapped what she saw to be a similar type of data collection and graphing activity from her prior teaching experience onto the investigation problem. This pedagogical strategy was more for the ease of data collection, efficiency, and student ownership, than carefully considering the nuances of a specific investigation, looking ahead to data analysis and interpretation, and connecting this back to the purpose of the investigation through the design (CBMS, 2001). Donna's extensive prior experience with statistics in the context of textbook problems and activities loosely connected to her curriculum appears to have influenced her ability to apply the process of statistical investigation in a context intended to teach ideas central to the curriculum. It is the juxtaposition of her background and experience with the statistical knowledge needed to implement purposeful statistical investigation connected to curriculum topics that gives rise to a number of implications regarding the statistics education of teachers.

IMPLICATIONS

To support planning and implementation of investigations connected to K–6 curriculum topics, teacher learning opportunities need to capitalize on occasions inside the statistical investigation process. Some basic ideas and concepts teachers need to learn are identified in this study. Examples include (a) how to define and create variables when none are inherent or obvious to an investigation, (b) how to do basic data manipulation like sorting, (c) how to gain the perspective to check and determine whether results of the analysis address the intended purpose of the investigation, and (d) how to discern when and what types of graphs to use in different situations. Others are enumerated in the section on data analysis and statistics in the CBMS (2001) document. As this study illustrates, the context in which the statistical concepts like data and distribution arise and are applied matters. Teachers need opportunities to construct understanding and recognize use of statistical concepts like data and distribution as they appear holistically in the context of conducting purposeful applied statistical investigation with children.

This suggestion raises a question to consider in the data-driven, activity-based trend and recommendations for teaching statistics (Cobb, 1993). The findings of this study lead one to ask whether that formula for learning statistics might be valuable for some teachers but detrimental to others wishing to use statistical investigation for teaching other content. Using statistical investigation as a tool for this purpose requires teachers to be able to reason in ways that require recognizing when and how context matters across all tasks of an investigation and making preplanned and spur-of-the-moment teaching decisions accordingly. People who apply statistical reasoning in real-world problems must be able to frame the problem and use their statistical knowledge in the framed context to solve it. Learning statistics through predeveloped or canned activities does not necessarily require the teacher to recognize the structure of the problem and to know how or when statistical knowledge and reasoning comes into play when student learning about a curriculum topic hinges on the outcome of the statistical investigation. As Lehrer and Schauble (2000) state, "When students work with data sets handed to them, they may lose the sense that data result from a constructive process" (p. 2). It is this constructive process that teachers must appreciate and understand themselves, in deep and sophisticated ways, in order to make decisions that guide and help children appreciate and understand the same.

Another implication is that teachers need to feel that they can and will learn more about statistics through the act of teaching statistical processes and content, and that it is acceptable to be simultaneously a learner and a teacher. Playing this dual role of teacher and learner is not without risk (Heaton, 2000) and needs to be supported and encouraged by statistics educators in their work with teachers. It is impossible to learn all the statistical content one needs to learn prior to teaching. Taking on the role of learner while teaching requires both confidence and a willingness to cope with uncertainty. One way for teachers to learn this is to see this disposition toward teaching, as well as learning, openly modeled by the statistics educators with whom they work.

Finally, a study such as this one could become an important tool in teacher education in the area of statistics, complementing the use of images of real classrooms and interactions and products of student work as a means of constructing knowledge for teaching in other areas of teacher education (Lampert & Ball, 2001; Merseth, 1996). Researchers have used studies like this to represent and help others understand the complexity of teaching and teacher knowledge while constructing their own knowledge for teaching. They offer a blend of statistical knowledge and practice such that teachers can see not only examples of statistical knowledge informing pedagogical decision making but also how particular pedagogical decisions can both positively and negatively affect data collection, summarization, and interpretation.

The development of more vignettes focused on statistical concepts and the process of statistical investigation would enable teachers to see statistical concepts as they appear in context and the ways statistical knowledge is used, or could be used, by teachers in investigative work with children. Furthermore, the development of a collection of examples of practice—situated in real classrooms around specific statistical concepts arising or deliberately taught while doing statistical investigations—offers a direction for creating usable knowledge for teaching from research. Additionally, using such examples from practice would continue to illustrate to teachers and teacher educators a key finding from this research: that in learning statistical knowledge for teaching, the context matters; and teachers need to learn where, when, why, and how it matters.

REFERENCES

Ball, D. L. (2002). *Mathematical proficiency for all students: Toward a strategic research and development program in mathematics education*. RAND Report. Washington, DC: Office of Education Research and Improvement, U.S. Department of Education.

Ball, D. L., Lubienski, S. T., & Mewborn, D. S. (2001). Mathematics. In V. Richardson (Ed.), *Handbook of research on teaching*, pp. 433–456.

Bright, G. W., & Friel, S. N. (1998). Graphical Representations: Helping students interpret data. In S. P. Lajoie (Ed.), *Reflections on statistics: Learning, teaching, and assessment in grades K–12*. Mahwah, NJ: Erlbaum.

Cobb, G. W. (1993). Reconsidering statistics education: A national science foundation conference, *Journal of Statistics Education [online], 1*(1), available e-mail: archive@jse.stat.ncsu.edu Message: Send jse/v1n1/cobb.

Cobb, P. (1999). Individual and collective mathematical development: The case of statistical data analysis, *Mathematical Thinking and Learning, 1*(1), 5–43.

Conference Board of Mathematical Sciences (March 2001). *Mathematical education of teachers project*, draft report. Washington, DC: Author.

Confrey, J., & Makar, K. (2001, August). Secondary teachers' inquiry into data. *Proceedings of the Second International Research Forum on Statistical Reasoning, Thinking, and Literacy*, University of New England, Armidale, NSW, Australia.

Creswell, J. W. (1998). *Qualitative inquiry and research design: Choosing among five traditions*. Thousand Oaks, CA: Sage Publications.

Fennema, E., & Nelson, B. S. (1997). *Mathematics teachers in transition*. Hillsdale, NJ: Erlbaum.

Friel, S. N., & Bright, G. W. (1998). Teach-Stat: A model for professional development in data analysis and statistics for teachers K–6. In S. P. Lajoie (Ed.), *Reflections on statistics: Learning, teaching, and assessment in grades K–12*. Mahwah, NJ: Erlbaum.

Friel, S. N., & Bright, G. W. (1997). A framework for assessing knowledge and learning in statistics (K–8). In J. Gal and J. B. Garfield (Ed.), *The assessment challenge in statistics education* (pp. 55–63). Amsterdam: IOS Press.

Graham, A. (1987). *Statistical investigations in the secondary school*. Cambridge, UK: Cambridge University Press.

Heaton, R. (2000). *Teaching mathematics to the new standards: Relearning the dance*. New York: Teachers College Press.

Heaton, R., & Mickelson, W. (2002). The learning and teaching of statistical investigation in teaching and teacher education. *Journal of Mathematics Teacher Education, 5*, 35–59.

Jaworski, B. (1998). The centrality of the researcher: Rigor in a constructivist inquiry into mathematics teaching. In A. Teppo (Ed.), *Qualitative research methods in mathematics education* (pp. 112–127). Reston, VA: National Council of Teachers of Mathematics (NCTM).

Lajoie, S. P. (1998). Reflections on a statistics agenda for K–12. In S.P. Lajoie (Ed.), *Reflections on statistics: Learning, teaching, and assessment in grades K–12*. Mahwah, NJ: Erlbaum.

Lajoie, S. P., & Romberg, T. A. (1998). Identifying an agenda for statistics instruction and assessment in K–12. In S. P. Lajoie (Ed.), *Reflections on statistics: Learning, teaching, and assessment in grades K–12*. Mahwah, NJ: Erlbaum.

Lampert, M., & Ball, D. L. (2001). *Teaching, multimedia, and mathematics: Investigations of real practice*. New York: Teachers College Press.

Lehrer, R., & Schauble, L. (2002). *Investigating real data in the classroom: Expanding children's understanding of math and science*. New York: Teachers College Press.

Lehrer, R., & Schauble, L. (2000). Inventing data structures for representational purposes: Elementary grade students' classification models, *Mathematical Thinking and Learning, 2*(1 & 2), 51–74.

Merseth, K. (1996). Cases and case methods in education. In J. Sikula (Ed.), *Handbook of research on teacher education* (2nd ed., pp. 722–744). New York: Macmillan.

National Council of Teachers of Mathematics (2000). *Principles & standards for school mathematics*. Reston, VA: Author.

National Council of Teachers of Mathematics (1989). *Curriculum and evaluation standards for school mathematics*. Reston, VA: Author.

Newmann, F. M., & Wehlage, G. G. (1997). *Successful school restructuring: A report to the public and educators*. Madison: Center on Organization and Restructuring of Schools, University of Wisconsin.

Russell, S. J., & Friel, S. N. (1989). Collecting and analyzing real data in the elementary school classroom. In P. R. Trafton & A. P. Shulte (Eds.), *New directions for elementary school mathematics* (pp. 134–148). Reston, VA: National Council of Teachers of Mathematics.

Schaeffer, R. L. (1988). Statistics in the schools: The past, present and future of the quantitative literacy project. *Proceedings of the American Statistical Association form the Section on Statistical Education* (pp. 71–78).

Scheaffer, R. L., Gnanadesikan, M., Watkins, A., & Witmer, J. A. (1996). *Activity-based statistics*. New York: Springer.

Scheaffer, R. L., Watkins, A. E., & Landwehr, J. M. (1998). What every high-school graduate should know about statistics. In S. P. Lajoie (Ed.), *Reflections on statistics: Learning, teaching, and assessment in grades K–12*. Mahwah, NJ: Erlbaum.

Schifter, D. (1996). *What's happening in math class?* (Vols. 1–2). New York: Teachers College Press.

Shaughnessy, J. M. (1992). Research in probability and statistics: Reflections and directions. In D. A. Grouws (Ed.), 1992, *Handbook of Research on Mathematics Teaching and Learning*. New York: MacMillan, pp. 465–494.

Shulman, L. S. (1986). Those who understand: Knowledge growth in teaching. *Educational Researcher*, 15(2), 4–14.

Utts, J. M. (1999). *Seeing Through Statistics* (2nd Ed). Pacific Grove, CA: Duxbury Press.

Watson, J. M. (2001). Profiling teachers' competence and confidence to teach particular mathematics topics: The case of chance and data. *Journal of Mathematics Teacher Education, 4*(4), 305–337.

Watson, J. M. (2000). Preservice mathematics teachers' understanding of sampling: Intuition or mathematics. *Mathematics Teacher Education Research Journal, 12*(2), 147–169.

Young, S. L. (1990). North American dinosaur data sheet; Graph the dinosaurs' lengths; Meat-eaters and Plant-eaters. *Arithmetic Teacher, 38*(1), 23–33.

Chapter 15

SECONDARY TEACHERS' STATISTICAL REASONING IN COMPARING TWO GROUPS[1]

Katie Makar[1] and Jere Confrey[2]
University of Texas at Austin, USA[1], and Washington University in St. Louis, USA[2]

OVERVIEW

The importance of distributions in understanding statistics has been well articulated in this book by other researchers (for example, Bakker & Gravemeijer, Chapter 7; Ben-Zvi, Chapter 6). The task of *comparing* two distributions provides further insight into this area of research, in particular that of variation, as well as to motivate other aspects of statistical reasoning. The research study described here was conducted at the end of a 6-month professional development sequence designed to assist secondary teachers in making sense of their students' results on a state-mandated academic test. In the United States, schools are currently under tremendous pressure to increase student test scores on state-developed academic tests.

This paper focuses on the statistical reasoning of four secondary teachers during interviews conducted at the end of the professional development sequence. The teachers conducted investigations using the software Fathom™ in addressing the research question: "How do you decide whether two groups are different?" Qualitative analysis examines the responses during these interviews, in which the teachers were asked to describe the relative performance of two groups of students in a school on their statewide mathematics test. Pre- and posttest quantitative analysis of statistical content knowledge provides triangulation (Stake, 1994), giving further insight into the teachers' understanding.

---

[1] This research was funded by the National Science Foundation (NSF) under ESR-9816023. The opinions expressed in this chapter do not necessarily reflect the views of NSF.

D. Ben-Zvi and J. Garfield (eds.),
The Challenge of Developing Statistical Literacy, Reasoning and Thinking, 353–373.

WHY STUDY TEACHERS' REASONING ABOUT COMPARING TWO GROUPS?

Statistics and data analysis are becoming increasingly important in our society for a literate citizenry. As such, many schools have begun to incorporate statistics and data analysis into their curriculum, beginning as early as Kindergarten (TERC, 1998). Although many schools are increasing their emphasis on statistics, very few are taking sufficient steps to help teachers master the statistics they are expected to teach. Professional development typically provided to teachers by their schools gives mathematics teachers little opportunity to improve their statistical content knowledge beyond evaluation of central tendency and simple interpretation of graphs and tables, while university statistics courses are rarely aimed at content teachers feel is relevant. Furthermore, U.S. teachers have little experience with data analysis and inferential statistics, yet in a time when teachers are under increasing pressure to improve student scores on state-mandated tests, teachers are required to make instructional decisions based on large quantities of data about their students' performance. Given that teachers are both the target and the vehicle of reform (Cohen & Ball, 1990), it is vital that we consider teachers' facility in statistical reasoning as well as possible vehicles for helping teachers improve their conceptual understanding of the statistics they are expected to teach. Enhanced understanding of teachers' statistical reasoning will help professional development leaders better design conceptual trajectories for advancing teacher reasoning in statistics, which should ultimately improve student understanding in probability and statistics.

Investigations involving comparing groups provide a motivational vehicle to learn statistics (see, for example, Konold & Pollatsek, 2002): They are steeped in context, necessitate a focus on both central tendency and distribution (for various aspects of distributions, see Chapter 7 this volume), and provide momentum for the conceptual development of hypothesis testing. Furthermore, tasks involving group comparisons are rich enough to be accessible to a broad array of learners at varying ages and levels of statistical understanding. Comparing distributions can be an interesting arena for researchers to gain insight into teachers' statistical reasoning, and in particular gave us an opportunity to understand teachers' reasoning about variation in a more sophisticated context.

Several curriculum projects make use of group comparisons as an avenue to teach statistical reasoning. At the elementary level, comparing two groups can be used to introduce the concepts of data and graphing, providing students with important early experiences in viewing and reasoning with distributions. For example, a first-grade curriculum (TERC, 1998) introduces primary students to distributions by having them compare and qualitatively describe the distribution of their classmates' ages to that of their classmates' siblings. Middle school students are able to build on earlier experiences with data and start to focus on descriptions of distributions: measures of center and spread, shapes of distributions, as well as gaps and outliers. For example, a sixth-grade curriculum puts these skills into a meaningful context for students by having students compare "typical" heights of males and females in their class, examining measures of center, describing the shapes of two distributions, and looking at gaps and outliers (Lappan, Fey, Fitzgerald, Friel, & Phillips, 1998). For older students, more open-ended designs and conventional descriptions of statistical variation can be introduced, which will help students build a foundation for inferential statistics or to inform debate of issues in light of available data.

At a wide variety of grade levels and settings, comparing groups has the potential for giving students authentic contexts to use data to answer meaningful questions, thus motivating the power of data in decision making. However, in order for teachers to provide these kinds of tasks for their students, they need to develop their own statistical understanding. Heaton and Mickelson (Chapter 14, this volume) described the experience of an elementary teacher's struggle to develop her own statistical reasoning as she worked to merge statistical investigations into the existing school curriculum. This chapter will examine statistical reasoning in secondary teachers as they build their statistical content knowledge through investigations of student assessment data, in particular the role of variation in considering what it means to compare two groups. (For additional discussions of the teachers' reasoning with data, see Confrey & Makar, 2002; Makar & Confrey, 2002.)

PREVIOUS RESEARCH ON COMPARING TWO GROUPS

Within the world of statistics, much concern is placed on making comparisons, either direct or implied. Whether the difference is between brands of peanut butter, or housing prices compared to last year, comparisons form the very fabric of research and of *principled* arguments (Abelson, 1995):

> The idea of *comparison* is crucial. To make a point that is at all meaningful, statistical presentations must refer to differences between observation and expectation, or differences among observations. Observed differences lead to why questions, which in turn trigger a search for explanatory factors ... When we expect a difference and don't find any, we may ask, "Why is there *not* a difference?" (p. 3)

Lehrer and Schauble (2000), in their work with children's graphical construction, indicate that young students "are often disconcerted when they find a discrepancy between the expected value of a measure and its observed value" (p. 104). Watson and Moritz (1999) argue that comparisons of data sets provide a meaningful backdrop for students to gain a deeper understanding of the arithmetic mean as well as strong intuitive approaches to compare groups through balancing and visual strategies, "hopefully avoiding the tendency to 'apply a formula' without first obtaining an intuitive feeling for the data sets involved" (p. 166).

The task of comparing groups appears in the literature as an impetus for students to begin to consider data as a distribution instead of focusing on individuals, in addition to motivate students to take into account measures of variation as well as center (Konold & Higgins, 2002). Lehrer and Schauble (2000) found that as older students solved problems in which they compared two distributions, they began to look at both centrality and dispersion. In their study, comparing groups served as an impetus for students to gain an appreciation for measures beyond center. For example, they report on a group of fifth graders who, when experimenting with different diets for hornworms, found that the hornworms in the two treatment groups showed differences not only in their typical lengths but also in the *variability* of their lengths. This caused the students to speculate and discuss reasons why the lengths in one group varied more, showing that "considerations of variability inspired the generation of explanations that linked observed patterns to mechanisms that might account for them" (p. 129).

Examining the context of a problem is critical for understanding group comparisons. Confrey & Makar (2002) discuss the role of context in statistical learning while examining the process of teachers' inquiry into data. In one activity they describe, teachers examined several pairs of graphs void of context and reasoned about comparisons between graphs in each pair at a very superficial level in a discussion that lasted only about 5 minutes. However, when the same graphs were examined again in light of a context relevant to the teachers (quiz scores), a much more in-depth analysis took place in a discussion lasting 40 minutes. This discussion was the first time in their study that the teachers articulated variation in a distribution as being useful. When the teachers could compare distributions in a personally meaningful context, they began to gain a more robust understanding of distribution. Similarly, Cobb (1999) found that by comparing the distributions in the context of judging the relative lifespan of two types of batteries, students were compelled to consider what it meant for one battery to be preferred over another—does one consider overall performance, or consistency? Here, students negotiated a purposeful reason to consider variation in the context of what constitutes a "better" battery.

Comparing two groups also becomes a powerful tool in light of its use toward a consideration of statistical inference. Watson & Moritz (1999) argue specifically that comparing two groups provides the groundwork "to the more sophisticated comparing of data sets which takes place when t-tests and ANOVAs are introduced later" (p. 166). Without first building an intuitive foundation, inferential reasoning can become recipe-like, encouraging black-and-white deterministic rather than probabilistic reasoning. "The accept-reject dichotomy has a seductive appeal in the context of making categorical statements" (p. 38, Abelson, 1995). Although formal methods of inference are not usually a topic in school-level statistics content, an ability to look "beyond the data" (Friel, Curcio, & Bright, 2001) is a desired skill. Basic conceptual development of statistical inference can lead to assistance in understanding one of the most difficult, but foundational concepts in university-level statistics: sampling distributions (delMas, Garfield, & Chance, 1999).

RESEARCH DESIGN AND METHODOLOGY

The research described in this chapter was part of an NSF-funded research project developed and carried out by a research team at the Systemic Research Collaborative for Education in Math, Science, and Technology at the University of Texas at Austin. Although this chapter focuses on the results of interviews taken at the end of the study, the experience of the participants in the research project is key to understanding their background knowledge and experience in statistical reasoning. It should be noted that research on these teachers' statistical reasoning was not the purpose of the workshop, which was to examine the effects of the professional development sequence within a larger systemic reform project (Confrey, in preparation). The authors saw an opportunity, after the workshops were planned, to examine the teachers' statistical reasoning through a set of clinical interviews. This chapter is the result.

The 6-month professional development research project took place in two phases: 18 contact hours of full-day and after-school meetings, followed by a 2-week summer institute. The project was conceived as a mathematical parallel of the National Writing Project, where teachers focus on their own writing rather than how to teach writing. A mission of the National Writing Project (2002), and a belief that was fundamental to our study, is that if teachers are given the opportunity to focus on their own learning of the content that they teach—to *do* writing, or mathematics, in an authentic context—they will better understand the learning process and hence teach with greater sensitivity to students' conceptual development (Lieberman & Wood, 2003). Our professional development sequence was designed under the assumption that if mathematics teachers are immersed in content beyond the level that they teach, and developed through their own investigations as statisticians within a context that they find compelling and useful, then they will teach statistics more authentically and their increased content knowledge will translate into improved practice.

During the professional development sequence, teachers learned a core of statistical content: descriptive statistics and graphing, correlation and regression, sampling distributions, the Central Limit Theorem, confidence intervals, and basic concepts of statistical inference. These concepts were not developed formally, as they would be in a university course; rather, teachers were given extensive experience with sampling distributions through simulations in order to (a) help them understand concepts of sampling variation that we thought was critical to their working with data and (b) give them access to powerful statistical ideas. Statistical concepts were introduced only as they were needed to make sense of the data; many of the teachers already had at least a working knowledge of descriptive statistics and graphing, as indicated by their statistics pretest.

During the workshops and summer institute, teachers conducted increasingly independent investigations focused on the analysis of their students' high-stakes state assessment data. For the teachers, this was a compelling context in which to learn statistics. In Texas, there is much emphasis on the Texas Assessment of Academic Skills (TAAS, www.tea.state.tx.us), the high-stakes state assessment where students and schools are held accountable for their performance on the battery of tests. Teachers felt they would be empowered if they were able to interpret TAAS data instead of having to rely on experts to tell them what the data meant and what actions the school needed to take in order to raise test scores. Because many of the "lessons" we wanted them to gain from working with data involved sampling variation, we felt it critical to give them enough experience to develop an intuition about this type of variation.

Many of the investigations were supported by the use of the statistical learning-based software, Fathom (Finzer, 2000), to examine data directly as well as to create simulations to test conjectures. The software allowed teachers to fluidly and informally investigate relationships in the data because of the ease with which Fathom creates graphs through a drag-and-drop process. Most statistical software tends to be like a "black box" with a purpose that supports a data-in, results-out mind-set that can work to encourage misconceptions in early learners who expect to find definitive answers in the data. Fathom insists that users build simulations in the same way that one would construct a sampling distribution: by creating a randomization process, defining and collecting measures from each randomization, and then iteratively collecting these measures. The levels of abstraction that make sampling distributions and hypothesis testing so difficult for university students (delMas et al., 1999) are not avoided in Fathom, but made transparent through creating visual structures that parallel these processes, allowing users to better visualize the concepts underlying the abstract nature of a sampling distribution. Fathom was also a powerful tool for analysis and supported the use of authentic data, even reading data directly from websites, thus empowering teachers to greater access to the many data sets that are available on the Internet.

During the workshops with the teachers, we often used sampling distributions to illustrate and investigate statistical concepts—properties of the normal distribution, the Central Limit Theorem, the effect of sample size on sampling variability, the null hypothesis, *p*-values, and hypothesis testing. In addition, these statistical concepts were applied during investigations of relationships in the data. It is important to note that we did not focus explicitly on group comparisons during the professional development workshops; we did not formally develop a list of procedures for comparing groups, nor had the teachers seen a task similar to the one we asked them to perform for the interview. During the workshops, the teachers did engage in two structured activities in which comparing two groups was central. The first activity took place in the early stages of the professional development program, when teachers were first learning the software and investigating descriptive statistics. In this activity (Erickson, 2001, p. 206), a sample of student scores on the Scholastic Aptitude Test (SAT; a national test many U.S. students are required to take as part of their college application) and the grade point averages of males and females were examined and informally compared. The second activity, *Orbital Express* (Erickson, 2001, p. 276), took place in the final week of the summer institute when investigating the concept of a null hypothesis and working with more advanced features of the software. In this activity, teachers dropped two types of wadded paper and attempted to hit a target 10 feet below. The distance each wad fell from the target was then entered into one column in Fathom, and the type of paper thrown was entered in a second column. Using the *scramble attribute* feature of the software, the values in the second column (type of paper) were randomized, simulating a null hypothesis, and the difference in the median of each group was calculated. This process was repeated 100 times to create a "null hypothesis" distribution, showing the amount of variation in the differences between the medians of the groups that might be expected just by chance. The difference found in the original sample was then examined in light of this "null hypothesis" distribution.

SUBJECTS AND DATA COLLECTION

This chapter focuses primarily on four secondary mathematics teachers from Texas who took part in the professional development program just described. Two of these participants joined the project later and were part of an abbreviated repeat of the first phase of the professional development sequence. One of the four subjects was a preservice teacher while the other three were experienced, credentialed secondary mathematics teachers who taught 13- to 16-year-old students. Two of the teachers had obtained a university degree in mathematics, the preservice teacher was working on her mathematics degree, and the remaining teacher had a degree in the social sciences. The two men and two women consisted of one Hispanic and three non-Hispanic whites. All but the preservice teacher had taken a traditional introductory statistics course 5–15 years previously during their university coursework. These were the only four teachers who took part in Phase II of the project (the summer institute), due partly to a change in the administration at the school and scheduling conflicts with teachers teaching summer school.

Data collected on the subjects included a pre-post test of statistical content knowledge, which was analyzed using a t-test in a repeated measures design. In addition, all of the sessions with the teachers were videotaped. Interviews were conducted at the end of the study in which participants were asked to compare the performances of two groups, and which will comprise the main source of data for this chapter. The interviews were videotaped, and major portions were transcribed and then analyzed using the qualitative methodology of grounded theory (Strauss & Corbin, 1998). Under this methodology, the transcripts were first subjected to open coding in the software NVivo (QSR, 1999) to capture the phenomenon observed in the teachers' own words and actions and to allow potential categories to emerge from the data that would describe strategies, mind-set, and other insights into how the teachers were comparing groups. Secondly, initial categories were organized into hierarchical trees and subjected to axial coding to begin to tie common themes into larger categories. Finally, the data were analyzed with selective coding to further investigate various dimensions of the categories and better describe the phenomenon observed.

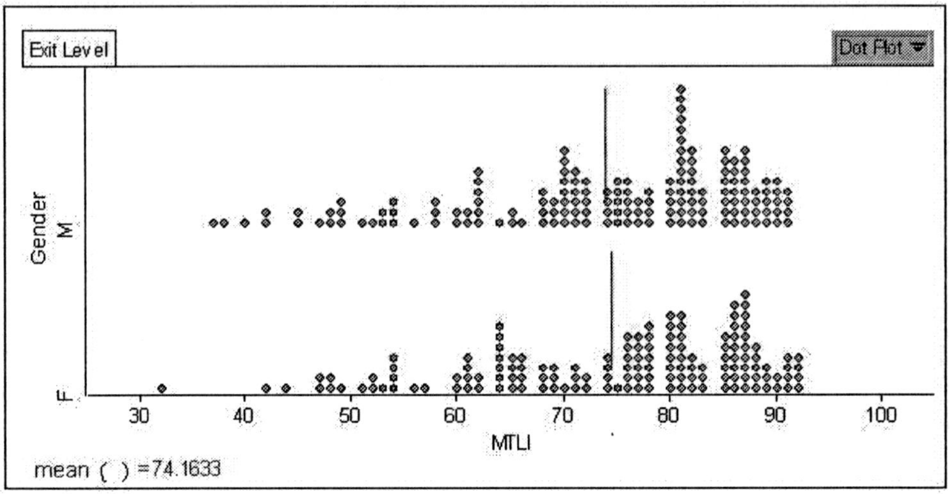

Figure 1. A dot plot of student test scores with means plotted in Fathom created by one teacher.

INTERVIEW TASK

In the interviews, which took place during the last two days of the summer institute, subjects were given a raw data set of student TAAS scores from a hypothetical low-performing high school and asked to use Fathom to compare the performance of males and females in the school. Although all of the data used in the example was not from a single school, it was in fact authentic student data created from a compilation of scores drawn from several schools in Texas. Figure 1 shows a graph similar to ones that each of the teachers initially created in Fathom from the given data. The MTLI on the horizontal axis of this graph is the Mathematics "Texas Learning Index" on TAAS; a MTLI score of 70 is considered passing. In the context in which the state high-stakes test exists, it is not just the means that are relevant to consider. Schools are held accountable for the proportion of students who pass the TAAS test, so it is also important to consider the proportion passing for each group. In a second task during the interview, subjects were asked to investigate the low-performing status of the school, based on analysis of the performance of the state-defined ethnic subgroups within the school and to make a campus-based program recommendation to the principal for the following year. The analysis in this chapter focuses on the first interview task.

RESULTS

In analyzing these teachers' reasoning about comparing two groups, we assumed that the professional development sequence had an impact on their content knowledge. Rather than examine teachers' reasoning about comparing two groups with teachers who had little experience with data or diverse backgrounds in statistical content knowledge, we chose to examine teachers who had developed their statistical understanding through rich experiences as investigators. We recognize that the reasoning ability of this group, therefore, is not *typical* of secondary teachers, but what *could* occur after a relatively short period of professional focus on building conceptual understanding and experience with powerful statistical ideas. The overarching purpose of the professional development was to give them rich experiences as investigators with school data. We did discuss many concepts in inferential statistics, but the majority of these more advanced concepts (e.g., t-tests, confidence intervals, null hypothesis, *p*-values) were experienced through simulations on a conceptual, not formal level.

To measure whether the content that was taught had an impact on teachers' understanding and to assess the level of statistical content knowledge at the time of the interviews, a pre-post test of content knowledge was given to teachers. The result of the analysis is given in Table 1. The data summary shows significant growth ($\alpha = 0.05$) in their overall content knowledge as well as for two individual areas (Sampling distributions and Inference), even though the number of teachers in the study was small ($n = 4$).

Table 1. Results of pre-post test of statistical content knowledge using a t-test and repeated measures design, n=4

Topic	Pretest Mean Percent Correct	Posttest Mean Percent Correct	Difference	t	p-value
Descriptive Statistics	61%	79%	18%	2.0	0.14
Graphical Representation	75%	83%	8%	0.5	0.63
Sampling Distribution	8%	75%	67%	4.9	< 0.01
Inference and Hypothesis Testing	6%	59%	53%	18.0	< 0.01
Overall	**35%**	**71%**	**36%**	**6.8**	**< 0.01**

While quantitative methods could be used to measure content knowledge, it was necessary to use qualitative methods to better understand teachers' *statistical reasoning* about comparing two groups. During the qualitative analysis, 20 initial categories were organized into four final categories—conjectures, context, variation, and conclusions—by collapsing and generalizing other categories. Finally, the researchers developed a preliminary framework for examining statistical reasoning (Makar & Confrey, 2002). This chapter focuses on elements that are specific to teachers' reasoning about comparing two distributions. Of special interest are teachers' conceptions of variation, which bring with them several issues that are unique to the task of comparing groups. In examining these teachers' descriptions of the two distributions, it will be interesting to note how they choose to compare the characteristics of the each distribution. For example, do they see these measures as absolute, or do they recognize the possibility of inherent error? That is, do they view any small differences in these measures quantitatively as absolute differences, or do they indicate a tolerance for variation in these measures, so that if these students were tested again, they might expect to see slightly different results?

EXAMPLES OF TEACHERS' REASONING ABOUT COMPARING DISTRIBUTIONS

In this section, we discuss the data from interviews with the four subjects from the second phase of the study: Larry, Leesa, Natalie, and Toby. These four teachers were the only four participating in Phase II of the study, the 2-week summer institute.

The first transcript we examine is Larry's, who has taught middle school math for 6 years. He has an undergraduate major in mathematics and is certified to teach mathematics at the middle and high school levels. Larry's initial portrayal of his comparison of the two distributions began with a visual evaluation of the similarity of their dispersion, then a numerical description of the means and standard deviation of each of the two distributions. He finished with a comparison of these measures:

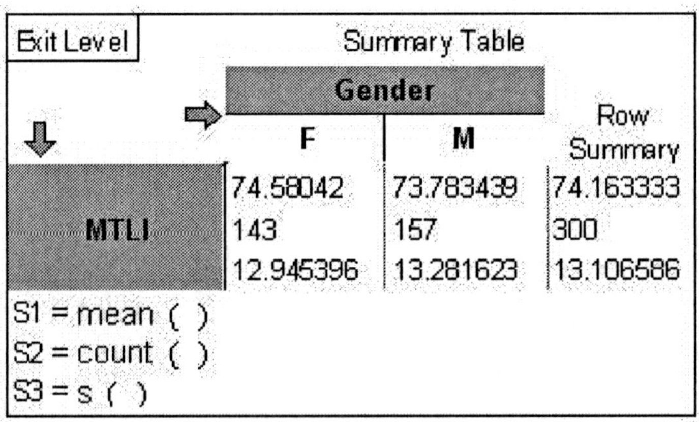

Figure 2. Larry's summary table in Fathom. The three rows correspond to the values of the mean, count, and standard deviation for the females, males, and then total group.

Larry: I'm just first dropping them, both of them in a graph (Figure 1), the math scores of the males and females. Um, both of them seem to be fairly equally distributed, maybe. I'm going to try and find the means of each one (mumbles). I'll just graph them, then. Hmm. So they're fairly close ... I'm pulling down a summary table (Figure 2) so I can actually find a number for each one. The, uh, so I actually get a real number. So it's giving me the count and mean of each one. Also, here I can find out, uh (mumbles), I can find the standard deviation of each one to see how close they are from the mean.

KM: And how, how will that help you?

Larry: Well, if, even if I didn't see the graph, I can look at the females are even a little tighter, around a higher mean.

KM: OK.

Larry: On both sides. As opposed to the men, also—that are a little more spread around, around a lower average.

Larry later considered the difference of the means more directly, estimating the difference from the figure:

Larry: Even though they're going to be very close, I, I think, I, I mean, there's not a great difference between the men and the women. But the women look like they scored maybe one or two points higher.

Larry here acknowledged that the difference between the means was very close, but did not interpret the difference as anything other than a 1- or 2-point difference. At the end of the first part of the interview, Larry informally compared the extreme values of the two distributions, as well as their means and proportion passing, to summarize his analysis:

KM: Just describe for me, if you were going to compare those two groups, the performances of those two groups. Describe the similarities and differences.

Larry: OK. The females have a larger range, because the lowest score and the highest score are—the lowest score of the females is lower than the lowest score of the males, and the highest score of the females is higher than the highest score of the males. Uh, so the, the range is higher. Yet, still the, the mean score is higher than the average score of each of—, the females is higher than the average score of the males.

Larry's comparisons consisted of independent descriptions of each distribution along with direct comparisons of center and dispersion. While he considered the variability of each distribution, he did not indicate a sense of the variation between the measures of the two distributions—that is, he compared the means and dispersions of the two distributions qualitatively or in absolute terms. He concluded that the mean of the females was higher than that of the males by observing an estimated 1- or 2-point difference in the means. While he asserted that these were close, Larry indicated no particular inclination to investigate whether the difference in the means was significant.

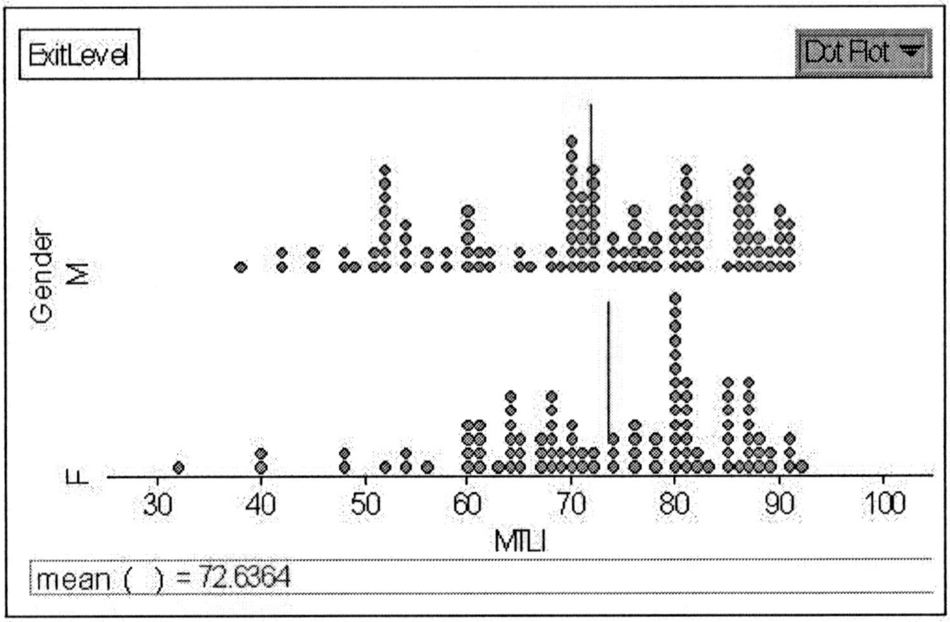

Figure 3. Leesa's initial dot plot of Gender vs. MTLI (TAAS math scores).

Leesa has taught middle school mathematics for 7 years, has undergraduate majors in the social sciences, and is the chair of her mathematics department. Her initial description included comparisons of the shape, range, maximums, and means of each distribution (Figure 3; note that Leesa's data set, and those of the other two teachers who follow, are slightly different than Larry's):

Leesa: OK, um, let's see. This looks skewed to the left [pointing to the top distribution]. Well, they both look skewed to the left. Uh, the range of the males looks like it goes from about nine[ty]—well, they're about the same. There's a bigger range in the female performance because of this, this one student right here who has got a 32.

KM: OK.

Leesa: Um. A high score is a 92 on female and on male it's a 91. Um, and then I can also, I can also go and find the mean. And then, [pause] the edit formula and plot the mean on both of those [Leesa selects the graph, chooses "edit formula" from the context menu, and plots the means on the graph in Fathom]. So for the females it looks like their line is about 72.6, no, 73 [Leesa moves the cursor close to the mean line for the females and reads the location of the cursor in the corner of the Fathom screen]. And then for the males, it looks like about 72.

KM: OK.

Leesa: So the average female score is just a little bit higher than, than the average male.

Leesa seemed to view the measures she stated as descriptions of each distribution separately, although she made some comparison in these measures, indicating some tolerance for variation in her qualitative description of the range of the distributions as "about the same." She did not hold on to this view, however, when she moved from qualitatively to quantitatively comparing the distributions. For example, while she found that the mean for females was higher than for males, she did not indicate whether she interpreted this 1-point difference as their centers being about the same. In the interview, Leesa went on to compare the proportion of each group that passed [63% of the females passed compared to 68% of the males]. She noted that alternative measures were giving her seemingly conflicting information about the relative performance of the two groups, stating, "More boys passed than girls when you look at percentages, but, and the mean score of the girls is higher."

When asked to sum up whether she saw any difference between the performance of males and females at the school, Leesa considered using a sampling distribution this time to provide evidence that the difference between the two groups was not significant. However, her attempt to do so included a laundry list of methods we had used in the summer institute to examine variability, including a reference to "what we did this morning," which was a procedure in Fathom called *stacking* that probably would not have been useful in this situation:

KM: So can you, you say whether one performed better than the other?
Leesa: No.
KM: What evidence could you give me to, that there wasn't any difference, what would you say?
Leesa: Um, I can do that test hypothesis thing. Um, I could do one of those, um, like what we did this morning, the sample and see if there was any—How many students were there? 231?
KM: Uh-huh.
Leesa: I could do a smaller sample and just kind of test and see if, see what the means look like each time ... OK, then when you do standard deviation—is that really going to help me here? Because, let's plot it and see what it looks like [Leesa plots a marker on her graph in Fathom, one standard deviation above the means].
KM: OK, why do you think that might give you something, or are you just going to see—
Leesa: Um. I just want to see if this, if this mean, if this mean [pointing to the females in Figure 3]—
KM: Uh-huh?
Leesa: —falls within one standard deviation of the top mean [the males].
KM: Do you think it will?
Leesa: Yes. (pause) So it's not like it's a huge difference, I guess.
KM: So what does checking where the standard deviation, what does that tell you? What does that measure? Try and think out loud.
Leesa: Um, OK. Standard deviation means that most of the scores are going to fall within there [the interval within one standard deviation of the mean]. So, I don't really see how that—OK, I understand what we were doing yesterday when we had the standard deviation and then, you know, when we had, uh,

when we looked to see if that would, if that was really weird. And if it fell outside the standard deviations, when we looked at z-scores and they were really high, if it fell way out here, then we know that was something, not typical.

KM: OK.

Leesa: OK, but since this, these are so close together, and it falls within, you know, that that's pretty typical and, it might go either way.

Unlike Larry, Leesa indicated a tolerance for variation between the measures she used to compare the two groups. Even though the means of the groups *were* different, she acknowledged that the difference was not enough for her to decide whether one group performed better than the other. She struggled, however, with providing evidence that the difference was not meaningful. Her explanation contained a hybrid of concepts related to the distribution of scores and that of a sampling distribution, together with a list of procedures she might try.

Natalie, a preservice teacher and mathematics major with no previous statistical coursework, immediately took a less deterministic stance in her comparison of the performances of males and females on the TAAS test at the hypothetical school. Natalie initially created a dot plot of the data (similar to the one in Figure 3), then changed it to a histogram. She then created a summary table in Fathom to calculate the means and standard deviations of the MTLI score for each gender:

Natalie: It looks like the mean for the females is a couple of points higher than the mean for the males [pointing to the summary table], but whether or not that's significant, I don't know yet ... I don't think they're very different. It just happens to come up a little bit higher, but the standard deviation is 13 points, so 2-point difference isn't all that much ... The, the range looks about the same to me, I mean, there's a few extra down there in the females, but I don't think that's very significant. They look pretty similar ... I don't think they're, they're very different.

Natalie immediately considered whether the difference she was seeing in the means was significant and went on to conclude that the 2-point difference in the means of the two groups was probably not significant, relative to the distribution of scores. She compared the 2-point difference in means to the standard deviation rather than to their standard error, since she did not consider the size of the group in her interpretation of significance. It's possible that she was considering not *statistical significance,* but a more informal notion of a *meaningful difference* relative to the distribution of scores.

The final interview was with Toby, an experienced high school teacher who has been teaching for over 10 years. Toby's initial comparison between the two groups (creating a graph similar to Figure 3) was based on a visual interpretation, before considering a numerical comparison:

KM: Describe to me what you see, compare those two groups.

Toby: Well, just by looking at that I would say that the, the men scored better than the women. Um, then I would probably drop, um, means in there. Um, probably get an idea of what that was. Uh, 74, closer to 74, and that was 72. Not, not that much difference. They're about the same.

KM: The same?

Toby: Yes.

KM: And you're basing that on?

Toby: Uh, that the means are pretty close together and that, there's about, uh, there, there are no real outliers ... The females averaged higher, um, there's one kind of low one out there, but there's not that much, they're a pretty close group, pretty closely grouped. If we had to go farther, we might, now I don't know how big this set is but I used all of the data, so.

KM: So if somebody said, you know, is there any difference between these two groups?

Toby: Well, to get, well, we could do those things like what we've been doing. Uh. How, how many is this? Uh, only 230. Well, uh. And they're all there. We can do one of those things about, you know, pick 50 of them at a time, find that average, pick 50 at a time, find that average, pick 50 at a time, and then look at that, uh, the average of those.

KM: Uh-huh.

Toby: OK. And, uh, that's going to tend to squish the data together, and, towards whatever the real mean of that data is, but it would also give me a, uh, idea of, of the spread or the vari—how, how the highs and lows were.

KM: OK.

Toby: Of that spread.

Toby also interpreted the difference that he found in the means as being "about the same," indicating he, too, possessed an expectation of variation between the measures of the two groups. Toby also recognized that a sampling distribution of some kind would help support his assertion that the difference between the two groups was not significant, but he had similar difficulties determining how to set up a sampling distribution or how to incorporate the sizes of the groups.

DISCUSSION

In examining teachers' reasoning about comparing distributions, we found that teachers were generally comfortable working with and examining traditional descriptive statistical measures as a means of informal comparison. An interesting contrast occurs, however, when we consider teachers' conceptions of *variability* when reasoning about comparing two distributions. As indicated in the literature, variability is an under-researched area of statistical thinking (Meletiou, 2000). Yet attitude toward variability could provide an important indication of statistical mind-set (Wild & Pfannkuch, 1999). Having an understanding and tolerance of variability encompasses a broad range of ideas. In examining the concept of variability with only one distribution, one considers the variation of values *within* that distribution. However, descriptive statistics for a single distribution are often viewed without regard to variability of the statistical measures themselves. With one distribution, there is little motivation to consider or investigate possible sources of variation in the measures drawn. Comparing distributions creates a situation where one is pushed to consider measures less deterministically. Depending on the measure that dominates the comparison (often a mean), how does one interpret differences found in measures *between* groups? That is, how does one determine whether any difference between the dominant measures is meaningful or significant? Further, how do teachers manage the distinction between these two kinds of variation? By considering variation between distributions, we are encouraged to consider sources of variation in these measures. In this chapter, we discuss three different ways that teachers considered issues of variability when reasoning about comparing two distributions: (1) how teachers interpreted variation *within a group*—the variability of data; (2) how teachers interpreted variation *between groups*—the variability of measures; and (3) how teachers *distinguished* between these two types of variation.

In the interviews, all four teachers knew that scores within each distribution of scores would possess variability—that is, they did not expect the data in the distribution of scores would all have the same value. Teachers' conceptions of this *within-group variation* were heard in their descriptions of shape, distribution, outliers, standard deviation, range, "domain" (maximum and minimum values), and "whiskers" on a box plot (not included in the preceding excerpts, but used by two of the teachers). Additional qualitative descriptions included statements about a distribution being "tighter" or "more spread out." Commonly, teachers calculated the standard deviation of each set almost immediately and somewhat automatically.

While all of the teachers clearly recognized variation *within* a single distribution, they articulated a variety of meanings about variation *between* two distributions. From our interaction with them in the workshops, we anticipated they would demonstrate their view of between-group variation by acting in one of four ways: (a) by calculating descriptive statistics for each group without making any comparisons; (b) by comparing descriptive statistics (e.g., indicating a difference in magnitude or that one was greater then the other); (c) by first comparing the descriptive measures of the two distributions as described earlier, then indicating whether they considered the difference to be meaningful by relying on informal techniques or intuition; or (d) by investigating whether the differences they found in the measures to be statistically significant using a formal test, such as the randomization test the teachers carried out during the *Orbital Express* activity (Erickson, 2001, p. 276) using the *scramble attribute* feature in Fathom, which randomizes one attribute of the data.

In addition to describing the variation *within* each distribution separately, the teachers typically reported some aspect of the similarity or differences in the measure of dispersion between the two distributions, by comparing range or standard deviation. They may also have compared shapes or means, for example, by noting that the mean of the females' scores was 2 points higher than that of the males. In some cases, teachers indicated an intuition about variation between measures, but struggled to quantify the evidence for their observations. One reason for our perception that teachers had difficulty in quantifying variation between distributions may be that the participants felt they were being pushed to provide evidence of what seemed to them to be an obvious example of two distributions that were "about the same." Perhaps to the teachers, the sameness could be seen visually, and they would not feel compelled to provide evidence of this observation under less test-like circumstances.

Two of the teachers, Leesa and Natalie, attempted to formally test whether the difference in the means of the two distributions was significant using some form of a standard deviation taken from the data distributions. Furthermore, Toby, as well as Leesa, checked the size of the population to see if it was "large enough" to draw samples from, perhaps recalling that several times during the workshop they had created sampling distributions by drawing random samples from a state data set of 10,000 student test scores. Neither of them, however, used the size of the data set in determining whether the difference in means between the males and females was significant. Overall, the three who considered using a sampling distribution struggled to understand the circumstances under which using one would be helpful nor were they able to separate the variability in the distributions of the data sets from that of the related sampling distribution, confirming that this is a very difficult concept to understand in statistics, consistent with the findings of delMas, Garfield, and Chance (1999).

Using Confrey's (1991, 1998) concept of *voice and perspective,* the authors brought to the research their own *perspective* of statistical reasoning surrounding the task of comparing distributions. By listening to teacher *voice* we were able to gain further insight into our own understanding of variation as we worked to understand the teachers' reasoning. Although the literature clearly points to sampling distributions as a stumbling point for students in inferential statistics, we had thought that abundant experience with simulations involving sampling distributions within meaningful problems that would demonstrate their power would be sufficient to help teachers overcome this difficulty. In fact, the conflicts teachers had in using sampling distributions may have been compounded by the way in which sampling distributions and simulations were introduced together without providing sufficiently motivating tasks for teachers to create a need for them. We learned that a wealth of experience with sampling distributions to solve interesting problems was not sufficient for their understanding. We believe, given our analysis of teachers' reasoning in this area, that sampling distribution concepts need to be developed more slowly, allowing teachers to conceptually construct the notion of a sampling distribution rather than have it presented as part of a "good way" to solve the problem at hand.

Comparing distributions raises another important issue about variation—which variation are we referring to when we compare two distributions? With a single distribution, discussions of variation are meant to describe variation *within* the distribution at hand. Having two distributions to compare provides a motivation to compare variation *between* the distributions. For example, if we observe that the performance of males and females on a test differs by 2 points, what does this 2-point difference tell us? Could this difference just be due to random variation, or could it indicate a more meaningful underlying phenomenon? When comparing groups and considering variation between distributions, it is important to consider whether the data being compared is that of a *sample* or a *population.* Traditional introductory instruction in significance testing often uses sampling distributions as a way to generalize our findings from a sample to some larger, unknown population.

Whether data should be considered as a population or a sample is somewhat problematic in the context of a school's student assessment data and indicates that these distinctions are not always clear-cut (Chance, 2002). On one hand, it makes sense to consider a set of student test data from a school as its own population. When comparing two groups, however, sampling distributions can inform us as to whether the difference between groups is *meaningful,* hence pushing us to consider measures beyond descriptive statistics. Simulations can be used to support a broader, inference-like view of a difference even though we are not necessarily trying to generalize to a larger population. In this case, we can investigate the difference in means between male and female performance through the use of a randomization test. That is, under the null hypothesis that there is no difference between the performance of males and females on a test, if we were to randomize the students' genders and then compare the means of the two groups, how likely is a difference of 2 points to occur between males and females *just by chance*? On the other hand, we might want to conceptualize the two groups as samples in a larger population of all students who pass through a school over many years to make inferences about the school itself, even though the samples are not randomly selected, assuming one is willing to accept these as representative samples.

In working with teachers, we found that capturing and influencing teachers' statistical reasoning is much more complex than trying to understand and describe students' reasoning. Firstly, students are expected to be learners, but teachers consider themselves experts. Therefore, it is very difficult for most experienced teachers to admit what they do not know and be open to learning and discussing their reasoning. Fortunately, statistics is a content area in which few teachers are expected to have knowledge, making it a viable entrance for teachers to reexperience being learners. Secondly, unless experienced teachers are enrolled in a masters program, they are usually not an easily accessible group for the kind of long-term study that can affect teachers' thinking. The study described here began with an agreement between a school principal and our research group to commit the entire mathematics department of seven teachers to the research project, including a 2-week summer institute. By the end of the study however, only the two strongest of the seven original teachers remained. This raises both an important question and limitation of the study. First, how one can engage experienced secondary teachers in research that hopes to both influence and study teacher learning and practice? Second, the four teachers in the study likely had higher mathematical content knowledge than might be considered typical. In addition, they were very committed to improving their own practice, were highly engaged during activities and discussion, and were more open than most to consider weaknesses in their own understanding.

Comparing two groups provides a rich context in which to build statistical reasoning. At a very early age in school, group comparisons can provide an impetus to collect data and later, to view data as a distribution. At an advanced level, an interesting problem involving comparing distributions can stimulate learners to consider not only measures of dispersion within each group, but comparisons of measures between groups, and hence to consider variation within the measures themselves. Just as algebra and calculus are considered to be gatekeepers to higher mathematics, understanding sampling distributions may be a gatekeeper to advanced statistical reasoning. However, simply presenting sampling distributions as a precursor to hypothesis testing may aggravate the difficulty learners have with its underlying concepts.

Further work is needed in better understanding reasoning about sampling distributions as well as ways to think about facilitating learners' conceptual development of variation within a distribution with an eye toward developing a tolerance and expectation for variation in statistical measures. Understanding sampling distributions is by no means a cure for the difficulty of understanding variation of any sort, or toward loosening a deterministic view of statistics and data analysis. It is the authors' hope, however, that better understanding of teachers' reasoning about comparing groups will open further discussion of building an intuition of variation in data and statistics for teachers as well as students.

IMPLICATIONS

We ascertained that comparing distributions holds great potential for encouraging learners to broaden their view of statistics and data. As researchers, we found comparing distributions to be a fruitful arena for expanding teachers' understanding of distribution and conceptions of variability as well as a motivating reason to introduce sampling distributions. However, we found it important to specify which kind of variation we are discussing when comparing two distributions. Teachers' reasoning about variation in the context of group comparisons was examined in three areas: variation *within* a distribution, variation *between* groups (variation of measures), and the struggle to interpret the difference between these two types of variation. The importance of making this distinction surprised us, and motivated us to consider both our own understanding and the way in which we planned our conjectured learning trajectory. This study implies that sources of variation in both data and in measures need to be discussed frequently when working with data, and again as measures are compared between distributions, to engender a tolerance for variation both within and between distributions.

At a more advanced level of statistical content, our study supports the findings of delMas et al. (1999) about the difficulty in understanding sampling distributions and implies that the teaching of sampling distributions needs to be done more carefully. Furthermore, traditional teaching of hypothesis and significance testing and the overreliance on computer simulations may actually promote misconceptions rather than advance understanding of sampling distributions. In addition, discussion about the distinctions and ambiguities between considering data as a sample or a population need to occur in the teaching of significance testing and among the research community.

H. G. Wells predicted decades ago that "Statistical thinking will one day be as necessary for efficient citizenship as the ability to read and write" (quoted in Snee, 1990, p. 117). If our goal is to promote statistical reasoning in our students, we must better understand and engender the statistical thinking and reasoning of teachers.

Snee (1990) highlights in his definition of statistical thinking in the quality control industry the importance of a recognition that "variation is all around us, present in everything we do" (p. 118). The concept of variation needs to be engendered early and continuously when teaching statistical reasoning. The teaching of statistics throughout schooling, with an emphasis on distribution and variation, may provide a way to loosen the deterministic stance of teachers, students, and the public toward data and statistics. More research is needed in this area.

REFERENCES

Abelson, R. (1995). *Statistics as principled argument.* Hillsdale, NJ: Erlbaum.

Chance, B. L. (2002). Personal communication (email: April 11, 2002).

Cobb, P. (1999). Individual and collective mathematical development. The case of statistical data analysis. *Mathematical Thinking and Learning, 1*(1), 5–43.

Cohen, D. K., & Ball, D. L. (1990). Relations between policy and practice: A commentary. *Educational Evaluation and Policy Analysis, 12*(3), 249–256.

Confrey, J. (1991). Learning to listen: A student's understanding of powers of ten. In E. von Glasersfeld (Ed.), *Radical constructivism in mathematics education* (pp. 111–138). Dordrecht, The Netherlands: Kluwer Academic Publishers.

Confrey, J. (1998). Voice and perspective: Hearing epistemological innovation in students' words. In M. Larochelle & N. Bednarz & J. Garrison (Eds.), *Constructivism and education* (pp. 104–120). New York: Cambridge University Press.

Confrey, J. (in preparation). *Systemic crossfire.* Unpublished manuscript.

Confrey, J., & Makar, K. (2002). *Developing secondary teachers' statistical inquiry through immersion in high-stakes accountability data.* Paper presented at the Twenty-fourth Annual Meeting of the North American Chapter of the International Group for the Psychology of Mathematics Education (PME-NA), Athens, GA.

delMas, R. C., Garfield, J., & Chance, B. L. (1999). A model of classroom research in action: Developing simulation activities to improve students' statistical reasoning. *Journal of Statistics Education, 7*(3).

Erickson, T. (2001). *Data in depth: Exploring mathematics with Fathom.* Emeryville, CA: Key Curriculum Press.

Finzer, W. (2000). Fathom (Version 1.1). Emeryville, CA: Key Curriculum Press.

Friel, S., Curcio, F., & Bright, G. (2001). Making sense of graphs: Critical factors influencing comprehension and instructional implications. *Journal for Research in Mathematics Education, 32*(2), 124–158.

Konold, C., & Higgins, T. (2002). Highlights of related research. In S. J. Russell, D. Schifter, & V. Bastable (Eds.), *Developing mathematical ideas: Working with data*, (pp. 165-201). Parsippany, NJ: Seymour Publications.

Konold, C., & Pollatsek, A. (2002). Data analysis as the search for signals in noisy processes. *Journal for Research in Mathematics Education, 33*(4), 259–289.

Lappan, G., Fey, J. T., Fitzgerald, W. M., Friel, S. N., & Phillips, E. D. (1998). *Connected Mathematics: Data about us.* White Plains, NY: Seymour.

Lehrer, R., & Schauble, L. (2000). Modeling in mathematics and science. In R. Glaser (Ed.), *Advances in instructional psychology: Educational design and cognitive science* (Vol. 5, pp. 101–159). Mahwah, NJ: Erlbaum.

Lieberman, A., & Wood, D. R. (2003). *Inside the National Writing Project: Connecting network learning and classroom teaching.* New York: Teachers College Press.

Makar, K., & Confrey, J. (2002). *Comparing two distributions: Investigating secondary teachers' statistical thinking.* Paper presented at the Sixth International Conference on Teaching Statistics (ICOTS-6), Cape Town, South Africa.

Meletiou, M. (2000). *Developing students' conceptions of variation: An untapped well in statistical reasoning.* Unpublished dissertation, University of Texas, Austin.

National Writing Project. (2002, April). *National Writing Project Mission.* Author. Retrieved April 28, 2002, from www.writingproject.org

QSR. (1999). NVivo (Version 1.1). Melbourne, Australia: Qualitative Solutions and Research Pty. Ltd.

Snee, R. (1990). Statistical thinking and its contribution to total quality. *The American Statistician, 44*(2), 116–121.

Stake, R. E. (1994). Case studies. In N. K. Denzin & Y. S. Lincoln (Eds.), *Handbook of qualitative research*. Thousand Oaks, CA: Sage.

Strauss, A., & Corbin, J. (1998). *Basics of qualitative research: Techniques and procedures for developing grounded theory*. Thousand Oaks, CA: Sage.

TERC. (1998). *Investigations in number, data, and space.* White Plains, NY: Seymour.

Watson, J., & Moritz, J. (1999). The beginning of statistical inference: Comparing two data sets. *Educational Studies in Mathematics, 37,* 145–168.

Wild, C., & Pfannkuch, M. (1999). Statistical thinking in empirical enquiry. *International Statistical Review, 67*(3), 223–265.

Chapter 16

PRINCIPLES OF INSTRUCTIONAL DESIGN FOR SUPPORTING THE DEVELOPMENT OF STUDENTS' STATISTICAL REASONING

Paul Cobb and Kay McClain
Vanderbilt University, USA

OVERVIEW

This chapter proposes design principles for developing statistical reasoning in elementary school. In doing so, we will draw on a classroom design experiment that we conducted several years ago in the United States with 12-year-old students that focused on the analysis of univariate data. Experiments of this type involve tightly integrated cycles of instructional design and the analysis of students' learning that feeds back to inform the revision of the design. However, before giving an overview of the experiment and discussing specific principles for supporting students' development of statistical reasoning, we need to clarify that we take a relatively broad view of statistics. The approach that we followed in the classroom design experiment is consistent with G. Cobb and Moore's (1997) argument that data analysis comprises three main aspects: data generation, exploratory data analysis (EDA), and statistical inference. Although Cobb and Moore are primarily concerned with the teaching and learning of statistics at the college level, we contend that the major aspects of their argument also apply to the middle and high school levels.

EDA involves the investigation of the specific data at hand (Shaughnessey, Garfield, & Greer, 1996). Cobb and Moore (1997) argue that EDA should be the initial focus of statistics instruction since it is concerned with trends and patterns in data sets and does not involve an explicit consideration of sample-population relations. In such an approach, students therefore do not initially need to support their conclusions with probabilistic statements of confidence. Instead, conclusions are informal and are based on meaningful patterns identified in specific data sets.

Cobb and Moore's (1997) proposal reflects their contention that EDA is a necessary precursor to statistical inference. Statistical inference is probabilistic in that the intent is to assess the likelihood that patterns identified in a sample are not

D. Ben-Zvi and J. Garfield (eds.),
The Challenge of Developing Statistical Literacy, Reasoning and Thinking, 375–395.

specific to that batch of data, but indicate trends in the larger population from which the data were generated. As Cobb and Moore indicate, the key idea underpinning statistical inference—that of sampling distribution—is relatively challenging even at the college level.

Cobb and Moore also argue that students are not in a position to appreciate the relevance of the third aspect of statistics, a carefully designed data generation process, until they become familiar with data analysis. They observe that "if you teach design before data analysis, it is harder for students to understand why design matters" (Cobb & Moore, 1997, p. 816). However, although they recommend introducing design after EDA, they also recognize that the crucial understandings that students should develop center on the relationship between the legitimacy of conclusions drawn from the data and the soundness of the process by which the data were generated.

Cobb and Moore's (1997) observations provide an initial framework for instructional design within which to develop specific design principles. In the design experiment that we conducted with the 12-year-old students, we focused primarily on EDA and on the process of generating data. We did, however, also explore the possibilities for statistical inference in both this experiment and in a follow-up design experiment we conducted with some of the same students that emphasized the analysis of bivariate data[1]. As Bakker and Gravemeijer (Chapter 7) illustrate, productive instructional activities for supporting the development of the students' reasoning about statistical inference include those in which students describe the characteristics of a data set that they anticipate will be relatively stable if the data generation process is repeated or if the size of the sample is increased. In our view, students' initial intuitions about the relative stability of the shape of both univariate and bivariate data sets constitute a potential starting point for an instructional sequence that culminates with students' development of a relatively deep understanding of the crucial idea of sampling distribution (Cobb, McClain, & Gravemeijer, 2003; Saldanha and Thompson, 2001). We introduce this conjecture to situate our discussion of design principles in a broader instructional context, since our focus in the remainder of this chapter will be on supporting the development of students' reasoning about data in the contexts of EDA and data generation. To ground the proposed design principles, we first give a short overview of the classroom design experiment and then frame it as a paradigm case in which to tease out design principles that address five aspects of the classroom environment that proved critical in supporting the students' statistical learning:

- The focus on central statistical ideas
- The instructional activities
- The classroom activity structure
- The computer-based tools the students used
- The classroom discourse

OVERVIEW OF THE CLASSROOM DESIGN EXPERIMENT

Initial Assessments of the Students' Reasoning

In preparing for the design experiment, we conducted interviews and whole-class performance assessments with a group of seventh graders from the same school in which we planned to work. These assessments indicated that data analysis for most of these students involved "doing something with the numbers" (McGatha, Cobb, & McClain, 1999). In other words, they did not view data as measures of aspects or features of a situation that had been generated in order to understand a phenomenon or make a decision or judgment (e.g., the points that a player scores in a series of basketball games as a measure of her skill at the game). In a very real sense, rather than analyzing data, the students were simply manipulating numbers in a relatively procedural manner. Further, when the students compared two data sets (e.g., the points scored by two basketball players in a series of games), they typically calculated the means without considering whether this would enable them to address the question or issue at hand. For example, in the case of the points scored by the two basketball players, simply calculating the means would not necessarily be a good way to select a player for an important game because it ignores possible differences in the range and variability of the players' scores (i.e., the player with a slightly lower mean could be much more consistent).

In interpreting these findings, we did not view ourselves as documenting an inherent psychological stage in seventh graders' reasoning about data. Instead, we were documenting the consequences of the students' prior instruction in statistics. They had, for example, previously studied measures of center (i.e., mean, mode, and median) as well as several types of statistical graphs (e.g., bar graphs, histograms, and pie charts). Our assessments of the students' reasoning at the beginning of the experiment tell us something about not just the content but also the quality of their prior instruction. The assessments indicate, for example, that classroom activities had emphasized calculational procedures and conventions for drawing graphs rather than the creation and manipulation of graphs to detect trends and patterns in the data. This view of the students' reasoning as situated with respect to prior instruction was useful in that it enabled us to clarify the starting points for the design experiment. For example, we concluded from the assessments that our immediate goal was not one of merely remediating certain competencies and skills. Instead, the challenge was to influence the students' beliefs about what it means to do statistics in school. In doing so, it would be essential that they actually begin to analyze data in order to address a significant question rather than simply manipulate numbers and draw specific types of graphs.

Concluding Assessments of the Students' Reasoning

The students' reasoning in these initial assessments contrasts sharply with the ways in which they analyzed data at the end of the 10-week experiment. As an illustration, in one instructional activity, the students compared two treatment protocols for AIDS patients by analyzing the T-cell counts of people who had enrolled in one of the two protocols. Their task was to assess whether a new experimental protocol in which 46 people had enrolled was more successful in raising T-cell counts than a standard protocol in which 186 people had enrolled. The data the students analyzed is shown in Figure 1 as it was displayed in the second of two computer-based tools that they used. All 29 students in the class concluded from their analyses that the experimental treatment protocol was more effective. Nonetheless, the subsequent whole-class discussion lasted for over an hour and focused on both the adequacy of the reports the students had written for a chief medical officer and the soundness of their arguments.

For example, one group of students had partitioned the two data sets at T-cell counts of 525 by using one of the options on the computer tool as shown in Figure 1. In the course of the discussion, it became clear that their choice of 525 was not arbitrary. Instead, they had observed that what they referred to as the "hill" in the experimental treatment data was above 525, whereas the "hill" in the standard treatment data was below 525. It was also apparent from the discussion that both they and the other students who contributed to the discussion reasoned about the display shown in Figure 1 in terms of relative rather than absolute frequencies (i.e., they focused on the *proportion* rather than the number of the patients in each treatment protocol whose T-cell counts were above and below 525). This was indicated by explanations in which students argued that most of the T-cell counts in the experimental treatment were above 525, but most of the T-cell counts in the traditional treatment were below 525.

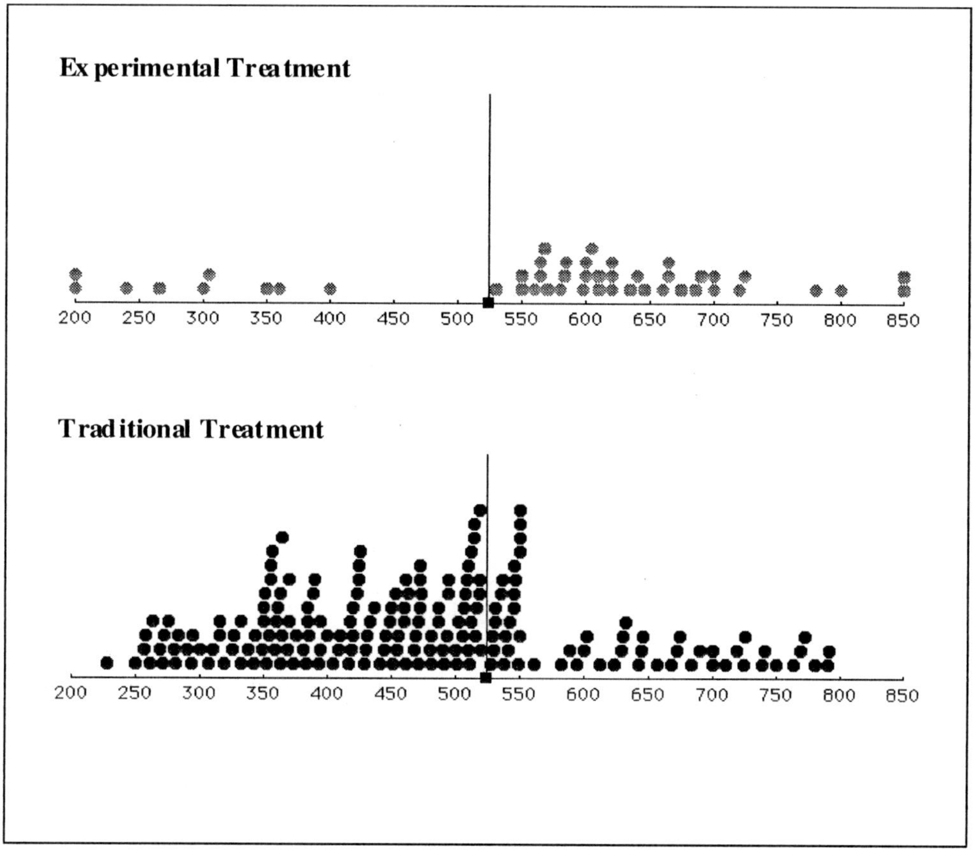

Figure 1. The AIDS protocol data partitioned at T-cell counts of 525.

This analysis was one of the most elementary that the students produced on this instructional activity. As a point of comparison, another group of students had used an option on the computer tool that enabled them to hide the dots that represented the individual data values and had then used another option on the tool to partition the two data sets into four groups, each of which contained one-fourth of the data points (see Figure 2). In this option, 25 percent of the data in each data set are located in each of the four intervals bounded by the vertical bars (similar to a box plot). As one student explained, these graphs show that the experimental treatment is more effective because the T-cell counts of 75 percent of the patients in this treatment were above 550, whereas the T-cell counts of only 25 percent of the patients who had enrolled in the standard treatment were above 550. This student's argument was representative in that he, like the other students who contributed to the discussion, was actually reasoning about data rather than attempting to recall procedures for manipulating numerical values.

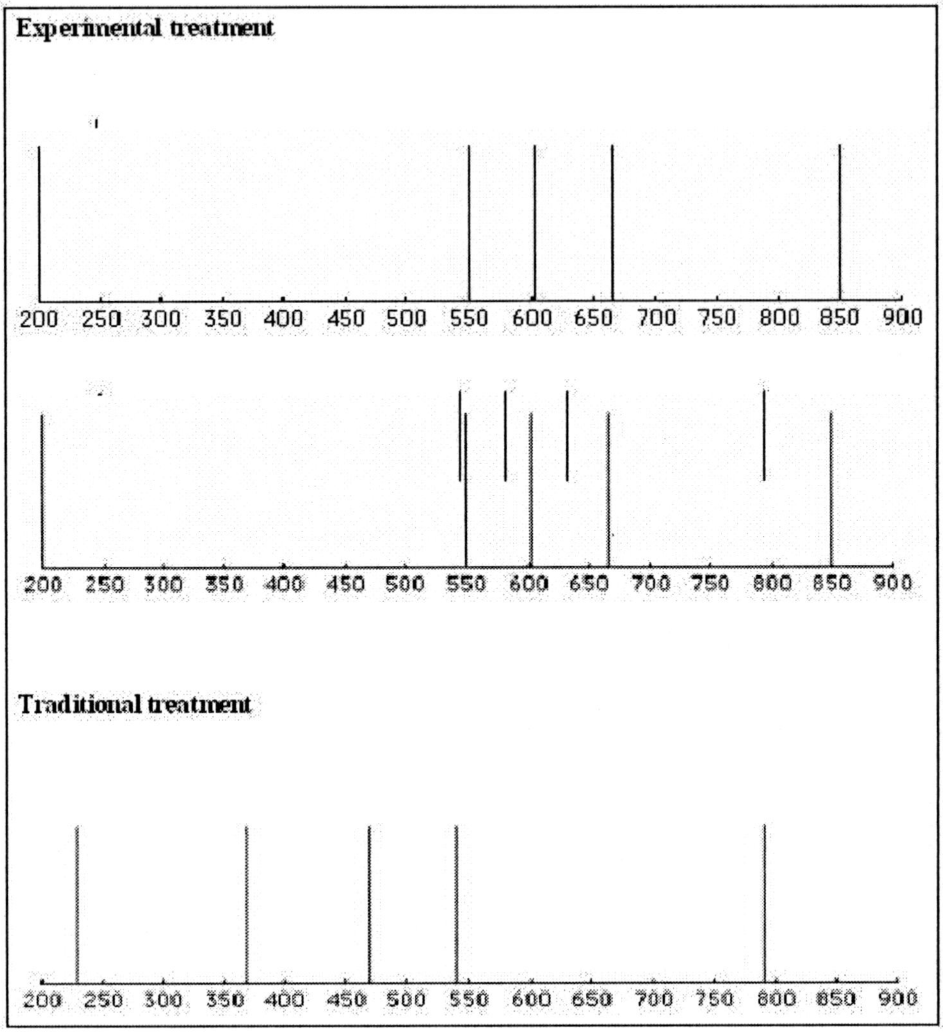

Figure 2. The AIDS protocol data organized into four equal groups with the individual data points hidden.

We have described the design experiment in some detail elsewhere and have documented the major phases in the development of the students' reasoning about data (Cobb, 1999; McClain, Cobb, & Gravemeijer, 2000). In addition, Bakker and Gravemeijer (Chapter 7) report on a series of four classroom design experiments that they conducted in the Netherlands in which students used the same two computer tools. For our present purposes, it therefore suffices to note that our classroom observations were corroborated by individual interviews that we conducted to document the students' reasoning at the end of the experiment. The analysis of these interviews indicates that a significant majority of the students could readily interpret graphs of two *unequal* data sets organized either into equal interval widths (an analogue of histograms) or into four equal groups (an analogue of box plots) in terms of patterns in how the data were distributed. In this regard, Konold, Pollatsek, Well, & Gagnon (1996) argue that a focus on the rate of occurrence (i.e., the proportion) of data within a range of values (e.g., above or below T-cell counts of 525) is at the heart of what they term a statistical perspective. Because discussions in the latter part of the experiment involved a concern for the proportion of data within various ranges of values, the students appeared to be developing this statistical perspective. It is also worth noting that when we began the follow-up design experiment with some of the same students nine months later, there was no regression in their statistical reasoning (Cobb et al., 2003). The students' progress at the beginning of this follow-up experiment was in fact such that they could *all* interpret univariate data sets organized into equal interval widths and into four equal groups in these relatively sophisticated ways within the first three or four class sessions.

This overview gives some indication of how the students' reasoning about data changed during the 10-week experiment. We now turn our attention to the process of that change and the design principles inherent in the means by which it was supported and organized.

CENTRAL STATISTICAL IDEAS

Distribution as an Overarching Statistical Idea

In their discussion of instructional design, Wiggins and McTighe (1998) emphasize the importance of beginning the design process by identifying the "big ideas" that are at the heart of the discipline, that have enduring value beyond the classroom, and that offer potential for engaging students. This design principle is particularly important in the case of elementary statistics instruction given that curricula frequently reduce the domain to a collection of at best loosely related concepts (e.g., mean, mode, median) together with conventions for making various types of graphs. McGatha (2000) documents the actual process by which we prepared for the design experiment. As she describes, our proposal of *distribution* as an overarching statistical idea emerged as we attempted to synthesize the research literature and analyzed the interviews and classroom performance assessments that we conducted as part of our pilot work. One of the primary goals for the design experiment was therefore that the students would come to reason about data sets as entities that are distributed within a space of possible values (Konold et al., 1996; Hancock, Kaput, & Goldsmith, 1992; Konold & Higgins, in press; Wilensky, 1997). Bakker and Gravemeijer (Chapter 7, Table 1) clarify the central, organizing role of distribution, thereby illustrating that notions such as center, spread, skewness, and relative density can then be viewed as ways of characterizing how specific data sets are distributed within this space of values. We would only add to their account that various statistical graphs or inscriptions then become ways of structuring data distributions in order to identify relevant trends or patterns. As an illustration, the students who analyzed the AIDS treatment data by organizing the two data sets into four equal groups used this precursor of the box plot in order to identify patterns that were relevant in determining which of the two treatments was more effective. More generally, in the approach that we took in the design experiment, the students' development of increasingly sophisticated ways of reasoning about data was inextricably bound up with their development of increasingly sophisticated ways of inscribing data (Biehler, 1993; de Lange, van Reeuwijk, Burrill, & Romberg, 1993; Lehrer & Romberg, 1996).

Bivariate Data Sets as Distributions

We can illustrate the importance of explicating central statistical ideas as a basic design principle by extending this focus on distribution to the analysis of bivariate data. Because statistical covariation involves coordinating the variation of two sets of measures, the characteristics of directionality and strength are sometimes viewed as being relatively transparent in two-dimensional inscriptions such as scatter plots. However, a focus on the way that bivariate data are distributed reveals that proficient statistical analysts' imagery of covariation is no more two-dimensional than their imagery of univariate distributions is one-dimensional. This is clearer in the case of univariate data in that inscriptions such as line plots involve, for the proficient user, a second dimension that indicates relative frequency. In the case of bivariate data, however, scatter plots do not provide such direct perceptual support for a third dimension corresponding to relative frequency. Instead, it appears that proficient analysts read this third dimension from the relative density of the data points[2]. This analysis of the types of reasoning that are involved in viewing bivariate data sets as distributions serves to clarify both the overall instructional goal and the primary challenge facing the instructional designer, that of enabling students to read this implicit third dimension into two-dimensional inscriptions such as scatter plots and thus to see the distributional shape of the data.

It should be clear from the illustrations we have given as well as from Bakker and Gravemeijer's (Chapter 7) discussion of their design experiments that a focus on overarching ideas can lead to a far-reaching reconceptualization of the statistics curriculum. This design principle therefore contrasts sharply with research that focuses on standard topics in current curricula in isolation. The benefit of adhering to the principle of identifying central statistical ideas is that it contributes to the development of relatively coherent instructional designs. The development of the students' statistical reasoning in the design experiment that we conducted can in fact be viewed as the first phase of a long-term learning trajectory that extends to the university level and encompasses the key idea of sampling distribution.

INSTRUCTIONAL ACTIVITIES

The Investigative Spirit of Data Analysis

As we have indicated, our primary focus in the design experiment was on exploratory data analysis and the process of generating data rather than on statistical inference. We found Biehler and Steinbring's (1991) characterization of EDA as detective work particularly helpful in that it emphasizes that the purpose is to search for evidence. In contrast, statistical inference plays the role of the jury that decides whether this evidence is sufficient to make claims about the population from which the data were drawn. Biehler and Steinbring's metaphor of detective makes it clear that an exploratory or investigative orientation is central to data analysis and constitutes an important instructional goal in its own right. From this, we concluded as a basic design principle for elementary statistics instruction that students' activity in the classroom should involve the investigative spirit of data analysis from the outset. This in turn implied that the instructional activities should all involve analyzing data sets that students view as realistic for a purpose that they consider legitimate.

The instructional activities that we developed in the course of the design experiment involved either (a) analyzing a single data set in order to understand a phenomenon, or (b) comparing two data sets in order to make a decision or judgment. The example of the AIDS treatment activity illustrates the second of the two types of instructional activities. In describing this activity, we also noted that the students were required to write a report of their analyses for a chief medical officer. This requirement supported the students' engagement in what might be termed genuine data analysis by orienting them to take account of a specific audience to either understand a phenomenon or to make a decision based on their analyses. In this regard, we note that data are typically analyzed with a particular audience in mind almost everywhere except in school (cf. Noss, Pozzi, & Hoyles, 1999).

Focusing on Significant Statistical Ideas

In addition to ensuring that the students' activity was imbued with the investigative spirit of data analysis, we also had to make certain that significant statistical ideas emerged as the focus of conversations during whole-class discussions of the students' analyses (cf. Hancock, Kaput, & Goldsmith, 1992). The challenge for us as instructional designers was therefore to transcend what Dewey (1981) termed the *dichotomy between process and content* by systematically supporting the emergence of key statistical ideas while simultaneously ensuring that the analyses the students conducted involved an investigative orientation. This is a nontrivial issue in that inquiry-based instructional approaches have sometimes been criticized for emphasizing the process of inquiry at the expense of substantive disciplinary ideas.

In approaching this challenge, we viewed the various data-based arguments that the students produced as they completed the instructional activities as a primary resource on which the teacher could draw to initiate and guide whole-class discussions that focused on significant statistical ideas. As a basic instructional design principle, our goal when developing specific instructional activities was therefore to ensure that the students' analyses constituted such a resource for the teacher. This would enable the teacher to initiate and guide the direction of whole-class discussions that furthered her instructional agenda by capitalizing on the diverse ways in which the students had organized and interpreted the data sets. In the case of the AIDS instructional activity, for example, the issues that emerged as explicit topics of conversation during the subsequent whole-class discussion included the contrast between absolute and relative frequency, the interpretation of data organized into four equal groups, and the use of percentages to quantify the proportion of the data located in particular intervals (Cobb, 1999; McClain et al., 2000).

The enactment of this design principle required extremely detailed instructional planning, in the course of which we attempted to anticipate the range of data-based arguments the students might produce as they completed specific instructional activities. Our discussions of seemingly inconsequential features of task scenarios and of the particular characteristics of data sets were therefore quite lengthy since minor modifications to an instructional activity could significantly influence the types of analyses the students would produce and thus the resources on which the teacher could draw to further her instructional agenda.

As an illustration, we purposefully constructed data sets with a significantly different number of data points when we developed the AIDS activity, so that the contrast between absolute and relative frequency might become explicit. This in turn required a task scenario in which the inequality in the size of the data sets would seem reasonable to the students and in which they would view the issue under investigation to be worthy of their engagement. Although the AIDS activity proved to be productive, on several occasions our conjectures about either the level of the students' engagement in an activity or the types of analyses they would produce turned out to be ill founded. In these situations, our immediate task was to analyze the classroom session in order to understand why the instructional activity had proven to be inadequate and thus revise our conjectures and develop a new instructional activity. In our view, this cyclic process of testing and revising conjectures about the seemingly minor features of instructional activities is essential if we are to develop relatively long-term instructional sequences in which teachers can support students' development of significant statistical ideas by drawing on their inquiry-oriented reasoning as a primary resource.

THE CLASSROOM ACTIVITY STRUCTURE

Talking through the Data Generation Process

As we have indicated, one of our concerns at the beginning of the design experiment was that the students would view data not merely as numbers, but as measures of an aspect of a situation that were relevant to the question under investigation. To this end, the teacher introduced each instructional activity by talking through the data generation process with the students. These conversations often involved protracted discussions during which the teacher and students together framed the particular phenomenon under investigation (e.g., AIDS), clarified its significance (e.g., the importance of developing more effective treatments), delineated relevant aspects of the situation that should be measured (e.g., T-cell counts), and considered how they might be measured (e.g., taking blood samples). The teacher then introduced the data the students were to analyze as being generated by this process. The resulting structure of classroom activities, which often spanned two or more class sessions, was therefore (a) a whole-class discussion of the data generation process, (b) an individual or small-group activity in which the students usually worked at computers to analyze data, and (c) a whole-class discussion of the students' analyses.

In developing this classroom activity structure, we conjectured that as a result of participating in discussions of the data generation process, data sets would come to have a history for the students such that they reflected the interests and purposes for which they were generated (cf. Latour, 1987; Lehrer & Romberg, 1996; Roth, 1997). This conjecture proved to be well founded. For example, we have clear indications that within a week of the beginning of the design experiment, doing statistics in the project classroom actually involved analyzing data for the students (Cobb, 1999; McClain et al., 2000). In addition, changes in the way that the students contributed to discussions of the data generation process as the design experiment progressed indicate that there was a gradual transfer of responsibility from the teacher to the students.

Initially, the teacher had to take an extremely proactive role. However, later in the experiment the students increasingly initiated shifts in these discussions, in the course of which they raised concerns about sampling processes as well as the control of extraneous variables. We have documented the process by which the students learned about data generation and the means by which that learning was supported elsewhere (Cobb & Tzou, 2000). For our current purposes, it suffices to note that the issues the students raised in the latter part of the experiment indicate that most if not all had come to realize that the legitimacy of the conclusions drawn from data depends crucially on the data generation process.

We should clarify that the teacher did not attempt to teach the students how to generate sound data directly. Instead, she guided the development of a classroom culture in which a premium was placed on the development of data-based arguments. It was against this background that the students gradually became able to anticipate the implications of the data generation process for the conclusions that they would be able to draw from data.

Data Collection and Data Generation

Given that our focus in this chapter is on design principles, it is important to note that design decisions relating to data generation are frequently reduced to the question of whether students should collect the data that they analyze. We decided that the students would for the most part not collect data during the design experiment, for two reasons. First, we had a limited number of classroom sessions available in which to conduct the design experiment; and second, we wanted to ensure that the data sets the students analyzed had particular characteristics so that the teacher could guide the emergence of issues that would further her instructional agenda. However, an interpretation of the design experiment as merely a case of students coming to reason meaningfully about data that they have not generated themselves misses the larger point. As a design principle for elementary statistics instruction, we contend on the basis of our findings that it is important for students to talk through the data generation process whether or not they actually collect data.

Our rationale for this claim becomes apparent when we note that data collection is but one phase in the data generation process, one that involves making measurements. The science education literature is relevant in this regard since it indicates that students who are involved in collecting their own data often do not understand the fundamental reasons for doing so and are primarily concerned with following methodological procedures and getting "the right data." In our view, such cases are predictable consequences of instructional designs that fail to engage students in the phases of the data generation process that precede data collection. These preceding phases involve clarifying the significance of the phenomenon under investigation, delineating relevant aspects of the phenomenon that should be measured, and considering how they might be measured. A primary purpose for engaging students in these phases is to enable them to remain cognizant of the purposes underpinning their inquiries and, eventually, to appreciate the influence of data generation on the legitimacy of the conclusions they can draw from the data they collect. In an approach of this type, the series of methodological decisions that make the collection of data possible are not assumed to be transparent to students, but instead become an explicit focus of discussion in the course of which students engage in all phases of the data generation process.

TOOL USE

As we have noted, the use of computer-based tools to create and manipulate graphical representations of data is central to exploratory data analysis (EDA). In the design experiment, the students used two computer tools that were explicitly designed to support the development of their statistical reasoning. We described the second of these tools when we discussed students' analyses of the AIDS treatment data. Bakker and Gravemeijer (Chapter 7) illustrate the range of options available for structuring data on both this tool and the first tool, the interface of which is shown in Figure 3. As Bakker and Gravemeijer also clarify, students could use this first tool to order, partition, and otherwise organize sets of up to 40 data points in a relatively immediate way. When data are entered, each data point is inscribed as a horizontal bar. Figure 3 shows data on the life spans of ten batteries of each of two different brands that were generated to investigate which of the two brands is superior in this respect.

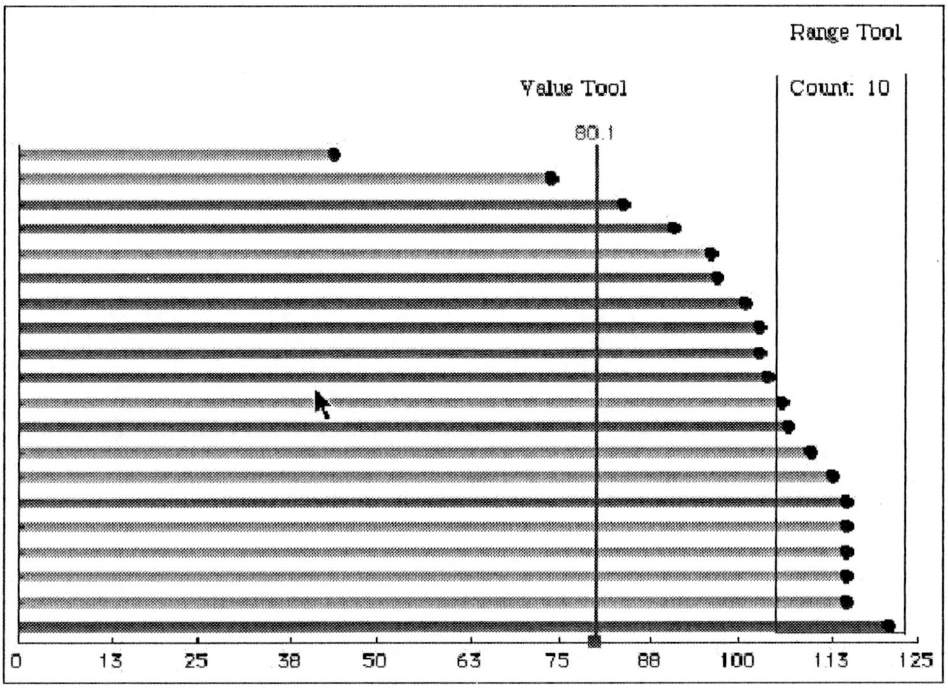

Figure 3. The first computer Minitool.

Compatibility with Students' Current Reasoning

A design principle that guided the development of the two computer tools was that they should fit with students' reasoning at a particular point in the instructional sequence (cf. Gravemeijer, 1994). It was apparent from our classroom observations that the tools did fit with the students' reasoning since they could use them to investigate trends and patterns in data with only a brief introduction. These observations indicate that when they were first introduced in the design experiment, the ways in which data were inscribed in the tools were transparent to the students.

In the case of the first tool, we have noted that one of our concerns at the beginning of the experiment was that the students would actually analyze data rather than merely manipulate numbers. It was for this reason that we decided to inscribe individual data values as horizontal bars. In addition, the initial data sets that the students analyzed when this tool was introduced within the first week of the experiment were selected so that the measurements made when generating the data had a sense of linearity and thus lent themselves to this type of inscription (e.g., the braking distances of cars, the life spans of batteries). As we have indicated, the choice of this inscription together with the approach of talking through the data generation process proved to be effective in that the teacher was able to initiate a shift in classroom discourse such that all the students actually began to reason about data as they completed the second instructional activity involving the first tool.

Supporting the Development of Students' Reasoning

A second design principle that guided the development of the two computer tools was that the students would come to reason about data in increasingly sophisticated ways as they used the tools and participated in the subsequent whole-class discussions of their analyses. We therefore viewed the design of the tools that the students would use as a primary means of supporting the reorganization of their statistical reasoning (cf. Dorfler, 1993; Kaput, 1991; Meira, 1998; Pea, 1993). In the case of the first tool, the students dragged the vertical value bar along the axis to either partition data sets or find the value of specific data points. In addition, they used the range tool to isolate a particular interval and compare the number of data points of each data set that were in that interval. In Figure 3, the range tool has been used to bound the 10 longest lasting batteries. It was as the students used the computer tool in these ways that they began to reason about (a) the maximum and minimum values and the range of data sets, (b) the number of data points above or below a particular value or within a specified interval, and (c) the median and its relation to the mean. Against the background of these developments, the teacher introduced the second tool in which data points were inscribed as dots in an axis plot (see Figure 1).

Sequencing the Use of Tools

Our intention in designing the second tool was to build on the ways of reasoning about data that the students had developed as they used the first tool. As Bakker and Gravemeijer note, the dots at the end of the bars in the first tool have, in effect, been collapsed down onto the axis in the second tool. The teacher in fact introduced this new way of inscribing data first by showing a data set inscribed as horizontal bars, and then by removing the bars to leave only the dots, and finally by transposing the dots onto the horizontal axis. As we had conjectured, the students were able to use the second tool to analyze data with little additional guidance, and it was apparent that the axis plot inscription signified a set of data values rather than merely a collection of dots spaced along a line. However, this development cannot be explained solely by the teacher's careful introduction of the new tool. Instead, we have to take account of a further aspect of the students' activity as they used the first tool in order to explain why the second tool fit with their reasoning.

We can tease out this aspect of the students' learning by focusing on their reasoning as they used the first tool to compare data sets in terms of the number of data points either within a particular interval or above or below a particular value. To illustrate, one student explained that he had analyzed the battery data by using the value bar to partition the data at 80 hours as shown in Figure 3. He then argued that some of the batteries of one brand were below 80 hours, whereas all those of the other brand lasted more than 80 hours. He judged this latter brand to be superior because, as he put it, he wanted a consistent battery. The crucial point to note is that in making arguments of this type, the students focused on the location of the dots at the end of the bars with respect to the axis. In other words, a subtle but important shift occurred as the students used the first tool. Originally, the individual data values were represented by the lengths of the bars. However, in the very process of using the tool, these values came to be signified by the endpoints of the bars.

As a result of this development, the second tool fit with the students' reasoning when it was introduced; they could readily understand the teacher's explanation of collapsing the dots at the end of the bars down onto the axis. Further, because the options in this new tool all involved partitioning data sets in various ways, the students could use it immediately because they had routinely partitioned data sets when they used range and value bar options on the first tool. This in turn made it possible for the second tool to serve as a means of supporting the development of their statistical reasoning. As our discussion of the AIDS treatment activity illustrates, students came to view data sets as holistic distributions that have shape rather than as amorphous collections of individual data points, to reason about these shapes in terms of relative rather than absolute frequencies, and to structure data sets in increasingly sophisticated ways.

It is almost impossible to deduce this subtle but important aspect of the students' learning by inspecting the physical characteristics of the first tool. As a third principle for the design of tools, we did not attempt to build the statistical ideas we wanted students to learn into the two computer tools and then hope that they might come to see them in some mysterious and unexplained way. Instead, when we designed the tools, we focused squarely on how the students might actually use them and what they might learn as they did so. Although this principle is relatively general, it is particularly important in the case of statistical data analysis given the central role of computer-based analysis tools and graphical representations in the discipline. The value of this principle is that it orients the designer to consider how the students' use of a proposed tool will change the nature of their activity as they analyze data and thus the types of reasoning that they might develop. In the case of the design experiment, a focus on data sets as holistic distributions rather than as collections of individual data points might not have become routine had the design of the tools been significantly different.

CLASSROOM DISCOURSE

The frequent references we have made to the whole-class discussions in which the students shared and critiqued their analyses indicates the value we attribute to this discourse as a means of supporting the students' learning. To this point, we have emphasized that these discussions should focus on significant statistical ideas that advance the teachers' instructional agenda. In further clarifying the importance of the whole-class discussions, we consider norms or standards for what counts as an acceptable data-based argument and then return to our goal of ensuring that significant statistical ideas emerge as topics of conversation.

Norms for Statistical Argumentation

Bakker and Gravemeijer (Chapter 7) report that the establishment of productive classroom norms is as important in supporting students' learning as the use of suitable computer tools, the careful planning of instructional activities, and the skills of the teacher in managing whole-class discussions. We can illustrate the significance of a key classroom norm—that of what counts as an acceptable data-based argument—by returning to the students' analyses of the battery data.

The first student who explained her reasoning said that she had focused on the 10 highest data values (i.e., those bounded by the range tool as shown in Figure 3). She went on to note that 7 of the 10 longest lasting batteries were of one brand and concluded that this brand was better. However, during the ensuing discussion, it became apparent that her decision to focus on the 10 rather than, say, the 14 longest lasting batteries was relatively arbitrary. In contrast, the next student who presented an analysis explained that he had partitioned the data at 80 hours because he wanted a consistent battery that lasted at least 80 hours. In doing so, he clarified why his approach to organizing the data was relevant to the question at hand—that of deciding which of the two brands was superior.

As the classroom discussion continued, the obligation that the students should give a justification of this type became increasingly explicit. For example, a third student compared the two analyses by commenting that although 7 of the 10 longest lasting batteries were of one brand, the 2 lowest batteries were also of this brand, and "if you were using the batteries for something important, you could end up with one of those bad batteries." Because of exchanges like this, the teacher and students established relatively early in the design experiment that to be acceptable, an argument had to justify why the method of structuring the data was relevant to the question under investigation. In our view, the establishment of this norm of argumentation constitutes an important design principle for statistics instruction. On the one hand, it serves to delegitimize analyses in which students simply produce a collection of statistics (e.g., mean, median, range) rather than attempt to identify trends and patterns in the data that are relevant to the issue they are investigating. On the other hand, it serves as a means of inducting students into an important disciplinary norm—namely, that the appropriateness of the statistics used when conducting an analysis has to be justified with respect to the question being addressed.

Focusing on Significant Statistical Ideas

Returning to the previously stated goal of ensuring that classroom discussions focus on significant statistical ideas, it is helpful if we outline the approach the teacher took when planning for the whole-class discussions. In the latter part of the design experiment, we organized instructional activities so that the students conducted their analyses and wrote their reports in one classroom session, and then the teacher conducted the whole-class discussion with them in the following classroom session. The teacher found this arrangement productive because she could review the students' reports prior to the whole-class discussion to gain a sense of the various ways in which students had reasoned about the data. This in turn enabled her to develop conjectures about statistically significant issues that might emerge as topics of conversation. Her intent in planning for discussions in this way was to capitalize on the students' reasoning by identifying data analyses that, when compared and contrasted, might give rise to substantive statistical conversations (McClain, 2002). In the case of the AIDS treatment data, for example, the teacher selected a sequence of four analyses for discussion, so that the issues of reasoning proportionally about data and of interpreting data organized into four equal groups might come to the fore.

Our purpose in describing this planning process is to emphasize that although the design of instructional activities and tools is important, the expertise of a knowledgeable teacher in guiding productive discussions by capitalizing on students' reasoning is also critical. Earlier in this chapter, we noted that the challenge of transcending what Dewey (1981) termed the dichotomy between process and content is especially pressing in the case of statistical data analysis, given that an investigative orientation is integral to the discipline. Thus, in contrast to attempts to make curricula teacher-proof, our final design principle attributes a central role to the teacher. We in fact find it useful to view the teacher as a designer who is responsible for organizing substantive classroom discussions that can serve as primary means of supporting students' induction into the values, beliefs, and ways of knowing of the discipline. The final design principle is therefore that our task in developing instructional activities and tools is to take account of the mediating role of the teacher rather than to view ourselves as supporting the students' statistical learning directly. The challenge is then to make it possible for the teacher to organize productive learning experiences for students by capitalizing on the diverse ways in which they use tools to complete specific instructional activities.

DISCUSSION

In this chapter, we have framed a classroom design experiment as a paradigm case in which to propose a number of design principles for supporting the development of students' statistical reasoning. These principles involve formulating and testing conjectures about:

1. Central statistical ideas, such as distribution, that can serve to orient the development of an instructional design
2. The characteristics of instructional activities that
   a) Make it possible for students' classroom activity to be imbued with the investigative spirit of data analysis
   b) Enable teachers to achieve their instructional agendas by building on the range of data-based arguments that students produce
3. Classroom activity structures that support the development of students' reasoning about data generation as well as data analysis
4. The characteristics of data analysis tools that
   a) Fit with students' reasoning when they are first introduced in an instructional sequence
   b) Serve as a primary means of supporting students' development of increasingly sophisticated forms of statistical reasoning
5. The characteristics of classroom discourse in which
   a) Statistical arguments explain why the way in which the data have been organized gives rise to insights into the phenomenon under investigation
   b) Students engage in sustained exchanges that focus on significant statistical ideas

Because we have discussed the principles in separate sections of the chapter, they might appear to be somewhat independent. We therefore need to stress that they are in fact highly interrelated. For example, the instructional activities as they were actually realized in the classroom depended on:

- The overall goal for doing statistics (i.e., to identify patterns in data that are relevant to the question or issue at hand)
- The structure of classroom activities (e.g., talking through the data generation process)
- The computer tools that the students used to conduct their analyses
- The nature of the of the classroom discourse (e.g., engaging in discussion in which significant statistical issues emerge as topics of conversation)

It is relatively easy to imagine how the instructional activities might have been realized very differently in a classroom where the overall goal is to apply prescribed methods to data, or where there are no whole-class discussions and the teacher simply grades students' analyses.

Given the interdependencies, it is reasonable to view the various principles we have discussed as serving to orient the design of productive classroom *activity systems*. The intent of instructional design from this perspective is to provide teachers with the resources necessary to guide the development of their classrooms as activity systems in which students develop significant statistical ideas as they participate in them and contribute to their evolution. They are, in short, systems designed to produce the learning of significant statistical ideas.

The comprehensive nature of a classroom activity system indicates that the approach we take to instructional design extends far beyond the traditional focus on curriculum while simultaneously acknowledging the vital, mediating role of the teacher. Because this perspective might seem unorthodox, we close by illustrating that it is in fact highly consistent with current research in the learning sciences. Bransford, Brown, and Cocking (2000) synthesize this research in the highly influential book, *How People Learn,* and propose a framework that consists of four overlapping lenses for examining learning environments. The first of these lenses focuses on the extent to which learning environments are *knowledge centered* in the sense of being based on a careful analysis of what we want people to know and be able to do as a result of instruction. In this regard, we discussed the importance of organizing instruction around overarching statistical ideas such as distribution, of ensuring that classroom discussions focus on significant statistical ideas, and of designing tools as a means of supporting the development of students' statistical reasoning. The second lens is *learner centered* and examines the extent to which a learning environment builds on the strengths, interests, and preconceptions of learners. We illustrated this focus when we discussed (a) the initial data generation discussions and the importance of cultivating students' interests in the issue under investigation, (b) the approach of designing tools that fit with students' current statistical reasoning, and (c) the process of planning whole-class discussions by building on students' analyses.

The third lens of the How People Learn Framework is *assessment centered* and examines the extent to which students' thinking is made visible, so that teachers can adjust instruction to their students' reasoning and students have multiple opportunities to test and revise their ideas. This lens was evident when we discussed the value of whole-class discussions in which students shared their analyses and received feedback, and when we indicated how the reports the students wrote enabled the teacher to assess their statistical reasoning. The final lens in the Framework is *community centered* and examines the extent to which the classroom is an environment in which students not only feel safe to ask questions but also can learn to work collaboratively. Our discussion of the AIDS and batteries instructional activities served to illustrate these general features of the classroom, and we also stressed the importance of the discipline specific norm of what counts as an acceptable data-based argument.

The broad compatibility between the instructional design principles we have proposed for elementary statistics instruction and the How People Learn Framework gives the principles some credibility. In addition, the grounding of the Framework in an extensive, multidisciplinary research base adds weight to our claim that it is productive for our purposes as statistics educators to view classrooms as activity systems that are designed to support students' learning of significant statistical ideas. As a result, although the set of principles that we have proposed might appear unduly wide ranging, we contend that approaches considering only the design of instructional activities and computer tools are in fact overly narrow.

NOTES

1 The second author served as the teacher in both this and the prior design experiment that focused on the analysis of univariate data and was assisted by the first author.

2 This notion of an implicit third dimension in bivariate data was first brought to our attention by Patrick Thompson (personal communication, August 1998).

REFERENCES

Biehler, R. (1993). Software tools and mathematics education: The case of statistics. In C. Keitel & K. Ruthven (Eds.), *Learning from computers: Mathematics education and technology* (pp. 68–100). Berlin: Springer.

Biehler, R., & Steinbring, H. (1991). Entdeckende Statistik, Strenget-und-Blatter, Boxplots: Konzepte, Begrundungen und Enfahrungen eines Unterrichtsversuches [Explorations in statistics, stem-and-leaf, boxplots: Concepts, justifications, and experience in a teaching experiment]. *Der Mathematikunterricht, 37*(6), 5–32.

Bransford, J., Brown, A. L., & Cocking, R. R. (Eds.) (2000). *How people learn: Brain, mind, experience, and school.* Washington, DC: National Academy Press.

Cobb, G. W., & Moore, D. S. (1997). Mathematics, statistics, and teaching. *American Mathematical Monthly, 104,* 801–823.

Cobb, P. (1999). Individual and collective mathematical learning: The case of statistical data analysis. *Mathematical Thinking and Learning, 1,* 5–44.

Cobb, P., McClain, K., & Gravemeijer, K. P. E. (2003). Learning about statistical covariation. *Cognition and Instruction, 21,* 1–78.

Cobb, P., & Tzou, C. (2000). *Learning about data creation.* Paper presented at the annual meeting of the American Educational Research Association, New Orleans.

de Lange, J., van Reeuwijk, M., Burrill, G., & Romberg, T. (1993). *Learning and testing mathematics in context. The case: Data visualization.* Madison: University of Wisconsin, National Center for Research in Mathematical Sciences Education.

Dewey, J. (1981). Experience and nature. In J. A. Boydston (Ed.), *John Dewey: The later works, 1925–1953* (Vol. 1). Carbondale: Southern Illinois University Press.

Dorfler, W. (1993). Computer use and views of the mind. In C. Keitel & K. Ruthven (Eds.), *Learning from computers: Mathematics education and technology* (pp. 159–186). Berlin: Springer-Verlag.

Gravemeijer, K. E. P. (1994) *Developing realistic mathematics education.* Utrecht, The Netherlands: CD-B Press.

Hancock, C., Kaput, J. J., & Goldsmith, L. T. (1992). Authentic inquiry with data: Critical barriers to classroom implementation. *Educational Psychologist, 27,* 337–364.

Kaput, J. J. (1991). Notations and representations as mediators of constructive processes. In E. von Glasersfeld (Ed.), *Constructivism in mathematics education* (pp. 53–74). Dordrecht, The Netherlands: Kluwer.

Konold, C., & Higgins, T. (in press). Working with Data. In S. J. Russell & D. Schifter & V. Bastable (Eds.), *Developing Mathematical Ideas: Collecting, Representing, and Analyzing Data.* Parsippany, NJ: Seymour.

Konold, C., Pollatsek, A., Well, A., & Gagnon, A. (1996, July). *Students' analyzing data: Research of critical barriers.* Paper presented the Roundtable Conference of the International Association for Statistics Education, Granada, Spain.

Latour, B. (1987). *Science in action.* Cambridge, MA: Harvard University Press.

Lehrer, R., & Romberg, T. (1996). Exploring children's data modeling. *Cognition and Instruction, 14,* 69–108.

McClain, K. (2002). Teacher's and students' understanding: The role of tools and inscriptions in supporting effective communication. *Journal of the Learning Sciences, 11,* 216–241.

McClain, K., Cobb, P., & Gravemeijer, K. (2000). Supporting students' ways of reasoning about data. In M. Burke (Ed.), *Learning mathematics for a new century* (2001 Yearbook of the National Council of Teachers of Mathematics, pp. 174–187). Reston, VA: National Council of Teachers of Mathematics.

McGatha, M. (2000). *Instructional design in the context of classroom-based research: Documenting the learning of a research team as it engaged in a mathematics design experiment.* Unpublished dissertation, Vanderbilt University, Nashville, TN.

McGatha, M., Cobb, P., & McClain K. (1999, April). *An analysis of student's initial statistical understandings.* Paper presented at the annual meeting of the American Educational Research Association, Montreal.

Meira, L. (1998). Making sense of instructional devices: The emergence of transparency in mathematical activity. *Journal for Research in Mathematics Education, 29,* 121–142.

Noss, R., Pozzi, S., & Hoyles, C. (1999). Touching epistemologies: Statistics in practice. *Educational Studies in Mathematics, 40,* 25–51.

Pea, R. D. (1993). Practices of distributed intelligence and designs for education. In G. Salomon (Ed.), *Distributed cognitions* (pp. 47–87). New York: Cambridge University Press.

Roth, W. M. (1997). Where is the context in contextual word problems? Mathematical practices and products in grade 8 students' answers to story problems. *Cognition and Instruction, 14,* 487–527.

Saldanha, L. A., & Thompson, P. W. (2001). Students' reasoning about sampling distributions and statistical inference. In R. Speiser & C. Maher (Eds.), *Proceedings of the Twenty Third Annual Meeting of the North American Chapter of the International Group for the Psychology of Mathematics Education* (vol. 1, pp. 449–454), Snowbird, Utah. ERIC Clearinghouse for Science, Mathematics, and Environmental Education, Columbus, OH.

Shaughnessey, J. M., Garfield, J., & Greer, B. (1996). Data handling. In A. J. Bishop, K. Clements, C. Keitel, J. Kilpatrick, & C. Laborde (Eds.), *International handbook of mathematics education* (part 1, pp. 205–237). Dordrecht, The Netherlands: Kluwer.

Wiggins, G., & McTighe, J. (1998). *Understanding by design.* Alexandria, VA: Association for Curriculum and Supervision.

Wilensky, U. (1997). What is normal anyway? Therapy for epistemological anxiety. *Educational Studies in Mathematics, 33,* 171–202.

Chapter 17

RESEARCH ON STATISTICAL LITERACY, REASONING, AND THINKING: ISSUES, CHALLENGES, AND IMPLICATIONS

Joan Garfield[1] and Dani Ben-Zvi[2]
University of Minnesota, USA[1], and University of Haifa, Israel[2]

INTRODUCTION

The collection of studies in this book represents cutting-edge research on statistical literacy, reasoning, and thinking in the emerging area of statistics education. This chapter describes some of the main issues and challenges, as well as implications for teaching and assessing students, raised by these studies. Because statistics education is a new field, taking on its own place in educational research, this chapter begins with some comments on statistics education as an emerging research area, and then concentrates on various issues related to research on statistical literacy, reasoning, and thinking. Some of the topics discussed are the need to focus research, instruction, and assessment on the big ideas of statistics; the role of technology in developing statistical reasoning; addressing the diversity of learners (e.g., students at different educational levels as well as their teachers); and research methodologies for studying statistical reasoning. Finally, we consider implications for teaching and assessing students and suggest future research directions.

STATISTICS EDUCATION AS AN EMERGING RESEARCH AREA

Statistics and statistics education are relatively new disciplines. Statistics has only recently been introduced into school curricula (e.g., NCTM, 2000) and is a new academic major at the college level (Bryce, 2002). In the United States, the NCTM standards (2000) recommend that instructional programs from pre-kindergarten through grade 12 focus more on statistical reasoning. The goals of their suggested statistics curriculum include

- Enable all students to formulate questions that can be addressed with data and collect, organize, and display relevant data to answer them.

D. Ben-Zvi and J. Garfield (eds.),
The Challenge of Developing Statistical Literacy, Reasoning and Thinking, 397–409.

- Select and use appropriate statistical methods to analyze data.
- Develop and evaluate inferences and predictions that are based on data.
- Understand and apply basic concepts of probability.

At the university level, statistics is taught at undergraduate as well as graduate levels across many disciplines. The students taking statistics at these levels may be preparing to be future "users" or "producers" of statistics in different fields of application (e.g., sciences, technology, industry, and medicine), or future statisticians or statistics teachers. Over the last 20 years there has been a steady increase in the numbers of statistics courses taught, to fulfill the growing demand for students and professionals who can use and understand statistical information.

Although the amount of statistics instruction at all levels is growing at a fast pace, the research to support statistics instruction is proceeding at a much slower rate. The research literature in statistics education is not well known; therefore, it is not often valued or utilized by statisticians, schools, or the immense number of other fields that use statistics (Joliffe, 1998). In fact, researchers in this area argue that the field still needs to define what research in statistics education is—not only to achieve academic recognition, but to convince others of its validity as a research discipline (Batanero, Garfield, Ottaviani, & Truran, 2000).

Unlike other research areas, the research studies on teaching and learning statistics have been conducted in, and influenced by, several different disciplines, each with its own perspectives, literatures, methodology, and research questions. For example, much of the early research was conducted by psychologists, often focusing on conceptions of chance and randomness (e.g., Piaget & Inhelder, 1975; Fischbein, 1975; and Kahneman, Slovic, & Tversky, 1982). Psychologists' dominant effort was to identify, through observations or paper and pencil tests, ways in which people make judgments of chance. Many researchers (for example, Kahneman et al., 1982; Konold, 1989) identified widespread errors in reasoning, finding that people tend to use nonstatistical heuristics to make judgments or decisions regarding chance events. By the end of the 1980s, there was strong evidence that many adults are unable to deal competently with a range of questions that require probabilistic thinking.

In the 1980s and 1990s, many researchers in mathematics education, motivated by the inclusion of statistics and probability in the elementary and secondary mathematics curriculum, began to explore students' understanding of ideas related to statistics and data analysis (e.g., Russell & Mokros, 1996; Mokros & Russell, 1995; Rubin, Bruce, & Tenney, 1991; Shaughnessy, 1992). These researchers found the mathematics education theoretical frameworks and methodologies relevant for research in statistics education (see, for example, Kelly & Lesh, 2000). During this same time period, several educational psychologists explored students' attitudes and anxiety about statistics in an attempt to predict success in statistics courses (e.g., Wisenbaker & Scott, 1997; Schau & Mattern, 1997), while cognitive psychologists examined ways to help students and adults correctly use statistical reasoning (e.g., Fong, Krantz, & Nisbett, 1986; Nisbett, 1993; Sedlmeier, 1999). A more recent group of researchers is emerging from the growing number of statisticians who are focusing their scholarship on educational issues (e.g., Chance, 2002; Lee, Zeleke, & Wachtel, 2002; Wild, Triggs, & Pfannkuch, 1997). Some of these researchers looked at particular classroom interventions and their impact on learning outcomes or developed models for teaching and for experts' statistical thinking.

What has been missing in the past few decades is a coordination of the research across the different disciplines described earlier, and a convergence of methods and important research questions. Without this coherence, it is hard to move the field forward and to build on the results by linking research to teaching. One example of an effort to coordinate the research across the different disciplines is the book edited by Lajoie (1998), which addresses issues of statistical content, learner needs, instructional methods, and assessment goals. It was the outcome of the coordinated work of statisticians, mathematics educators, and psychologists, who focused on formulating a research agenda for K–12 statistics education.

The International Research Forums on Statistical Reasoning, Thinking, and Literacy (SRTL-1, Israel; SRTL-2, Australia; and SRTL-3, USA) have been another important effort to achieve this goal, by bringing together an international group of researchers from across these disciplines to share their findings, discuss their methods, and generate important issues and research questions. Another goal of these research forums has been to make explicit connections to teaching practice, something that researchers are often criticized for failing to address.

RESEARCH ON STATISTICAL LITERACY, REASONING, AND THINKING

Although statistics is now viewed as a unique discipline, statistical content is most often taught in the mathematics curriculum (K–12) and in departments of mathematics (tertiary level). This has led to exhortations by leading statisticians, such as Moore (1998), about the differences between statistics and mathematics (see Chapter 4). These arguments challenge statisticians and statistics educators to carefully define the unique characteristics of statistics, and in particular, the distinctions between statistical literacy, reasoning, and thinking. We provided summaries of these arguments and related research in the early chapters of this book (see Chapters 1 through 4).

The authors of chapters in this book represent the growing network of researchers from SRTL-1 and SRTL-2 who are interested in statistical literacy, reasoning, and thinking, and who have been trained in the different disciplines (e.g., mathematics education, cognitive and educational psychology, and statistics). Many of the chapters describe collaborative studies, some including researchers from different disciplines (e.g., Chapters 2, 13, and 14). It may seem strange, given the quantitative nature of statistics, that most of the studies in this book include analyses of qualitative data, particularly videotaped observations or interviews. We have found that sharing these videos and their associated transcripts allows us to better present and discuss the important aspects of our work, as well as to solicit useful feedback from colleagues. Further discussion of methodological issues is provided later in this chapter.

The topics of the research studies presented in this book reflect the shift in emphasis in statistics instruction, from statistical techniques, formulas, and procedures to developing statistical reasoning and thinking. The chapters on individual aspects of reasoning focus on some core ideas of statistics, often referred to as the 'big ideas." Increasing attention is being paid in the educational research community to the need to clearly define and focus both research and instruction, and therefore, assessment, on the big ideas of a discipline (Bransford, Brown, & Cocking, 2000; Wiggins, 1998). We offer a list and description of the big ideas of statistics in the following section.

FOCUSING ON THE BIG IDEAS OF STATISTICS

The topics of the chapters in this book (e.g., data, distribution, averages, etc.) focus on some of the big ideas in statistics that students encounter in their educational experiences in elementary, secondary, or tertiary classes. Although many statistics educators and researchers today agree that there should be a greater focus on the big ideas of statistics, little has been written about what these ideas are. Friel (in press) offers a list similar to the one we provide here:

- **Data**—the need for data; how data represent characteristics or values in the real world; how data are obtained; different types of data, such as numbers, words, and so forth.
- **Distribution**—a representation of quantitative data that can be examined and described in terms of shape, center, and spread, as well as unique features such as gaps, clusters, outliers, and so on.
- **Trend**—a signal or pattern we are interested in. It could be a mean for one group, the difference of means for comparing two groups, a straight line for bivariate data, or a pattern over time for time-series data.
- **Variability**—the variation or noise around a signal for a data set, such as measurement error. Variability may also be of interest in that it helps describe and explain a data set, reflecting natural variation in measurements such as head sizes of adult men.
- **Models—**an ideal that is sometimes useful in understanding, explaining, or making predictions from data. A model is useful if it 'fits" the data well. Some examples of models are the normal curve, a straight line, or a binomial random variable with probability of 0.5.
- **Association**—a particular kind of relationship between two variables; information on one variable helps us understand, explain, or predict values of the other variable. Association may be observed between quantitative or categorical variables. This also includes being able to distinguish correlation from causality.
- **Samples and sampling**—the process of taking samples and comparing samples to a larger group. The sampling process is important in obtaining a representative sample. Samples are also used to generate theory, such as simulating sampling distributions to illustrate the Central Limit Theorem.
- **Inference**—ways of estimating and drawing conclusions about larger groups based on samples. Utts (2003) elaborates that this includes being able to differentiate between practical and statistical significance as well as knowing

the difference between finding "no effect" versus finding "no significant effect."

When we examine much of statistics instruction, it is not always clear how these big ideas are supposed to be presented and developed. In most statistics classroom instruction, the emphasis is on individual concepts and skills, and the big ideas are obscured by the focus on procedures and computations. After one topic has been studied, there is little mention of it again, and students fail to see how the big ideas are actually providing a foundation for course content and that they underlie statistical reasoning. For example, students may focus on how to compute different measures of center or variability without fully understanding the ideas of center and spread and their relationships to other big ideas, such as data and distribution. Later in their studies, students may fail to connect the idea of center and spread of sampling distributions with the ideas of center and spread in descriptive statistics. Or, when studying association, students may lose track of how center and spread of each variable are embedded in looking at bivariate relationships.

Major challenges that teachers face include not only finding ways to go beyond the individual concepts and skills, but leading students to develop an understanding of the big ideas and the interrelations among them. Such an approach will enable teachers to make the big ideas explicit and visible, throughout the curriculum. For example, Cobb (1999) suggests that focusing on distribution as a multifaceted end goal of instruction in seventh grade might bring more coherence in the middle school statistics curriculum and empower students' statistical reasoning. Bakker and Gravemeijer (Chapter 7) propose to focus instruction on the informal aspects of shape. Other illustrations of the need to focus on the big ideas of statistics and how to do it can be found in various chapters of this book: data (Chapter 6), center (Chapter 8), variability (Chapter 9), covariation (Chapter 10), and sampling (Chapters 12 and 13). It has been suggested that the use of technology-assisted learning environments can support—in many ways—students' construction of meanings for the big ideas of statistics (e.g., Garfield & Burrill, 1997).

THE ROLE OF TECHNOLOGY IN DEVELOPING STATISTICAL REASONING

Many of the chapters in this book mention the use of technology in developing statistical reasoning. This is not surprising, given how the discipline of statistics has depended on technology and how technology has been driving change in the field of statistics. Although there are many technological tools available, including graphing calculators, computers, and the World Wide Web, there is still a lack of research on how to best use these tools and how they affect student learning.

The interaction of technology with efforts to redefine both content and instruction in statistics in the K–12 curriculum provides a variety of strategies for *teaching* statistics and, at the same time, offers new ways of *doing* statistics (Garfield & Burrill, 1997). Today, computers, software, and the Internet are essential tools for instruction in statistics (Friel, in press).

Ben-Zvi (2000) describes how technological tools may be used to help students actively construct knowledge, by "doing" and "seeing" statistics, as well as to give students opportunities to reflect on observed phenomena. He views computers as cognitive tools that help transcend the limitations of the human mind. Therefore, technology is not just an amplifier of students' statistical power, but rather a reorganizer of students' physical and mental work. The following types of software, which are described in this book, are good examples of such tools:

- *Commercial statistical packages* for analyzing data and constructing visual representations of data such as spreadsheets (Excel©, Chapter 6), or data analysis programs (Statgraphics©, Chapter 11) that offer a variety of simultaneous representations that are easily manipulated and modified, as well as simulation of different distributions.
- *Educational data analysis tools* (Fathom©, Chapter 15) are intended to help students develop an understanding of data and data exploration. They support in-depth inquiry in statistics and data analysis through powerful statistical and plotting capabilities that give the user greater overall control in structuring and representing data (Friel, in press). Fathom also allows plotting functions, creating animated simulations, and has a "dragging" facility that dynamically updates data representations. This helps reveal the invariant phenomenon and the relationships among representations.
- *Web- or computer-based applets* were developed to demonstrate and visualize statistical concepts. Applets are typically small, web-based computer programs that visually illustrate a statistical concept by letting the user manipulate and change various parameters. The Minitools (Chapters 7 and 16), a particular type of applet, were designed to support an explicit "learning trajectory" to develop an understanding of a particular graph and its link to the data on which it is based.
- *Stand-alone simulation software*, such as Sampling SIM (Chapter 13), which was developed to provide a simulation of sampling distributions, with many capabilities allowing students to see the connections between individual samples, distributions of sample means, confidence intervals, and *p*-values.

The last three tools on this list (Fathom, Minitools, and Sampling SIM) were designed based on ideas about what students need to see and do in order to develop a conceptual understanding of abstract statistical concepts as well as develop the kinds of attitudes and reasoning required for analyzing data. Although these three tools were developed to improve student learning, Bakker (2002) distinguished between route-type software—small applets and applications, such as the Minitools that fit in a particular learning trajectory; and landscape-type software—larger applications, such as Fathom and TinkerPlots, that provide an open landscape in which teachers and students may freely explore data.

The increasing use of Internet and computer-mediated communication (CMC) in education has also influenced statistics education. Although not the focus of chapters in this book, there are numerous Internet uses in statistics classes that support the development of students' statistical reasoning. For example, data sources in downloadable formats are available on the Web to support active learning of exploratory data analysis. They are electronically available from data-set archives, government and official agencies, textbook data, etc. An additional example is the use of CMC tools, such as online forums, e-mail, and so forth to create electronic communities that support students' learning in face-to-face or distance learning.

It is important to note that despite its role in helping students learn and do statistics, technology is not available in all parts of the world, and not even in all classrooms in the more affluent countries. The research studies in this book address different instructional settings with and without the use of technology, as well as diverse types of students who are learning statistics at all levels.

DIVERSITY OF STUDENTS AND TEACHERS

With the growing emphasis on statistical literacy, reasoning, and thinking, statistics education research must address the diversity of students in statistics courses by considering issues of continuity (when to teach what), pedagogy (how to approach the content and develop desired learning outcomes), priority (prioritizing and sequencing of topics), and diversity (students' educational preparation and background, grade and level). For example, little attention has been given to the issue of when and how a new statistical idea or concept can be presented to students, or to the question of sequencing statistical ideas and concepts along the educational life span of a student.

The individual research studies in this book partially address such issues, but as a group reflect the diversity of students (and teachers) who learn and know statistics. The widest "student" population is addressed in research about statistical literacy, which includes school students through adults. Gal (Chapter 3) underscores the importance of statistical literacy education for all present and future citizens to enable them to function effectively in an information-laden society. The goal of statistical literacy research is to identify the components of literacy, to find ways to equip all citizens with basic literacy skills—such as being able to critically read the newspaper or evaluate media reports.

The students observed by Ben-Zvi (Chapter 6) as well as Bakker and Gravemeijer (Chapter 7) were high-ability students. The forms of reasoning exhibited by some of these students are to some extent unique to the specific settings and circumstances. However, these studies describe some important teaching and learning issues and how the reasoning might develop in other types of students. They also suggest meaningful and engaging activities such as making predictions graphs without having data and using software tools that support specific statistical ways of reasoning. The instructional suggestions in some chapters require establishing certain socio-mathematical (statistical) norms and practices (Cobb & McClain, Chapter 16), use of suitable computer tools, carefully planned instructional activities, and skills of the teacher to orchestrate class discussions.

Ben-Zvi (Chapter 6) and Mickelson and Heaton (Chapter 14) describe the teachers in their studies as above average in pedagogical and statistical knowledge and skills. It is likely that the role of "average" elementary and middle school teachers, normally not trained in statistics instruction, would be quite different. Teachers need careful guidance to teach such a new and complex subject. Hence, more studies are needed that explore how to equip school teachers at all levels with appropriate content knowledge and pedagogical knowledge, and to determine what kind of guidance they need to successfully teach these topics.

RESEARCH METHODOLOGIES TO STUDY STATISTICAL REASONING

The chapters in this book reveal a variety of research methods used to study statistical literacy, reasoning, and thinking. Ben-Zvi (Chapter 6), and Mickelson and Heaton (Chapter 14) use a case study approach in natural classroom settings to study one or two cases in great detail. Batanero, Tauber, and Sánchez (Chapter 11) use a semiotic approach to analyze students' responses to open-ended questions on an exam. Chance, delMas, and Garfield (Chapter 13) use collaborative classroom research to develop software and build a model of statistical reasoning. Their research is implemented in their own classes and with their students, using an iterative cycle to study the impact of an activity on students' reasoning as they develop their model. Their method of classroom research is similar to the classroom teaching experiment used in the studies by Bakker and Gravemeijer (Chapter 7) and Cobb and McClain (Chapter 16), who refer to this method as design experiment (described by Lesh, 2002). Watson (Chapter 12) uses a longitudinal approach to study children's development of reasoning about samples.

As mentioned earlier, videotaped classroom observations and teacher or student interviews were included in most studies as a way to gather qualitative data. We have found in analyzing these videos, that observing students' verbal actions as well as their physical gestures helps us better understand students' reasoning and the socio-cultural processes of learning. Other sources of qualitative data were students' responses to open-ended questions, field notes of teachers and researchers, and samples of students' work (e.g., graphs constructed, statistics projects).

Makar and Confrey (Chapter 15) combine qualitative data with quantitative data on teachers' statistical reasoning. Pre- and posttests of statistical content knowledge provided the main source of quantitative data for their study, while videotaped interviews were transcribed and then analyzed using grounded theory (Strauss & Corbin, 1998). A few other studies also include some quantitative data in the context of student assessment, for example, Reading and Shaughnessy (Chapter 9), Moritz (Chapter 10), Batanero et al. (Chapter 11), and Watson (Chapter 12).

It may seem surprising that few statistical summaries are actually included in these studies, given that the subject being studied by students or teachers is statistics. And it may seem surprising that the research studies in this book are not traditional designed experiments, involving control groups compared to groups that have received experimental treatment, the gold standard of experimental design. However, statistics education tends to follow the tradition of mathematics and science education, in using mostly qualitative methods to develop an understanding of the nature of students' thinking and reasoning, and to explore how these develop (see Kelly & Lesh, 2000). Perhaps after more of this baseline information is gathered and analyzed, the field will later include some small, experimental studies that allow for comparisons of particular activities, instructional methods, curricular trajectories, types of technological tools, or assessments.

Before we reach this stage of research, we need to further study the long-lasting effects of instruction on students' reasoning, and to continue the exploration of models of conceptual change and development. These models will be based on careful examination and analyses of how reasoning changes, either over an extensive period of time (as in longitudinal studies) or during periods of significant transition (as in some clinical interviews or classroom episodes).

IMPLICATIONS FOR TEACHING AND ASSESSING STUDENTS

In the three chapters that focus on the topics of statistical thinking (Chapter 2), statistical literacy (Chapter 3), and statistical reasoning (Chapter 4), each author, or pair of authors, recommends that instruction be designed to explicitly lead students to develop these particular learning outcomes. For example, Pfannkuch and Wild (Chapter 2) discuss the areas to emphasize for developing statistical thinking, Gal (Chapter 3) describes the knowledge bases and dispositions needed for statistical literacy, and delMas (Chapter 4) describes the kinds of experiences with data that should lead to statistical reasoning.

One important goal of this book is provide suggestions for how teachers may build on the research studies described to improve student learning of statistics. Although most teachers do not typically read the research related to learning their subject matter content, we encourage teachers of statistics at the elementary, secondary, and tertiary level to refer to chapters in this book for a concise summary of research on the different areas of reasoning. These studies provide ideas not only about the types of difficulties students have when learning particular topics, so that teachers may be aware of where errors and misconceptions might occur, but also what to look for in their informal and formal assessments of students learning. In addition, these studies provide valuable information regarding the type of statistical reasoning that can be expected at different age levels. The models of cognitive development in statistical reasoning documented in Chapter 5 enable teachers to trace students' individual and collective development in statistical reasoning during instruction. Because the cognitive models offer a coherent picture of students' statistical reasoning, they can provide a knowledge base for teachers in designing and implementing instruction.

These research studies include details on the complexity of the different statistical topics, explaining why they are so difficult for students to learn. As several authors stress, it is important for teachers to move beyond a focus on skills and computations, and the role of teacher as the one who delivers the content. Instead, the role of teacher suggested by the authors of these chapters is one of providing a carefully designed learning environment, appropriate technological tools, and access to real and interesting data sets. The teacher should orchestrate class work and discussion, establish socio-statistical norms (see Cobb and McClain, Chapter 16) and provide timely and nondirective interventions by the teacher as representative of the discipline in the classroom (e.g., Voigt, 1995). The teacher should be aware not only of the complexities and difficulty of the concepts, but of the desired learning goals—such as what good statistical literacy, reasoning, and thinking look like—so that assessments can be examined and compared to these goals. The teachers need to be comfortable with both the content and tools, and with the process of data analysis.

The chapters in this book stress that students need to be exposed to the big ideas and their associated reasoning in a variety of settings, through a course or over several years of instruction. The authors make many suggestions about how technology can be used to help students develop their reasoning, and suggest that students be prodded to explain what they see and learn when using these tools as a way to develop their reasoning.

Many of the authors present some types of learning activities and data sets that teachers can use in their classes at different school levels. They suggest that regardless of the activity used, teachers can find ways to observe their students carefully to see how their reasoning is affected by the activities. Teachers should also avoid assuming that students have learned the material merely because they have completed an activity on that topic. Finally, teachers are encouraged to apply the research tools on their classes, and to use the information gathered to continually revise and improve their activities, materials, and methods. We believe that it is better to learn a few concepts in depth, rather than trying to cover every topic. If this can be done in a systematic way, then more topics might be covered over a span of grades, rather than in one single grade level.

We agree with the many educators who have called for classroom instruction to be aligned with appropriate methods of assessment, which are used as a way to make reasoning visible to teachers as well as to students. Assessment should be used for formative as well as summative purposes, and it should be aligned with learning goals. In most cases, a type of performance assessment seems to best capture the full extent of students' statistical reasoning and thinking (Gal & Garfield, 1997; Garfield & Gal, 1999).

We suggest that it is often helpful to start by considering the types of assessment that are appropriate to measure the desired learning outcomes, and to work backward, thinking about instruction and activities that will lead to these goals (see Wiggins, 1998). Then assessment data gathered from students can be used to evaluate the extent to which these important learning goals (e.g., developing statistical reasoning) have been achieved.

FUTURE DIRECTIONS AND CHALLENGES

Given the importance of the learning outcomes described in this book, statistical literacy, reasoning and thinking, it is crucial that people working in this area use the same language and definitions when discussing these terms. Similarly, some standard goals for each outcome should be agreed upon and used in developing educational materials and curricula, designing assessments, preparing teachers' courses, and conducting future research.

Because the field of statistics education research is so new, there is a need for more research in all of the areas represented in this book. Studies need to be conducted in different educational settings, with different-aged students worldwide, and involving different educational materials and technological tools. As we continue to learn more about how different types of reasoning in statistics develop, we need to continue to explore cognitive developmental models, seeing how these apply to the different settings. There is also a need to validate these models, and to investigate how they may be used to promote reasoning, thinking, and literacy through carefully designed instruction.

There is a great need for assessment instruments and materials that may be used to assess statistical literacy, reasoning, and thinking. A set of accessible, high-quality instruments could be used in future evaluation and research projects to allow more comparison of students who study with different curricula or in different educational settings.

SUMMARY

This book focuses on one aspect of the "infancy" of the field of statistics education research, by attempting to grapple with the definitions, distinctions, and development of statistical literacy, reasoning, and thinking. As this field grows, the research studies in this volume should help provide a strong foundation as well as a common research literature. This is an exciting time, given the newness of the research area and the energy and enthusiasm of the contributing researchers and educators who are helping to shape the discipline as well as the future teaching and learning of statistics. We point out that there is room for more participants to help define and construct the research agenda and contribute to results. We hope to see many new faces at future gatherings of the international research community, whether at SRTL-4, or 5, or other venues such as the International Conference on Teaching Statistics (ICOTS), International Congress on Mathematical Education (ICME), and the International Group for the Psychology of Mathematics Education (PME).

REFERENCES

Bakker, A. (2002). Route-type and landscape-type software for learning statistical data analysis. In B. Phillips (Chief Ed.), *Developing a Statistically Literate Society: Proceedings of the Sixth International Conference on Teaching Statistics*, Voorburg, The Netherlands (CD-ROM).

Batanero, C., Garfield, J., Ottaviani, M. G., and Truran, J. (2000). Research in statistical education: Some priority questions. *Statistical Education Research Newsletter, 1*(2), 2–6.

Ben-Zvi, D. (2000). Toward understanding the role of technological tools in statistical learning. *Mathematical Thinking and Learning, 2*(1&2), 127–155.

Bransford, J., Brown, A. L., & Cocking, R. R. (Eds.). (2000). *How people learn: Brain, mind, experience, and school.* Washington, DC: National Academy Press.

Bryce, G. B. (2002). Undergraduate statistics education: An introduction and review of selected literature. *Journal of Statistics Education, 10*(2). Retrieved June 23, 2003 from http://www.amstat.org/publications/jse/v10n2/bryce.html.

Chance, B. L. (2002). Components of statistical thinking and implications for instruction and assessment. *Journal of Statistics Education, 10*(3). Retrieved April 7, 2003, from http://www.amstat.org/publications/jse/.

Cobb, P. (1999). Individual and collective mathematical development: The case of statistical data analysis. *Mathematical Thinking and Learning, 1*(1), 5–43.

Fischbein, E. (1975). *The intuitive sources of probabilistic thinking in children.* Dordrecht, The Netherlands: D. Reidel.

Fong G. T., Krantz, D. H., & Nisbett, R. E. (1986). The effects of statistical training on thinking about everyday problems. *Cognitive Psychology, 18*, 253–292.

Friel, S. N. (in press). The research frontier: Where technology interacts with the teaching and learning of data analysis and statistics. In M. K. Heid & G. W. Blume (Eds.), *Research on technology and the teaching and learning of mathematics: Syntheses and perspectives, vol. 1.* Greenwich, CT: Information Age Publishing.

Gal, I., & Garfield, J. B. (Eds.). (1997). *The assessment challenge in statistics education.* Voorburg, The Netherlands: International Statistical Institute.

Garfield, J. B., & Burrill, G. (Eds.). (1997). *Research on the role of technology in teaching and learning statistics. In Proceedings of the 1996 IASE Round Table Conference*, Granada, Spain. Voorburg, The Netherlands: International Statistical Institute.

Garfield, J., & Gal. I. (1999). Teaching and assessing statistical reasoning. In L. V. Stiff (Ed.), *Developing mathematical reasoning in grades K–12* (NCTM 1999 Yearbook), pp. 207–219. Reston, VA: National Council of Teachers of Mathematics.

Joliffe, F. (1998). What is research in statistical education? In L. Pereira-Mendoza, L. Seu Kea, T. Wee Kee, & W. K. Wong (Eds.), *Proceedings of the Fifth International Conference on Teaching Statistics*, pp. 801–806. Singapore: International Statistical Institute.

Kahneman, D., Slovic, P., & Tversky, A. (Eds.). (1982). *Judgment under uncertainty: Heuristics and biases.* New York: Cambridge University Press.

Konold, C. (1989). Informal conceptions of probability. *Cognition and Instruction, 6*, 59–98.

Kelly, A. E., & Lesh, R. A. (Eds.). (2000). *Handbook of research design in mathematics and science education.* Mahwah, NJ: Erlbaum.

Lajoie, S. P. (Ed.). (1998). *Reflections on statistics: Learning, teaching, and assessment in grades K–12.* Mahwah, NJ: Erlbaum.

Lee, C., Zeleke, A., & Wachtel, H. (2002). Where do students get lost? The concept of variation. In B. Phillips (Chief Ed.), Developing a Statistically Literate Society: *Proceedings of the Sixth International Conference on Teaching Statistics.* Voorburg: The Netherlands (CD-ROM).

Lesh, R. (2002). Research design in mathematics education: Focusing on design experiments. In L. English (Ed.), *International Handbook of Research Design in Mathematics Education.* Hillsdale, NJ: Erlbaum.

Mokros, J., & Russell, S. J. (1995). Children's concepts of a verage and representativeness. *Journal for Research in Mathematics Education, 26*, 20–39.

Moore, D. (1998). Statistics among the liberal arts. *Journal of the American Statistical Association, 93*, 1253–1259.

National Council of Teachers of Mathematics (NCTM). (2000). *Principles and standards for school mathematics.* Reston, VA: NCTM.

Nisbett, R. (1993). *Rules for reasoning.* Hillsdale, NJ: Erlbaum.

Piaget, J., & Inhelder, B. (1975). *The origin of the idea of chance in children.* London: Routledge & Kegan Paul.

Rubin, A., Bruce, B., & Tenney, Y. (1991). Learning about sampling: Trouble at the core of statistics. In D. Vere-Jones (Ed.), *Proceedings of the Third International Conference on Teaching Statistics, vol. 1* (pp. 314–319). Voorburg, The Netherlands: International Statistical Institute.

Russell, S. J., & Mokros, J. (1996). What do children understand about average? *Teaching Children Mathematics, 2*, 360–364.

Schau, C., & Mattern, N. (1997). Assessing students' connected understanding of statistical relationships. In I. Gal & J. B. Garfield (Eds.), *The Assessment Challenge in Statistics Education* (pp. 91–104). Amsterdam, Netherlands: IOS Press.

Sedlmeier, P. (1999). *Improving statistical reasoning: Theoretical models and practical implications.* Hillsdale, NJ: Erlbaum.

Shaughnessy, J. M. (1992). Research in probability and statistics: Reflections and directions. In D. Grouws (Ed.), *Handbook of research on mathematics teaching and learning* (pp. 465–494). New York: Macmillan.

Strauss, A., & Corbin, J. (1998). *Basics of qualitative research: Techniques and procedures for developing grounded theory.* Thousand Oaks, CA: Sage.

Utts, J. (2003). What educated citizens should know about statistics and probability. *The American Statistician, 57*(2), 74–79.

Voigt, J. (1995). Thematic patterns of interaction and socio-mathematical norms. In P. Cobb & H. Bauersfeld (Eds.), *Emergence of mathematical meaning: Interaction in classroom cultures* (pp. 163–201). Hillsdale, NJ: Erlbaum.
Wiggins, G. (1998). *Understanding by design*. Alexandria, VA: ASCD Press.
Wisenbaker, J., & Scott, J. (1997). Modeling aspects of students' attitudes and achievement in introductory statistics courses. Paper presented at AERA Annual Meeting, Chicago.
Wild, C., Triggs, C., & Pfannkuch, M. (1997). Assessment on a budget: Using traditional methods imaginatively. In I. Gal & J. B. Garfield (Eds.), *The Assessment Challenge in Statistics Education*. Amsterdam, Netherlands: IOS Press.

Author Index

O

P

Q

R

S

T

U

Subject Index

Printed in the United Kingdom
by Lightning Source UK Ltd.
116434UKS00006B/89